Sustainable Synthesis of Pharmaceuticals
Using Transition Metal Complexes as Catalysts

Green Chemistry Series

Titles in the series:

1: The Future of Glycerol: New Uses of a Versatile Raw Material
2: Alternative Solvents for Green Chemistry
3: Eco-Friendly Synthesis of Fine Chemicals
4: Sustainable Solutions for Modern Economies
5: Chemical Reactions and Processes under Flow Conditions
6: Radical Reactions in Aqueous Media
7: Aqueous Microwave Chemistry
8: The Future of Glycerol: 2nd Edition
9: Transportation Biofuels: Novel Pathways for the Production of Ethanol, Biogas and Biodiesel
10: Alternatives to Conventional Food Processing
11: Green Trends in Insect Control
12: A Handbook of Applied Biopolymer Technology: Synthesis, Degradation and Applications
13: Challenges in Green Analytical Chemistry
14: Advanced Oil Crop Biorefineries
15: Enantioselective Homogeneous Supported Catalysis
16: Natural Polymers Volume 1: Composites
17: Natural Polymers Volume 2: Nanocomposites
18: Integrated Forest Biorefineries
19: Sustainable Preparation of Metal Nanoparticles: Methods and Applications
20: Alternative Solvents for Green Chemistry: 2nd Edition
21: Natural Product Extraction: Principles and Applications
22: Element Recovery and Sustainability
23: Green Materials for Sustainable Water Remediation and Treatment
24: The Economic Utilisation of Food Co-Products
25: Biomass for Sustainable Applications: Pollution Remediation and Energy
26: From C-H to C-C Bonds: Cross-Dehydrogenative-Coupling
27: Renewable Resources for Biorefineries
28: Transition Metal Catalysis in Aerobic Alcohol Oxidation
29: Green Materials from Plant Oils
30: Polyhydroxyalkanoates (PHAs) Based Blends, Composites and Nanocomposites

31: Ball Milling Towards Green Synthesis: Applications, Projects, Challenges
32: Porous Carbon Materials from Sustainable Precursors
33: Heterogeneous Catalysis for Today's Challenges: Synthesis, Characterization and Applications
34: Chemical Biotechnology and Bioengineering
35: Microwave-Assisted Polymerization
36: Ionic Liquids in the Biorefinery Concept: Challenges and Perspectives
37: Starch-based Blends, Composites and Nanocomposites
38: Sustainable Catalysis: With Non-endangered Metals, Part 1
39: Sustainable Catalysis: With Non-endangered Metals, Part 2
40: Sustainable Catalysis: Without Metals or Other Endangered Elements, Part 1
41: Sustainable Catalysis: Without Metals or Other Endangered Elements, Part 2
42: Green Photo-active Nanomaterials
43: Commercializing Biobased Products: Opportunities, Challenges, Benefits, and Risks
44: Biomass Sugars for Non-Fuel Applications
45: White Biotechnology for Sustainable Chemistry
46: Green and Sustainable Medicinal Chemistry: Methods, Tools and Strategies for the 21st Century Pharmaceutical Industry
47: Alternative Energy Sources for Green Chemistry
48: High Pressure Technologies in Biomass Conversion
49: Sustainable Solvents: Perspectives from Research, Business and International Policy
50: Fast Pyrolysis of Biomass: Advances in Science and Technology
51: Catalyst-free Organic Synthesis
52: Hazardous Reagent Substitution: A Pharmaceutical Perspective
53: Alternatives to Conventional Food Processing: 2nd Edition
54: Sustainable Synthesis of Pharmaceuticals: Using Transition Metal Complexes as Catalysts

How to obtain future titles on publication:
A standing order plan is available for this series. A standing order will bring delivery of each new volume immediately on publication.

For further information please contact:
Book Sales Department, Royal Society of Chemistry, Thomas Graham House, Science Park, Milton Road, Cambridge, CB4 0WF, UK
Telephone: +44 (0)1223 420066, Fax: +44 (0)1223 420247
Email: booksales@rsc.org
Visit our website at www.rsc.org/books

Sustainable Synthesis of Pharmaceuticals
Using Transition Metal Complexes as Catalysts

Edited by

Mariette M. Pereira
University of Coimbra, Portugal
Email: mmpereira@qui.uc.pt

and

Mário J. F. Calvete
University of Coimbra, Portugal
Email: mcalvete@qui.uc.pt

ROYAL SOCIETY OF CHEMISTRY

THE QUEEN'S AWARDS FOR ENTERPRISE: INTERNATIONAL TRADE 2013

Green Chemistry Series No. 54

Print ISBN: 978-1-78262-934-4
PDF ISBN: 978-1-78801-065-8
EPUB ISBN: 978-1-78801-441-0
ISSN: 1757-7039

A catalogue record for this book is available from the British Library

The Royal Society of Chemistry is a charity, registered in England and Wales, Number 207890, and a company incorporated in England by Royal Charter (Registered No. RC000524), registered office: Burlington House, Piccadilly, London W1J 0BA, UK, Telephone: +44 (0) 207 4378 6556.

For further information see our web site at www.rsc.org

Printed in the United Kingdom by CPI Group (UK) Ltd, Croydon, CR0 4YY, UK

Preface

The idea that led to the edition of this book *Sustainable Synthesis of Pharmaceuticals: Using Transition Metal Complexes as Catalysts* for the Royal Society of Chemistry's Green Chemistry series resulted from discussions between professors from several Portuguese universities, and also researchers from the pharmaceutical industry involved in Master and Inter-University doctoral plans. From the contact with the students, we realized how important it would be to edit a book that, besides presenting a set of selected examples of the application of catalysts in the synthesis of drugs/synthons, would also discuss their application, with emphasis on the principles of green chemistry because we consider that it is urgent to implement these principles in pedagogical courses in order to obtain sustainable practices at the industrial level in the near future. Considering that the book might be of interest to academia and industry, students and researchers, we decided to invite teachers of international merit to develop a high-level book and also to extend the scope of the book's dissemination.

The book is organized in 10 chapters and all start with the fundamental concepts of reaction mechanisms and the relevance of guiding the new synthesis of pharmaceuticals by catalytic and non-stoichiometric processes, followed by a set of selected relevant examples, using well-known reactions, ranging from hydroformylation to carbonylation, passing by the use of alternative sustainable synthetic methodologies. Although hydrogenation *via* molecular hydrogen activation is one of the paradigmatic examples of the application of catalysts in the synthesis of pharmaceuticals, this topic was not included because there are already numerous review articles and books in the field. The literature included in each chapter was not intended to be an exhaustive review of the subject but rather a concise selection made by the authors in order to review the historical landmarks on the topic, the

Green Chemistry Series No. 54
Sustainable Synthesis of Pharmaceuticals: Using Transition Metal Complexes as Catalysts
Edited by Mariette M. Pereira and Mário J. F. Calvete

Published by the Royal Society of Chemistry, www.rsc.org

overall mechanism and a selection of examples where organometallics have been used as catalysts in the synthesis of active pharmaceutical ingredients.

The editors thank the students of the Master's and Doctoral Plans in chemistry and medicinal chemistry at the University of Coimbra, the University of Lisbon and the New University of Lisbon for ideas and comments.

We are indebted to all the authors who have contributed to this book!

A special acknowledgement to Professor David Allen (University of Sheffield) and Professor James Clark (University of York), who accepted and supported this project, and to Janet Freshwater and Robin Driscoll for all the support in the final edition. Finally, we thank the Royal Society of Chemistry for supporting this edition.

Mariette M. Pereira
Mário J. F. Calvete

Contents

Green Chemistry Series No. 54
Sustainable Synthesis of Pharmaceuticals: Using Transition Metal Complexes as Catalysts
Edited by Mariette M. Pereira and Mário J. F. Calvete

Published by the Royal Society of Chemistry, www.rsc.org

CHAPTER 1

Introduction

M. M. PEREIRA* AND M. J. F. CALVETE*

University of Coimbra, CQC, Department of Chemistry, Rua Larga, Coimbra 3004-535, Portugal
*Email: mmpereira@qui.uc.pt; mcalvete@qui.uc.pt

1.1 Introduction

The development of sustainable processes for the synthesis of new active pharmaceutical ingredients (API) continues to be one of the great challenges for medicinal chemistry at universities and in the pharmaceutical industry.[1]

Given the benefits to public health, for decades, the pharmaceutical industry was more concerned with the end product than with the means of producing it. For decades, the synthetic methods of pharmaceutical products were the ones that led to greater waste and those that least respected the principles of green chemistry (*green chemistry preferentially utilizes raw materials, avoids toxic and hazardous reagents and solvents, eliminates waste and when possible reduces the energy consumption,* Chapter 3). However, in the last decade several companies have decided to adhere to the philosophy of green chemistry by modifying their production processes and especially by designing new processes where metrics (Chapter 3) have already been taken into account and the principles of waste and solvent reduction have been considered very important issues for the development of new API process development.

A key feature for improving the sustainability of the pharmaceutical industry is the design of new processes according to the principles of green chemistry. In the last few decades several companies changed their practices and adhered in particular to principle number 9: "*catalytic reagents*

Green Chemistry Series No. 54
Sustainable Synthesis of Pharmaceuticals: Using Transition Metal Complexes as Catalysts
Edited by Mariette M. Pereira and Mário J. F. Calvete

Published by the Royal Society of Chemistry, www.rsc.org

(as selective as possible) are superior to stoichiometric reagents". Nowadays, it has become clear to the pharmaceutical industry that the substitution of stoichiometric chemical reactions by catalytic processes may solve several industrial problems: (i) increasing selectivity for the desired product, particularly as regards the synthesis of enantiomerically pure pharmaceuticals; (ii) reducing the costs due to lower energy consumption; and (iii) reducing solvents and the overall process cost.

The great relevance of the use of organometallic reagents as catalysts for organic synthesis is clearly evidenced by the attribution of several Nobel Prizes in the field.[2]

This book is aimed at researchers or post-graduate students, both in academia and the pharmaceutical industry, who are interested in developing processes for synthesis of drug synthons or APIs, using transition metals catalysts as the tool for achieving green chemistry purposes. As Clark reports in Chapter 2, the substitution of toxic metals by less toxic ones like iron and the development of new processes for recovering catalysts is clearly a new paradigm that the pharmaceutical industry should consider when introducing metals as catalysts for the development of new drugs. This issue is focused on in the great majority of the chapters.

In Chapter 2, J. Clark elucidates on the availability of critical chemical elements, mostly metals, and the strong discouragement of their use since there is a growing appreciation that not only are resources limited, but also their recovery is very difficult, and this vital part of sustainability must be recognized within green chemistry. It is crucial that efforts are increased both to use less metals and to design catalysts and processes to maximize recovery of the metals. Heterogeneous catalysts can definitely play a major role in this endeavour.

In Chapter 3, M. Pineiro and M. Calvete discuss the success of the philosophy and principles of green chemistry in the active search for more sustainable drug synthesis processes. A collection of diverse approaches has been reported so far, including the use of alternative reaction media and alternative technologies, such as microwaves, mechanochemistry, and ultrasound, especially when combined with new catalysts and catalytic systems, where the sustainability "improvement" is measured and quantified by using green chemistry metrics integrated in the drug discovery and development process in the pharmaceutical industry.

In Chapter 4, R. Skoda-Földes and L. Kollár shed light on pharmaceuticals arising from one-pot carbonylation processes, as typically the production of pharmaceuticals usually involves multistep syntheses where the selectivity and yield of the individual steps are of utmost importance. Among highly efficient catalytic processes, carbonylation received special attention as it involves both new carbon–carbon bond formation and the introduction of a synthetically useful functionality in the synthesis of carbonyl compounds and carboxylic acid derivatives. To achieve widespread application, more efficient catalysts should be developed that ensure higher turnover numbers and make it possible to carry out carbonylations at atmospheric conditions as well as to replace the starting material aryl iodides with cheaper bromides or, still preferably, chlorides. This reaction was described as a greener approach to prepare pharmaceuticals or their precursors bearing carboxylic acids, amides or esters functionalities in just one-pot.

In Chapter 5, M. Pereira highlights several aspects concerning the mechanism of rhodium-catalysed hydroformylation (still considered the "metal of choice" owing to its high activity and selectivity), the development and evolution of new metal catalysts and phosphorus ligands from a historical perspective, some strategies related to the synthesis of reusable catalysts for use in alternative media and the utilization of less toxic solvents and alternative metals. A set of selected examples for the direct transformations of olefins into aldehydes, *via* one-step 100% atom economy process, for the sustainable preparation of pharmaceutical intermediates or APIs is also described.

In Chapter 6, B. Royo describes the application of metal-catalysed asymmetric transfer hydrogenation and borrowing hydrogen processes to the synthesis of pharmaceuticals, especially those using Earth-abundant catalysts, which can replace precious metals, enabling simple and safe synthetic strategies with high atom economy and efficiency, providing an outstanding input for the discovery of new bioactive molecules useful in the pharmaceutical industry.

In Chapter 7, C. Romão discusses the development of green metal asymmetric epoxidation and sulfoxidation catalysis, obeying the highest possible number of "green chemistry" rules, and exploring/expanding the control over the exquisite mechanisms of both epoxidation and sulfoxidation, posing a very high research target that could lead to waste-free fully sustainable oxidation processes. Their very existence proves the endeavour is worth pursuing.

In Chapter 8, L. Martins, A. Phillips and A. Pombeiro provide a discussion on C–C bond formation in the sustainable synthesis of pharmaceuticals (APIs or other drug components), whose current production includes transition metal-catalysed C–C cross coupling reactions as key steps of the synthetic processes in view of their mildness, functional group compatibility and the high turnover of used catalysts. Emphasis is put on modern protocols, which allow the purge of metal catalysts, whilst still providing high purity compounds, by continuing development of optimized catalysts, ligands, additives and reaction conditions in a green environment.

In Chapter 9, E dos Santos, A. Granato, and A. Santos elucidate on metal-catalysed olefin metathesis reactions for greener synthon/drug synthesis, providing detailed valuable information on catalysts and conditions to perform metathesis reactions. Various applications of metathesis in the synthesis of active pharmaceutical ingredients have been recently disclosed. It has also been very important for chemical diversity generation and drug discovery, giving rise to new applications in the industrial synthesis of pharmaceuticals.

In Chapter 10, Q. Meng, M. Wang, and M. Vicente highlight tetravalent boron-based therapeutics, particularly with anticancer, antiviral, antibacterial and antifungal activities. The number of boron-containing therapeutics is expected to continue to increase as new synthetic compounds, such as those based on BODIPYs and related systems, continue to be investigated for potential applications in diagnosis, and in the treatment of cancer and other diseases, including *via* photodynamic and boron-neutron capture therapies.

References

1. J. Verghese, C. J. Kong, D. Rivalti, E. C. Yu, R. Krack, J. Alcázar, J. B. Manley, D. T. McQuade, S. Ahmad, K. Belecki and B. F. Gupton, *Green Chem.*, 2017, **19**, 2986.
2. A. M. Thayer, *Chem. Eng. News*, 2013, **91**, 68.

CHAPTER 2

Transition Metals in Greener Pharmaceutical Chemistry

J. H. CLARK

Green Chemistry Center of Excellence, University of York, York YO105DD, United Kingdom
Email: james.clark@york.ac.uk

2.1 Transition Metals in Greener Pharmaceutical Chemistry

When the first concerns about the future availability of some critical elements—mostly metals—were being discussed in Europe, there was a school of thought that Brussels might become a "metal-free zone", in other words that the EU would try to discourage the use of (some) metals in (some) applications. The growing appreciation that we cannot continue to use resources as though they were unlimited, and in a way that makes their recovery very difficult, is a vital part of sustainability and must be recognized within green chemistry. Medium-term availability problems are made worse by the rapid growth of new technologies that are ironically often driven by the desire for low-carbon energy, which use elements at levels never previously seen. This includes wind turbines (which use dysprosium, neodymium and praseodymium) and electric vehicles (antimony, neodymium, dysprosium and gadolinium) as well as rapid developments in batteries, which are using more and more lithium.

Green Chemistry Series No. 54
Sustainable Synthesis of Pharmaceuticals: Using Transition Metal Complexes as Catalysts
Edited by Mariette M. Pereira and Mário J. F. Calvete

Published by the Royal Society of Chemistry, www.rsc.org

Figure 2.1 Critical impacts to be considered for all processes. LCA stands for Life Cycle Assessment.
Reproduced from ref. 1 with permission from The Royal Society of Chemistry.

The latest and more holistic green chemistry metrics for pharmaceutical (and other fine chemical) processes accommodate this new concern as one of the key impacts to take into consideration (Figure 2.1).[1]

In the associated online tool that assesses the green credentials of any new or existing process, a red flag is waved over any metal used that is considered to be "critical" although such classifications can change as new technologies grow while others fade and as a result of a new discovery of the associated ore. New ore discoveries will however, almost certainly come at higher economic and environmental cost—we have exploited most of the easy resources.

It would, however, be oversimplified and unwise to seek to avoid the use of all critical metals as catalytic materials. The benefits may outweigh concerns over the availability of some metals, at least until we find adequate replacements. Catalysis continues to be the most important green chemical technology and is of course one of the key impacts in Figure 2.1. Metals play a dominant role in catalysis. Palladium, for example, has become a highly valued catalyst in many cross-coupling reactions, including Heck, Negishi and Suzuki, reactions that are increasingly used in the manufacture of pharmaceuticals. Alternative Pd-free routes, where possible, are generally more laborious, involve several additional reagents and create substantial quantities of waste. Palladium is currently considered to be of medium availability concern, with known reserves and current rate of consumption giving 50–100 remaining years, although this could easily change either way. Other very useful catalytic metals including ruthenium and iridium are believed to be down to less than 50 years, making their sustainability of more immediate concern and their continued use may be more problematic.[2]

In all cases it is important that we increase our efforts both to use less metal and to design the catalyst and process to maximize recovery of the metal. Heterogeneous catalysts can make recovery easier as may other two-phase reaction systems [*e.g.* where the catalyst and substrate(s) are in different liquid phases] but there is always a concern about leaching from the catalyst phase, likely leading to loss of the metal and to possible contamination of the product (which is strictly controlled for pharmaceutical compounds). We do need to develop new technologies for recovering trace metals from product mixtures as well as from wastes more generally. As we become more concerned about the sustainability of many metals and some other important elements (*e.g.* phosphorus) we should seriously consider new sources of those elements that are now in the form of wastes: WEEE (Waste Electrical and Electronic Equipment), mine tailings, emissions from catalytic convertors and others. Here techniques including phytomining (the recovery of metals from land using plants) might have an important role.[3]

Of course, an important alternative strategy to finding solutions to the continued use of critical metals in catalysis is to try to find more sustainable alternatives and the success of this approach is such that it led to a recent two-volume book on the subject.[4]

References

1. C. R. McElroy, A. Constantinou, L. C. Jones, L. Summerton and J. H. Clark, *Green Chem.*, 2015, **17**, 3111.
2. *Element Recovery and Sustainability*, ed. A. Hunt, RSC Green Chemistry Book Series, RSC Publishing, Cambridge, 2013.
3. J. H. Clark, H. L. Parker, E. L. Rylott, A. J. Hunt, J. R. Dodson, A. F. Taylor and N. C. Bruce, *PLoS One*, 2014, **9**, e87192.
4. *Sustainable Catalysis with Non-endangered Metals*, ed. M. North, RSC Green Chemistry Book Series, RSC Publishing, Cambridge, 2015.

CHAPTER 3

Sustainable Synthesis of Pharmaceuticals Using Alternative Techniques: Microwave, Sonochemistry and Mechanochemistry

M. PINEIRO* AND M. J. F. CALVETE

University of Coimbra, CQC, Department of Chemistry, RuaLarga, Coimbra 3004-535, Portugal
*Email: mpineiro@qui.uc.pt

3.1 Introduction

Sustainability can be broadly defined as "meeting the needs of the present without compromising the ability of future generations to meet their own needs".[1] It is a very comprehensive concept that embraces every aspect of human behaviour such as sustainable economy, sustainable agriculture, sustainable tourism and, naturally, sustainable chemistry.

Sustainable chemistry is, again, a broad concept that includes remediation technologies, exposure controls, water purification, alternative energy, chemical policies and also green chemistry. Green chemistry is a subset of the sustainable chemistry area, focused on the design, development and implementation of chemical products and processes. Green chemistry aims at reducing pollution at its source by minimizing or eliminating the hazards

Green Chemistry Series No. 54
Sustainable Synthesis of Pharmaceuticals: Using Transition Metal Complexes as Catalysts
Edited by Mariette M. Pereira and Mário J. F. Calvete

Published by the Royal Society of Chemistry, www.rsc.org

of chemical feedstocks, reagents, solvents and products. It assesses the solvent used in the process, explores the potential use of catalysts, promotes the incorporation of renewable feedstocks, attempts to minimize energy utilization and identifies materials that do not persist or bioaccumulate. To accomplish the green chemistry goals, it is necessary to reduce or eliminate the use and/or generation of hazardous materials or processes. Over the years, this has generated several principles for the design, development and implementation of chemical products and/or processes, summarized as the 12 principles (Figure 3.1):[2] minimum waste generation, atom economy, less hazardous chemical synthesis design, design of safer chemicals with minimum toxicity, use of safer solvents and auxiliaries, design of energy efficiency, use of renewable feedstock, reduction of derivatization steps, use of selective catalysts, design of products degradable after use, real-time analysis for pollution prevention and inherent safer chemistry for accident prevention. The first principle advocates the basic tenet of green chemistry, pollution prevention, while the other 11 deal with more specific topics, namely, atom economy, toxicity or energy use. These principles were further extended by Winterton and others,[3] and with the publication of the Twelve Principles of Green Engineering.[4]

To accomplish the green chemistry goals, new materials and processes must be successfully deployed in commercial endeavours. The pharmaceutical industry is devoted to discovering, developing, producing and commercializing drugs to increase the life and life-quality of patients, and it is largely responsible for the almost doubling of life expectancy from 1900 to 2000. It was one of the industry sectors that first embraced the field of green chemistry.[5] Pfizer, in the late 1990s, led the world's first corporate program to expand and develop green chemistry practices inside the pharmaceutical industry. The pharmaceutical industry has some widely used examples of reduced manufacturing costs of active pharmaceutical ingredients (APIs), many of which are award winning green chemistry technologies. Since 1995, when USA President Clinton established the Presidential Green Chemistry

1 **Prevent Waste**
2 **Atom Economy**
3 **Less Hazardous Synthesis**
4 **Design Benign Chemicals**
5 **Benign Solvents & Auxiliaries**
6 **Design for Energy Efficiency**
7 **Use of Renewable Feedstocks**
8 **Reduce Derivatives**
9 **Catalysis (*vs* Stoichiometric)**
10 **Design for Degradation**
11 **Real-Time Analysis for Pollution Prevention**
12 **Inherently Benign Chemistry for Accident Prevention**

Figure 3.1 The twelve green chemistry principles.

Challenge program, eight awards were given to the pharmaceutical industry.[6] In 2005, the ACS Green Chemistry Institute® and several global pharmaceutical corporations founded the ACS GCI Pharmaceutical Roundtable, currently formed of 16 pharmaceutical companies, to encourage innovation while catalysing the integration of green chemistry and green engineering in the pharmaceutical industry. These industry competitors work collaboratively to address the environmental impact of their manufacturing processes.[7]

To minimize this impact, alternative activation methods and alternative reaction media are required, aiming at reducing cost, energy, risk, hazard and waste. From the point of view of sustainability, greener methodologies with higher energy efficiency, such as microwave, mechanochemistry, and ultrasound, are recommended.[8,9]

Once the major cause of waste has been recognized, the solution to the waste problem was evident: replacement of the classical synthesis that employs stoichiometric amounts of inorganic (or organic) reagents by cleaner, catalytic alternatives. The fundamental role of catalysis in green chemistry was identified earlier,[10] and no subject pervades modern chemistry as that of catalysis.[11] The design and application of new catalysts and catalytic systems are simultaneously achieving the goals of environmental protection and economic benefit.[12] Acid–base catalysis, catalysis by transition metals and biocatalysis have an important role in the increase of sustainability, especially using phase transfer catalysts and supported catalysts.

Several approaches can thus be used and combined to increase the sustainability of synthetic processes. From the point of view of reactivity, one-pot, multicomponent reactions and domino reactions, regio-, stereo- and enantio-selective synthetic procedures are desirable. From the point of view of new reaction conditions there are two paths to be considered: the use of alternative reaction media, such as supercritical fluids, alternative solvents, water or ionic liquids, and the use of non-conventional techniques, such as microwave, ultrasound and mechanochemistry. The combination of the two paths to deliver suitable and effective green methods should, especially, be considered.

3.2 Metrics

Developing green synthetic methodologies requires a significant behavioural change for both industry and academia, and to support and reinforce this change it is necessary to measure progress, resorting to suitable indicators that reflect the development of 'greener' processes. This requirement led to many different proposals to determine how "green" a process is from a chemical and engineering perspective.[13,14] Different measures are described and briefly addressed in the following.

The environmental Sheldon's factor, simply known as the E factor,[15] was developed in order to highlight the amount of waste generated to produce 1 kg of chemical product across different branches of the chemical industry.

The E factor is defined as the mass ratio of waste to product and is now widely quoted across many different chemical industries as it provides a simple benchmark guide for different sectors of the chemical industry and has been widely published and presented in multiple venues. In 2017 Sheldon reviewed the E factor concept, defining the complete E Factor (cEF) to include in the amount of waste the mass of solvent and the mass of water.[16]

$$\text{E Factor} = \frac{\text{Total waste (kg)}}{\text{Total weight of desired product (kg)}}$$

or

$$\text{E Factor} = \frac{\Sigma\ \text{mass of materials} - \text{mass of isolated product}}{\text{mass of isolated product}}$$

$$\text{cEF} = \frac{\Sigma\ \text{mass of materials} + \Sigma\ \text{mass of solvents} + \Sigma\ \text{mass of water} - \text{mass of isolated product}}{\text{mass of isolated product}}$$

For example, the comparison of E factors of the homogeneous and heterogeneous catalytic processes in the alkylation of benzene shows a 30-fold preference towards the heterogeneous method.[17]

Another indicator, mass intensity, measures the amount of material needed to synthesize the desired product. It takes into account yield, reaction stoichiometry, solvents, and reagents covering everything that is placed inside a reaction vessel. It also includes all mass used in acid, base and salt as well as organic solvents used in wash, extractions, crystallizations, or solvent switching,

$$\text{Mass Intensity} = \frac{\text{Mass of all materials excluding water (kg)}}{\text{Total weight of product (kg)}}$$

or

$$\text{Mass Intensity} = \text{E} - \text{Factor} + 1$$

Mass intensity, as defined by GlaxoSmithKline (GSK), did not include water used in the process, giving rise to a source of confusion in the original E factor determination,[15] and solved with the cEF calculation.[16]

Another limitation that can be assigned to the E factor is that it does not take into account the nature and environmental impact of the generated waste. In order to arrive at a more meaningful prediction, the environmental quotient (Q) was introduced, with the E factor being multiplied by this quotient.[18]

Mass productivity is obtained by taking the mass intensity reciprocal, represented as a percentage:

$$\text{Mass Productivity} = \frac{1}{\text{Mass Intensity}} \times 100$$

The atom economy concept was established by Barry Trost[19] and made him recipient of the Presidential Green Chemistry Challenge Award.[20] Atom economy, which addresses the second and eighth principles of green chemistry, is the most fundamental concept applied to estimate the greenness of an organic reaction at the molecular level. It considers that the atoms present in the starting materials should end up in the product rather than in the waste.

$$\%\,\text{Atom Economy} = \frac{\text{Formula weight of desired product}}{\Sigma\ \text{Formula weight of reactants}} \times 100$$

The higher the atom economy is, the greener the chemical reaction. However, this metric does not take into account reaction yields, reactant stoichiometry or solvents. Carbon efficiency is defined as the percentage of carbon in the reactants that remains in the final product. In this calculation, reaction yield and stoichiometry of reactants are included.

$$\%\,\text{Carbon Efficiency} = \frac{\text{mol of product} \times \text{nC in product}}{\Sigma\ \text{mol of reactant}_i \times \text{nC in reactant}_i} \times 100$$

where nC is the number of carbons.

The metrics described above are focused significantly on quantifying waste generated from chemical processes; however, scientists and green chemists are currently working on improved green metrics, such as EcoScale, Green Star and Life Cycle Analysis (LCA), that are more associated with the philosophy of the 12 green chemistry principles.

EcoScale is a semi-quantitative post-synthesis analysis tool that evaluates the quality of an organic preparation based on yield, cost, safety conditions and ease of workup/purification.[21] It is a powerful tool to compare several preparations of the same product based on safety, economical and ecological features. An ideal reaction "*Compound A (substrate) undergoes a reaction with (or in the presence of) inexpensive compound(s) B to give the desired compound C in 100% yield, at room temperature, with a minimal risk for the operator and a minimal impact for the environment*" has an EcoScale value of 100. The EcoScale score for a particular preparation of the product in a high purity state (>98%) is calculated by lowering the maximum value of 100 by any applicable penalty points related to yield, price of reaction components, safety, technical setup, temperature, time, workup and purification, according to Table 3.1. The use of EcoScale allows the classification of the reactions, in which the reaction conditions used in the preparation of a high purity (>98%) product is ranked on a scale from 0 to 100 using the following scores: >75, excellent;

Table 3.1 Penalty points for EcoScale value calculation.

Parameter		Penalty points
1	**Yield**	(100 – %yield)/2
2	**Price of reaction components** (to obtain 10 mmol of end product)	
	Inexpensive (<$10)	0
	Expensive (>$10 and <$50)	3
	Very expensive (>$50)	5
3	**Safety**[a]	
	N (dangerous for environment)	5
	T (toxic)	5
	F (highly flammable)	5
	E (explosive)	10
	F+ (extremely flammable)	10
	T+ (extremely toxic)	10
4	**Technical setup**	
	Common setup	0
	Instruments for controlled addition of chemicals[b]	1
	Unconventional activation technique[c]	2
	Pressure equipment, >1 atm[d]	3
	Any additional special glassware	1
	(Inert) gas atmosphere	1
	Glove box	3
5	**Temperature/time**	
	Room temperature, <1 h	0
	Room temperature, <24 h	1
	Heating, <1 h	2
	Heating, >1 h	3
	Cooling to 0 °C	4
	Cooling, <0 °C	5
6	**Workup and purification**	
	None 0	0
	Cooling to room temperature	0
	Adding solvent	0
	Simple filtration	0
	Removal of solvent with bp <150 °C	0
	Crystallization and filtration	1
	Removal of solvent with bp >150 °C	2
	Solid-phase extraction	2
	Distillation	3
	Sublimation	3
	Liquid–liquid extraction[e]	3
	Classical chromatography	10

[a]Based on the hazard warning symbols.
[b]Dropping funnel, syringe pump, gas pressure regulator, *etc.*
[c]Microwave irradiation, ultrasound or photochemical activation, *etc.*
[d]$scCO_2$, high-pressure hydrogenation equipment, *etc.*
[e]If applicable, the process includes drying of solvent with desiccant and filtration of desiccant.

>50, acceptable; and <50, inadequate. The EcoScale website allows automatic calculation of EcoScale, for which the program requires the quantities/qualities of all reaction components, followed by the workup procedure. On this basis, it calculates all other parameters (the data are extracted from the ChemExper database), providing the final EcoScale value.[21]

This metric allows scientists and students to assess the sustainability of a chemical process with a higher level of complexity than that of those described before. However, the drawbacks of EcoScale are the arbitrary assignment of categories and penalties, and not considering the amount and safety of the solvent use as well as the amount of waste generated, its environmental, safety and social implications, which are also very important, especially for industrial processes. Therefore, research on EcoScale has focused on defining a similar method with capability for evaluating the industrial process.[22]

The Green Star metric was established on the basis of criteria to assess the accomplishment of each of the 12 principles, and is relevant for the situation under analysis.[23–25] This metric is implemented assigning the scores of 1, 2, or 3 (the maximum value of greenness) to each of the 12 principles, following the criteria presented in Table 3.2. It is applied in teaching experiments and, therefore, the fourth (design safer chemicals) and the eleventh (real-time analysis for pollution prevention) principles are generally excluded because the teaching laboratory work does not include the preparation of a new product. Obviously, for application at the industrial level, and on the whole in the pharmaceutical industry, this metric should include the fourth and eleventh principles. It was included in this chapter owing to its usefulness for students when evaluating pharmaceutical processes.

To evaluate the hazards to human health and the environment, and of potential chemical accident, every substance is classified in a scale from 1 to 3 by criteria, as shown in Table 3.3. Information for identifying whether the substances are renewable and break down to innocuous degradation products is used for classifying the substances following the criteria shown in Table 3.4. The information above can be obtained from the security

Table 3.2 Data used for the calculation of the E factor for the Pfizer process.

	Mass (g)	Volume (mL)	Mw (g mol^{-1})	Density (g mL^{-1})
Sertraline tetralone	63.6		291.17	
methylamine	21.1		31.06	
Ethanol	197.25	250	46.07	0.789
Pd/$CaCO_3$ [1%(w/w)]	0.64		106.42	
C [3.5% (w/w)]	2.23		12	
H_2 (not quantified)				
Mass of product	Yield	Mw	mol = mol of reactant × yield	g = mol × Mw
Sertraline imine	0.4	304.21	0.09	26.58
	E factor = 9.71			

Table 3.3 Data used for the calculation of the E factor for the Welch process.

	Mass (g)	Volume (mL)	Mw ($g\,mol^{-1}$)	Density ($g\,mL^{-1}$)
Sertraline tetralone	13.5		291.17	
Methylamine	9.1	13	31.06	0.7
Toluene	164.73	190		0.867
$TiCl_4$	4.46		106.42	
Methanol	59.4	75		0.792
$NaBH_4$	1.7		37.83	
Mass of product	Yield	Mw	mol = mol of reactant × yield	g = mol × Mw
Sertraline imine	0.33	304.21	0.02	4.67
	E factor = 53.1			

Table 3.4 Ecoscale value for classical synthesis and Pfizer's synthesis of sertraline hydrochloride.

		Penalty points	
Parameter		Pfizer	Welch
1	Yield	30	33.5
2	Price of reaction components (to obtain 10 mmol of end product)	3	8
3	Safety	45	50
4	Technical setup	3	1
5	Temperature/time	7	8
6	Workup and purification	0	13
Ecoscale (100 − penalty points)		12	−13.5

safety sheets (SDS) available online in the sites of several manufacturers of chemistry products; when information is absent, the worst score is chosen.

In 2002 Pfizer was awarded the Presidential Green Chemistry Challenge in the category of Greener Synthetic Pathways for the redesign of the manufacturing process of sertraline, the active ingredient of Zoloft®.[26] Comparing with the classical three-step methodology (Scheme 3.1a),[27] the Pfizer process is a single-step method (Scheme 3.1b)[28] that doubles the overall product yield, reduces raw material use by 20 to 60%, eliminates the use or generation of hazardous materials, reduces energy and water consumption, and increases workers' safety.

The E factor, estimated from the available data in the literature, decreased from 53 to 10 (see Tables 3.2 and 3.3), and the EcoScale increased from a negative value (−13.5) to 12 (Table 3.4).

A visual inspection provides an indication of the global greenness, where the greener the chart appears, the higher the degree of greenness. For comparison of different Green Star values, a Green Star Area Index (GSAI) can be calculated as a percentage (100 × green area of the Green Star/area of the Green Star of maximum greenness) (Tables 3.5–3.7). GSAI varies between

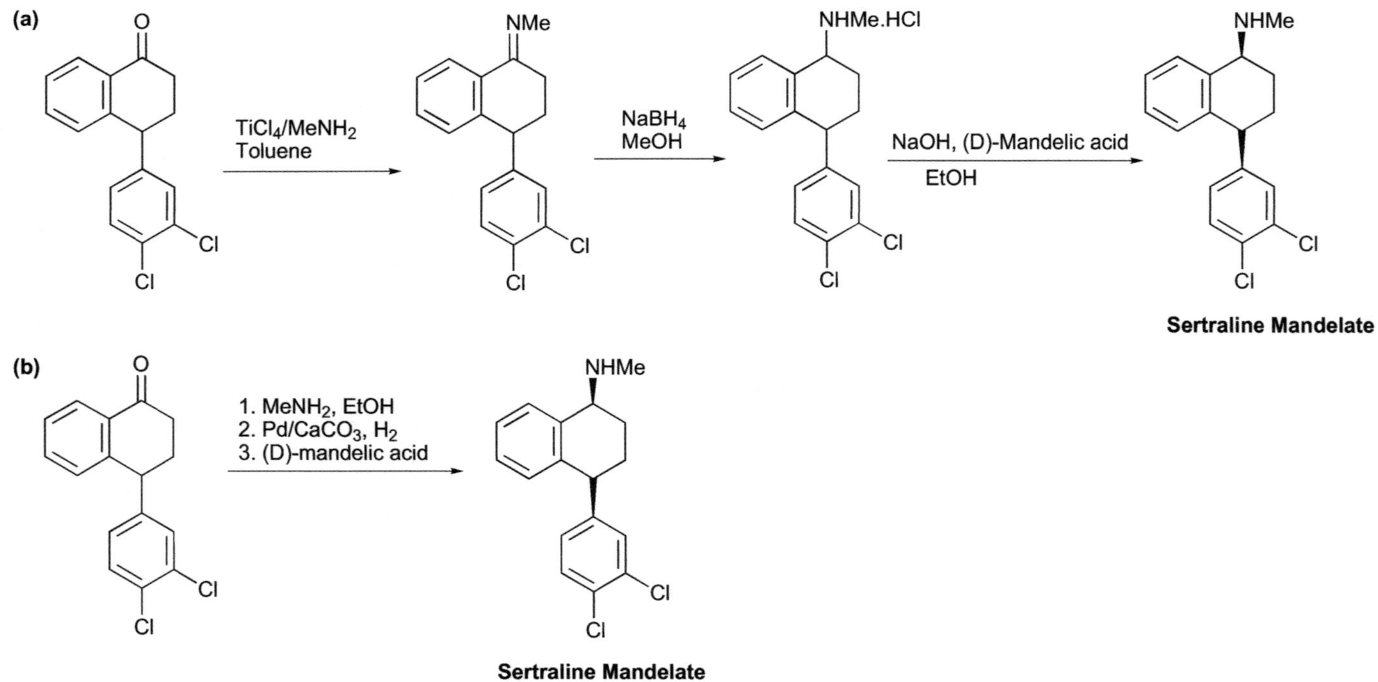

Scheme 3.1 (a) Classical synthesis of sertraline hydrochloride (b) as proposed by Pfizer.

Table 3.5 Score to construct the Green Star.

Green Chem. Principle	Criteria	Score
P1: Prevention[a]	Waste is innocuous (S = 1)	3
	Waste involves a moderate hazard to human health and the environment (S = 2 for at least one substance, no substances with S = 3)	2
	Waste involves a high hazard to human health and the environment (S = 3 for at least one substance)	1
P2: Atom Economy	Reactions without excess of reagents (≤10%) and without formation of byproducts	3
	Reactions without excess of reagents (≤10%) and with formation of byproducts	2
	Reactions with excess of reagents (>10%) and without formation of byproducts	2
	Reactions with excess of reagents (>10%) and with formation of byproducts	1
P3: Less Hazardous Chemical Synthesis[a]	All substances involved are innocuous (S = 1)	3
	Substances involved with moderate hazard to human health and the environment (S = 2, for at least one substance, no substances with S = 3)	2
	Substances involved with high hazard to human health and the environment (S = 3, for at least one substance)	1
P5: Safer Solvents and Auxiliary Substances[a]	Solvents and other auxiliary substances are not used, but if used are innocuous (S = 1)	3
	Solvents or other auxiliary substances are used with moderate hazard to human health and the environment (S = 2, for at least one substance, no substances with S = 3)	2
	Solvents or other auxiliary substances are used with high hazard to human health and environment (S = 3, for at least one substance)	1
P6: Increase Energy Efficiency	Room temperature and pressure	3
	Room pressure and temperature between 0 and 100 °C when cooling or heating is needed	2
	Pressure different from room pressure and/or temperature >100 °C or less than 0 °C	3
P7: Use Renewable Feedstocks[b]	All raw materials/feedstocks are renewable (S = 1)	3
	At least one raw material/feedstock is renewable, water is not considered (S = 1)	2
	None of the raw materials/feedstocks are renewable, water is not considered (S = 3)	1
P8: Reduce Derivatives	Derivatizations or similar operations are not used	3
	Only one derivatization or similar operation is used	2
	More than one derivatization or similar operation is used	1

Table 3.5 (*Continued*)

Green Chem. Principle	Criteria	Score
P9: Catalysts[a]	Catalysts are not used and if used are innocuous (S = 1)	3
	Catalysts are used with moderate hazard to human health and the environment (S = 2)	2
	Catalysts are used with high hazard to human health and the environment (S = 3)	1
P10: Design for Degradation[b]	All substances involved are degradable and break down to innocuous products (S = 1)	3
	All substances involved not degradable may be treated to render them degradable to innocuous products (S = 2)	2
	At least one substance is not degradable or it may not be treated to render degradable to innocuous products (S = 3)	1
P12: Safer Chemistry for Accident Prevention[a]	Substances used with low hazard to cause chemical accidents (S = 1, considering health and physical hazards)	3
	Substances used with moderate hazard to cause chemical accidents (S = 2, for at least one substance considering health and physical hazards, no substances with S = 3)	2
	Substances used with high hazard to cause chemical accidents (S = 3, for at least one substance considering health and physical hazards)	1

[a]According to Table 3.6.
[b]According to Table 3.7.

100 (maximum greenness, Figure 3.2b) and 0 (minimum greenness, Figure 3.2a). The improvement in the sustainability is also observable in the GSAI, which increased from 43 (Figure 3.2c) to 53 (Figure 3.2d).

Nowadays, apart from waste determination, green metrics focus on resource and mass efficiency, energy and waste, relying on the environmental, health, and safety profiles of the materials used and of the chemical process, overall LCA considerations, use of renewable feedstock, occupation hazard-risk and inherent safety using real analysis.

The LCA approach evaluates the "greenness" of a product or a process not only by the amount of waste generated and harmful materials released during the manufacturing process, but also considering the consumption of energy, the depletion of the raw materials that are used and the fate of chemical products in the environment.[29] Life Cycle Analysis (LCA) is a tool for the systematic evaluation of the environmental aspects of a product or service system through all stages of its life cycle. LCA provides an adequate instrument for environmental decision support, with reliable LCA performance being crucial to achieving life cycle economy. The International Organization

Table 3.6 Scores to classify the nature of substances, using Globally Harmonized System regulation.

	Hazards	Hazard codes	Score
Physical	Statements H	200, 201, 202, 203, 205, 220, 222, 224, 225, 228 (category 1), 230, 270, 271, 272, (category 2), 240, 241, 242 (type C and D), 250, 251, 260, 261 (category 2)	3
		204, 221, 223, 226, 227, 228, (category 2), 229, 231, 272 (category 3), 242 (type E and F), 252, 261 (category 3), 280, 281, 290	2
		No indication	1
	Statements EUH	001, 006, 014, 018, 019, 044, 209	3
		209A	2
		No indication	1
Health	Statements H	300, 301, 304, 310, 311, 314, 318, 330, 331, 334, 340, 341, 350, 351, 360, 361, 370, 371, 372, 373	3
		302, 305, 312, 315, 317, 319, 332, 335, 336, 362	2
		No indication	1
	Statements EUH	029, 031, 032, 070, 071, 201, 202, 206, 207	3
		066, 201A, 203, 204, 205, 208	2
		No indication	1
Environmental	Statements H	400, 401, 410, 411, 420	3
		402, 412, 413	2
		No indication	1
	Statements EUH	059	3
		No indication	

Table 3.7 Criteria to classify substances regarding degradability and renewability.

Characteristics	Criteria	Score
Degradability	Not degradable and may not be treated to render the substances degradable to innocuous products	3
	Not degradable but may be treated to render the substances degradable to innocuous products	2
	Degradable and breakable to innocuous products	1
Renewability	Not renewable	3
	Renewable	1

for Standardization (ISO), a world-wide federation of national standards bodies, has standardized this framework within the series ISO 14040 on LCA.[30] An effective LCA allows analysts to calculate the environmental impact of a product, identify the positive or negative environmental impact of a process or product, find opportunities for process and product improvement, compare and analyse several processes based on their environmental impacts, and quantitatively justify a change in a process or product.

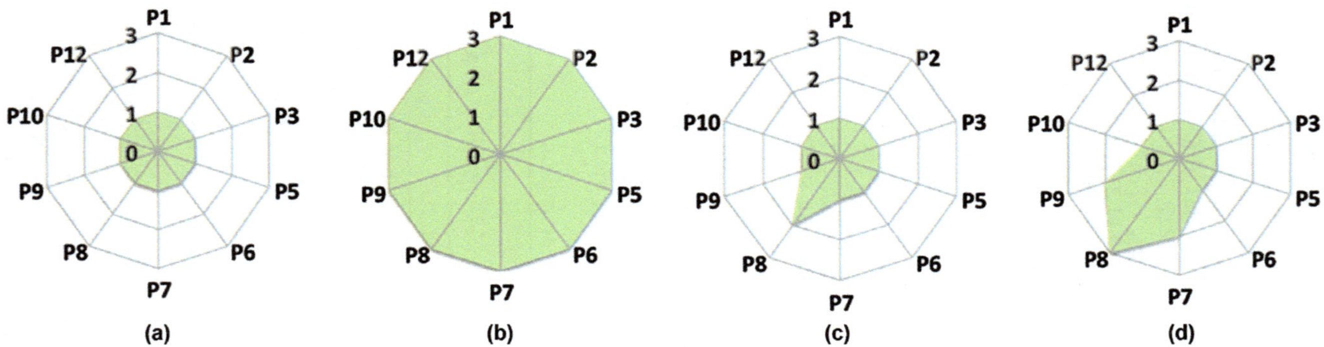

Figure 3.2 Green Star metrics: (a) minimum greenness; (b) maximum greenness; (c) Green Star for the classical Sertraline synthesis; (d) Green Star for the Pfizer Sertraline Synthesis, applied according to Table 3.5.[26]

The LCA method provides researchers or companies with quantitative data for their current products. By looking at a product's life, from raw material extraction to disposal, the environmental impact of each process and material can be analysed. The LCA allows analysts to determine and analyse the technological, economic, environmental, and social aspects of a product or process necessary to manage the complete life cycle. With this quantitative data, desired changes can be justified with respect to the cost and environmental impacts of a product or process. As a very important metric for pharmaceutical industrial processes, LCA is not easy to use because, among other complications, obtaining the necessary data for an analysis can be a difficult task as it is important not only to consider the data source, but also its validity. Data availability also varies by region, country, and continent, and it is mandatory to clearly define the inputs and outputs of a process or product. The difficulty of getting all this data undermines its usefulness for assessing the sustainability of chemical processes in the laboratory or in the classroom. An example of a flow chart to complete when initiating a LCA is presented[31] in Figure 3.3.

Microwave, ultrasound and mechanochemistry are tools with the potential to improve the sustainability and efficiency of synthetic processes, especially when combined with catalysis, alternative reaction media and solvent-free conditions. Herein we present a brief introduction of each technique and some examples of their use in the synthesis of pharmaceuticals, with examples of application of the metrics previously described, so as to assess the degree of improvement.

3.3 Microwave

Microwaves are electromagnetic waves in the frequency range of 0.3 to 300 GHz, which corresponds to wavelengths of ~1 m to 1 cm, respectively. Among the several frequency bands that are available, a frequency of 2.45 GHz is commonly used for domestic and industrial microwave reactors (Figure 3.4).

Microwaves, being of an electromagnetic nature, consist of time-varying electric and magnetic fields, and propagate through space at the speed of light. However, the magnetic part of the electromagnetic waves does not interact with organic media and, thus, will not participate in microwave heating for most chemical transformations. Radiation of this frequency is not strong enough to break chemical bonds, affecting only molecular rotation. The capability of a compound to convert microwave irradiation to heat is given by the interaction with the electric field and is therefore related to the polarity of the molecules and, more specifically, to the loss tangent, $\tan(\delta)$.

$$\tan(\delta) = \varepsilon_1(\omega)/\varepsilon_2(\omega)$$

in which ε_1 is the permittivity, ε_2 is the dielectric loss factor, and ω is the frequency. The higher $\tan(\delta)$ is, the better the compound will absorb

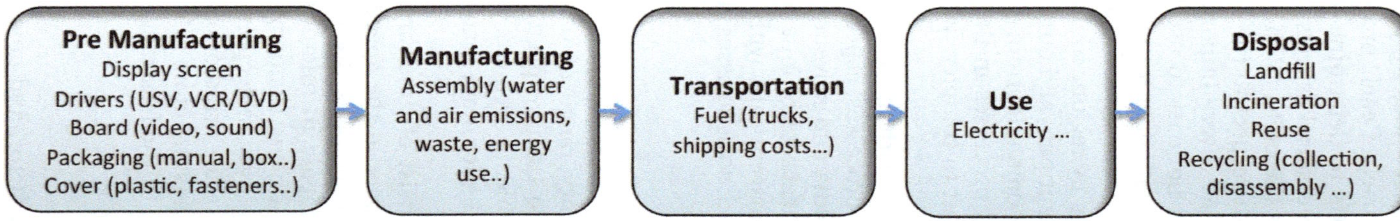

Figure 3.3 Flow chart for LCA calculation.

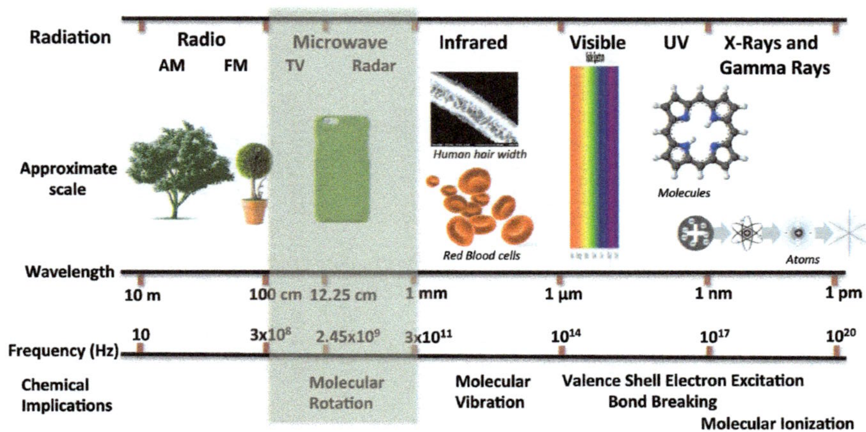

Figure 3.4 Comparison of the size of real objects with electromagnetic wavelengths.

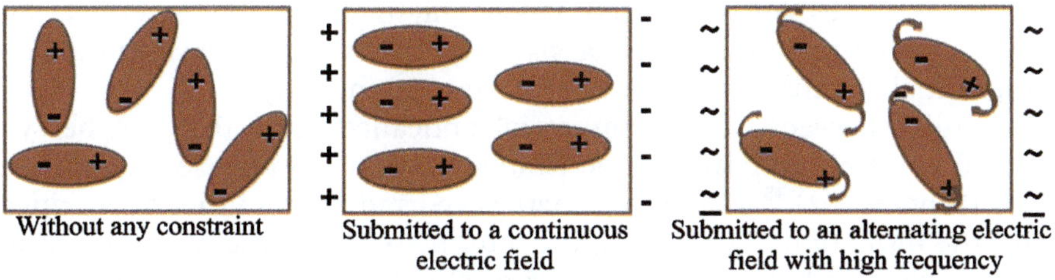

Figure 3.5 Effects of the surrounding electric field on the mutual orientation of dipoles.

microwaves, resulting in more effective heating. Consequently, polar substances are expected to heat up more efficiently than non-polar or less polar counterparts. Thus, microwave heating can result from dipolar polarization as a consequence of dipole–dipole interactions between polar molecules and the electric field. The dissipation of energy into heat is an outcome of agitation and intermolecular friction of molecules when dipoles change their mutual orientation at each alternation of the electric field and at very high frequencies (Figure 3.5).

This energy dissipation in the core of the material allows a much more regular temperature distribution, when compared to classical heating. Classical thermal phenomena (conduction, convection, radiation, *etc.*) only play a secondary role in the *a posteriori* equilibration of temperature. For liquid compounds (solvents), only polar molecules selectively absorb microwaves, with non-polar molecules being inert to the microwave dielectric loss. When heating fairly magnetic solid particles, the charge space polarization can be of prime importance since semiconductors contain free conduction electrons. Ionic species can be heated under microwave irradiation by an ionic heating mechanism. When a solution containing ions is placed in an electric field, this causes an ionic current, which gives rise to solution heating, proportionally to the ionic conductivity of the material. This latter property depends on ion concentration and is, in general, also dependent on temperature and frequency.[32,33]

The use of microwave irradiation often leads to an increase in reaction yield, considerably reducing the formation of secondary products, and therefore increasing reactivity, selectivity and productivity. Microwave irradiation also decreases the reaction time, thus promoting energy and time savings. Additionally, the efficiency of microwave heating through molecular interactions, particularly of those molecules with large dipole moments, allows the reduction or elimination of reaction solvents and the use of alternative and greener reaction media such as ionic liquids or water. Consequently, microwave chemistry fits many of the paradigms of green chemistry and contributes to the swift optimization of reaction conditions, allowing higher sustainability and feasibility for small-scale synthesis, having an important role in screening and target discovery processes, as well as in the lead optimization phases of drug development.[34]

Some examples are presented below, highlighting the potential and unique capabilities of microwave-assisted synthesis and catalysis in the field of medicinal chemistry and the pharmaceutical industry.

Imatinib (Gleevec®) is a commercial anticancer drug that can be synthesized by combining microwave irradiation and solid-phase organic synthesis (Scheme 3.2).[35] The use of microwave irradiation in the more critical steps, the formation of the guanidine intermediate and the final cyclization reaction, allowed the synthesis of the final compound in higher yield and purity than when using conventional heating methods and a considerable reaction time reduction. The guanidinylation of aniline was performed in just 10 min, a very short reaction time when compared to the 15 h needed under conventional heating, while the final cyclization step was performed in 50 min, also favourably compared to 20 h under conventional heating.[35]

The decrease in reaction time that can be achieved under microwave irradiation also plays an important role in the synthesis of radiolabelled derivatives.[36] For instance, when performing carbon-11 radiolabelling, for each 10 min taken by the procedure, the amount of radioactive material and the specific activity of the product will decrease by 29% simply due to the decay of the radionuclide. The most commonly used method for introducing a carbon-11 label into an organic molecule is by nucleophilic reaction of an alcohol, amine, amide or thiol with a labelled alkyl halide, such as $^{11}CH_3I$. Large reduction in reaction times have been achieved by using mono- and multi-modal microwave techniques, such as in the alkylation of an amide to give the benzodiazepine receptor antagonist, [*N*-methyl-^{11}C]flumazenil or [isopropyl-^{11}C]nimodipine (Scheme 3.3).[37]

Aripiprazole (Abilify®) is an anti-psychotic drug that was successfully synthetized under microwave irradiation. 7-Hydroxy-2(1*H*)-quinolinone and 1,4-dibromobutane react under microwave irradiation to yield 78.5% of 7-(4-bromobutoxy)-2(1*H*)-quinolinone in 1 min, which reacts with 1-(3-chlorophenyl)-piperazine hydrochloride (also successfully obtained under microwave irradiation with 94.5% yield) to yield in 2 minutes 95.2% of aripiprazole. Comparing with the conventional synthesis performed by the same author, the use of microwaves allows the reduction of the reaction time (from 35 h to 5 min), the increase of the overall reaction yield (from 54% to 75%) and a 35% reduction of the solvent quantity in the final step (Scheme 3.4).

The complete E factor (cEF) of 2.2 of the last step, under microwave irradiation, compares favourably with that of the conventional methodology (cEF of 3.6), only reflecting the increase in yield (from 84 to 95.2%) and the reduction in the quantity of solvent (from 75 to 50 mL).

When bringing together microwaves and catalysis, several synthetic routes to biologically active compounds have also been unequivocally improved. Botta and co-workers synthesized a series of dihydro-alkylthio-benzyl-oxopyrimidines *S*-aryl-*S*-DABO derivatives, as HIV-1 inhibitors, through C–S coupling between thiols and aryl boronic acids, promoted by Cu(II) in an Ullmann type reaction.[38] The use of microwave heating allowed very short reaction times; the reactions were also successful at room temperature, albeit with longer times and in lower yields (Scheme 3.5).

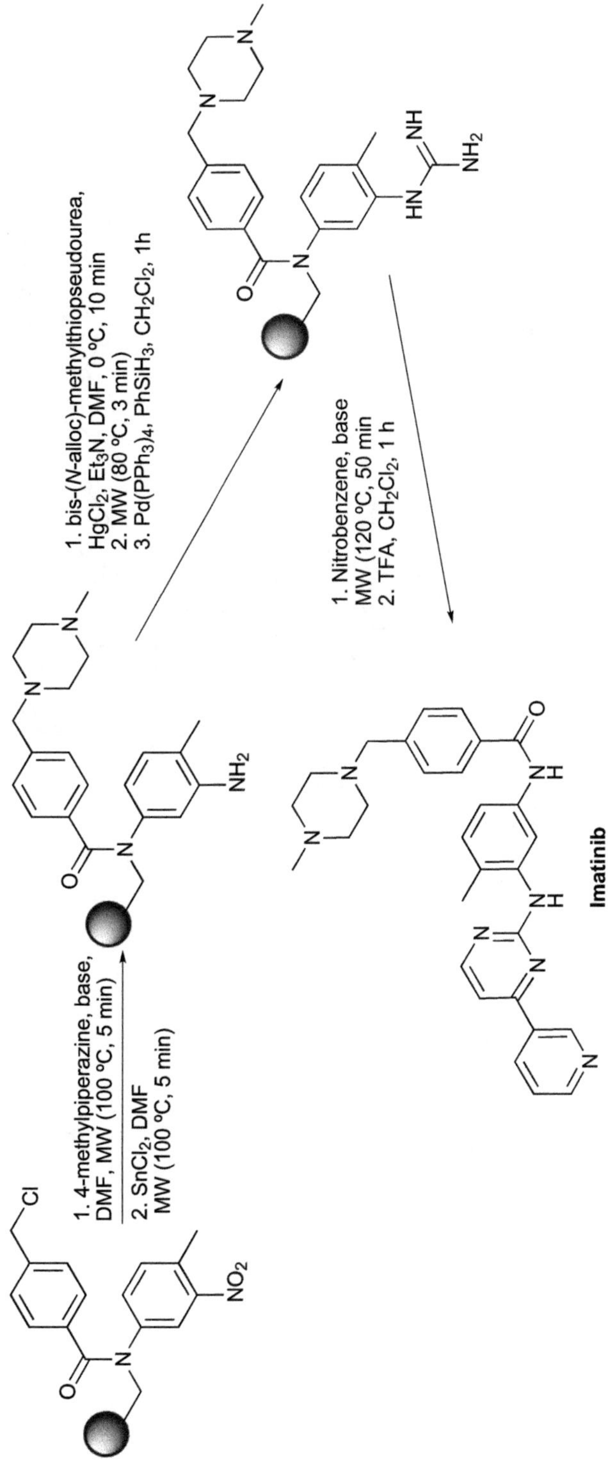

Scheme 3.2 Final steps of the synthesis of imatinib using microwave irradiation.

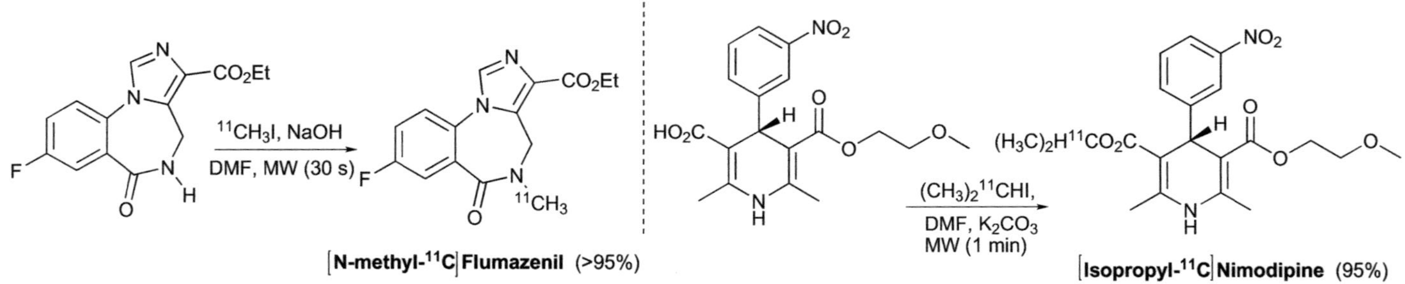

Scheme 3.3 Synthesis of radiolabelled compounds under microwave irradiation.

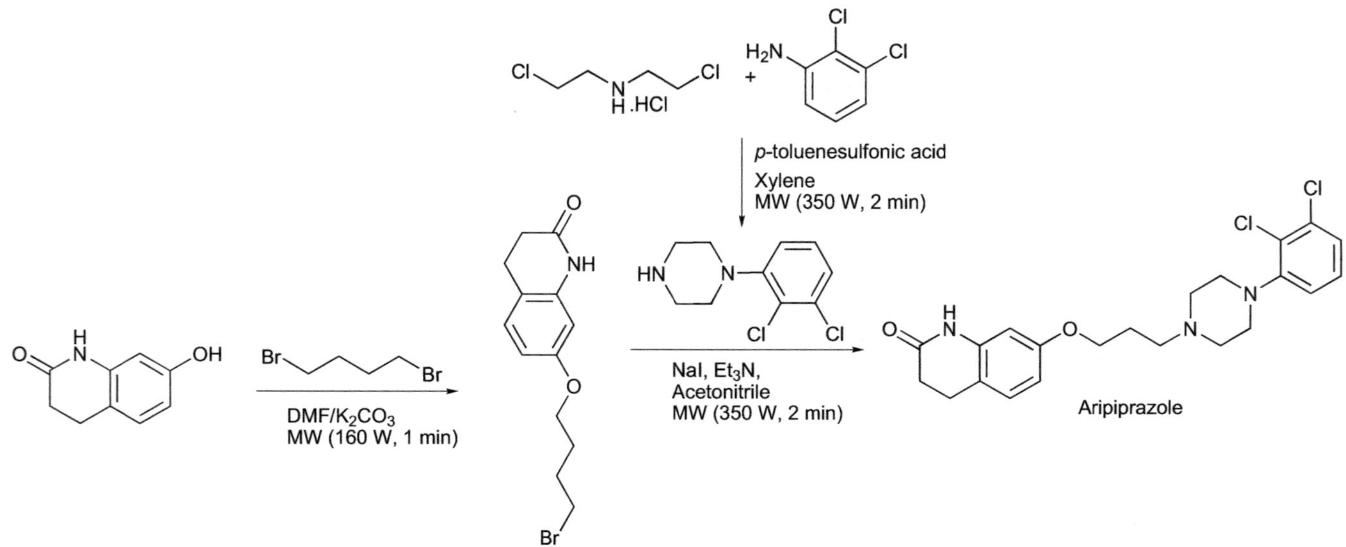

Scheme 3.4 Microwave-assisted synthesis of aripiprazole.[67]

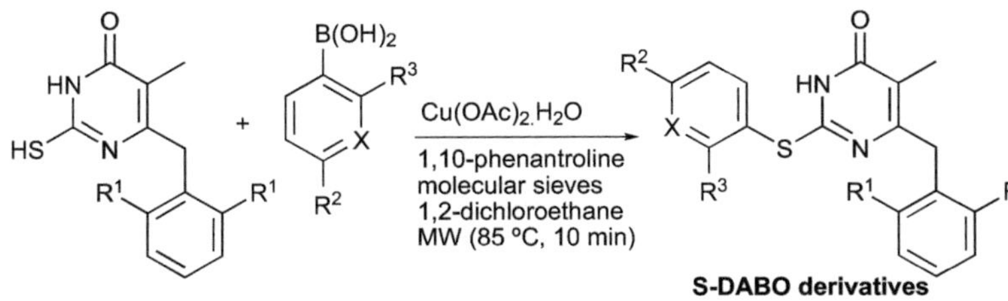

Scheme 3.5 Ullman-type reaction under microwave irradiation for the synthesis of *S*-aryl-*S*-DABO derivatives.[38]

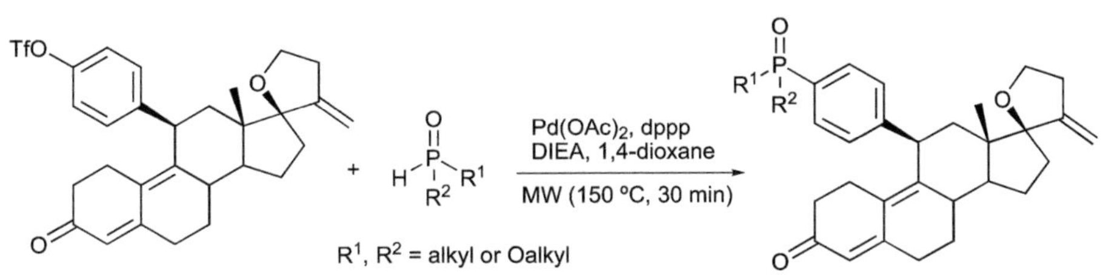

Scheme 3.6 Synthesis of steroidal progesterone receptor antagonists through Pd-catalysed C–P coupling under microwave irradiation.[39]

At Johnson & Johnson, a number of phosphorus-containing 11β-aryl-substituted steroids were synthesized as new progesterone receptor antagonists through Pd-catalysed C–P coupling under microwave irradiation.[39] The dialkylphosphites or dialkylphosphine oxides were coupled with an aryl triflate using 1,3-bis(diphneylphosphino)propane (dppp) as the ligand and microwave heating (Scheme 3.6). The authors found that, under microwave irradiation conditions, the desired coupling product was constantly provided with higher yield and purity than when prepared under conventional heating (100 °C for 16 h).[39]

Convolutamydine-A is an oxindole alkaloid, isolated in low yields from *Amathia convolute,* that reduces the differentiation of HL-60 human promyelocytic leukemia cells. Convolutamydine-A could be synthetized from the aldol reaction of 4,6-dibromoisatin with acetone. 4,6-Dibromoisatin is not commercially available but it could be obtained through a five-step sequence starting from *p*-nitroaniline (Scheme 3.7). The use of microwave irradiation in two steps, the synthesis of isonitrosoacetanilide and the cyclization to isatin,[40] allows the reduction of the reaction time from hours to minutes and also avoids the formation of resinous materials that often reduce the yield of the conventional procedure.[41] The cEF for the conventional procedure of the isonitrosoacetanilide synthesis step is 23.1, decreasing to 14.4 when performed under microwave irradiation, thus underlining the increase in sustainability for the microwave-assisted reaction.

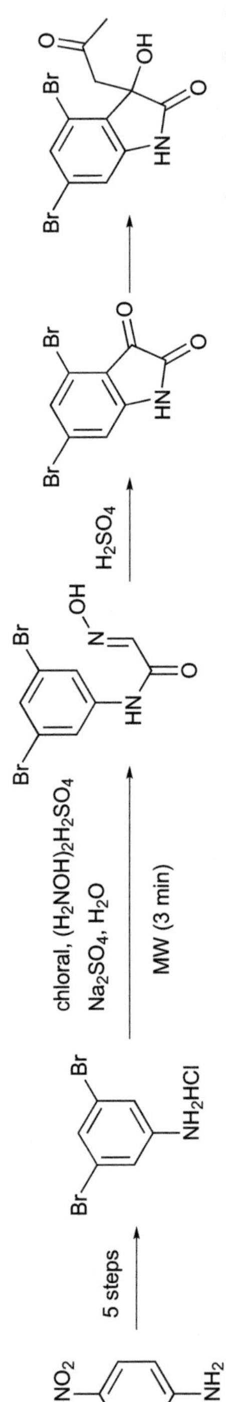

Convolutamydine A

Scheme 3.7 Microwave-assisted synthesis of convolutamydine-A.[40]

3.4 Sonochemistry

Sonochemistry is the set of techniques in chemistry that deals with the chemical and mechanical effects of ultrasound. Sound is a pressure wave transmitted through a gas, liquid or solid in compression and expansion cycles. Ultrasound comprises frequencies beyond human hearing, *i.e.* above 18 kHz, and is usually divided into two regions: conventional ultrasound, up to 100 kHz, where most of the devices operate, and diagnostic ultrasound, around 1–10 MHz. The propagation of a pressure wave, of adequate energy, through a liquid occurs when negative pressure exceeds the tensile strength of the liquid, by formation of vapour bubbles. This physical process creates, enlarges and implodes or collapses gaseous and vaporous bubble "cavities" in an irradiated liquid, and is thus called "cavitation". Cavitation induces very high local temperatures and pressures inside the cavities, leading to turbulent liquid flows, enhanced mass transfer and kinetic energy release, creating conditions to drive chemical reactions. Experimental results have shown that the cavities could attain temperatures of around 5000 K, pressures of 1000 atm and heating and cooling rates above 10^{10} K s^{-1}, meaning that cavities behave as microreactors where high-energy species, ions, radicals and excited states may be involved in reactions.[42,43] Although the effect of ultrasound is customarily interpreted in terms of cavitation, it is a complex process and often follows a non-linear behaviour that could be affected by several parameters, including solvent and solute characteristics, viscosity, surface tension, vapour pressure, temperature and frequency. Therefore, the interpretation of sonochemical reactions can be based on different models, depending on the reaction conditions. In homogeneous conditions, in the presence of highly volatile molecules, these enter into the cavities, being submitted to high temperatures and pressures, leading to bond scission and formation of chemical species with short life-times. These short-lived species then return to the liquid state at room temperature, reacting with other species. In homogeneous conditions, in the presence of low volatile compounds, it is not expected that these compounds enter the cavities; however, they may be affected by pressure waves generated from cavity collapse. On the other hand, in the case of heterogeneous conditions, the mechanical effects of cavitation have a predominant role and surface phenomena such as surface cleaning, erosion and also size reduction, dispersion and emulsification clearly influence the reaction outcome.[42,43]

In general, radical reactions can be enhanced by ultrasound irradiation in situations where radical and ionic mechanisms coexist, switching the conventional pathway and leading to a change in the nature or ratio of the reaction products. Together with reaction rate acceleration (decreasing energy input) and improved reaction yields, these properties make sonochemistry a tool to be considered for improving the sustainability of a chemical process.

Some selected examples among the relevant applications of sonochemistry in the synthesis of pharmaceuticals are presented below.

The pyrazole ring is present in numerous biologically active compounds, including prominent drug molecules such as Viagra® or Celebrex®. Pyrazolo[3,4-b]quinolin-5-ones, pyrazolo[5,1-b]quinazolin-8-ones, and pyrazolo[4,3-c]quinolizin-9-ones could be selectively obtained by three-component condensation of 5-aminopyrazoles, aromatic aldehydes, and cyclic 1,3-diketones in moderate to good isolated yields by specific variation of both the catalyst type and the reaction temperature, employing either rapid microwave flash heating or sonication as alternative energy sources (Scheme 3.8).[44]

The Diels–Alder $(4+2)$ π-electron cycloaddition reaction between conjugated dienes and reactive alkenes (dienophiles) is one of the most important reactions in synthetic organic chemistry. For instance, it has been used in the synthesis of lonapalene, a non-steroidal agent for the treatment of psoriasis.[45] The use of ultrasonic conditions led to a yield enhancement and a total reaction time reduction of the Diels–Alder reaction and also of the hydrolysis to obtain hydroquinone derivatives by an average of 90% (Scheme 3.9).[45,46] This methodology displays an EcoScale value of 79, a much higher value when compared to the classical methodology reported in 1987[45] for which we have calculated an EcoScale value of 33. The increase in sustainability is also indicated by the cEF value, which decreased from 35.4 in the classical conditions to 12.5 using sonochemistry.

Sonochemistry was also applied to the hydrogenation of cinnamaldehyde using hydrogen gas, Pd-black and Raney-Ni as catalysts at room temperature.[47] The typical conversion of an unsaturated aldehyde into an intermediate saturated aldehyde and final product benzenepropanol was observed. Ultrasound enhanced intermediate formation by 28% for Pd-black and 14% for RANEY-Ni when compared to conventionally stirred experiments. In addition, ultrasound increased activity by nine-fold for Pd-black and 20-fold for RANEY-Ni (Scheme 3.10).[47]

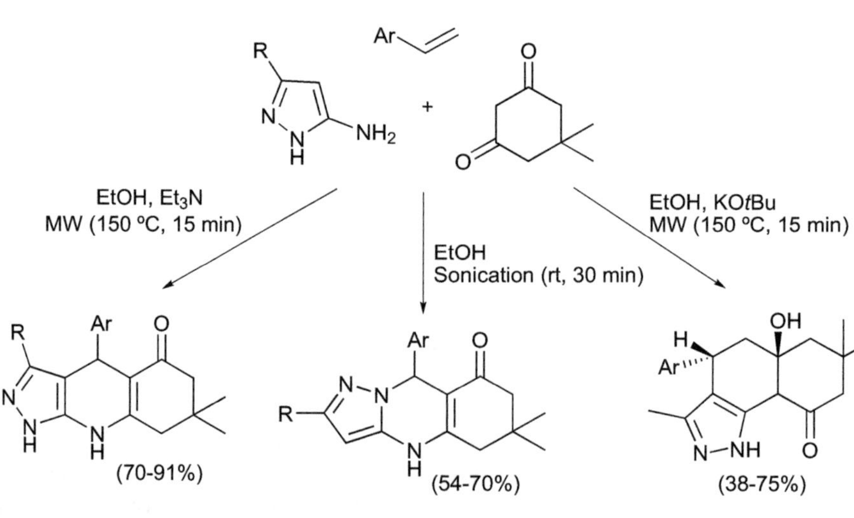

Scheme 3.8 Selective synthesis of pyrazolo[3,4-b]quinolin-5-ones, pyrazolo[5,1-b]-quinazolin-8-ones, and pyrazolo[4,3-c]quinolizin-9-ones under microwave or ultrasound irradiation.[44]

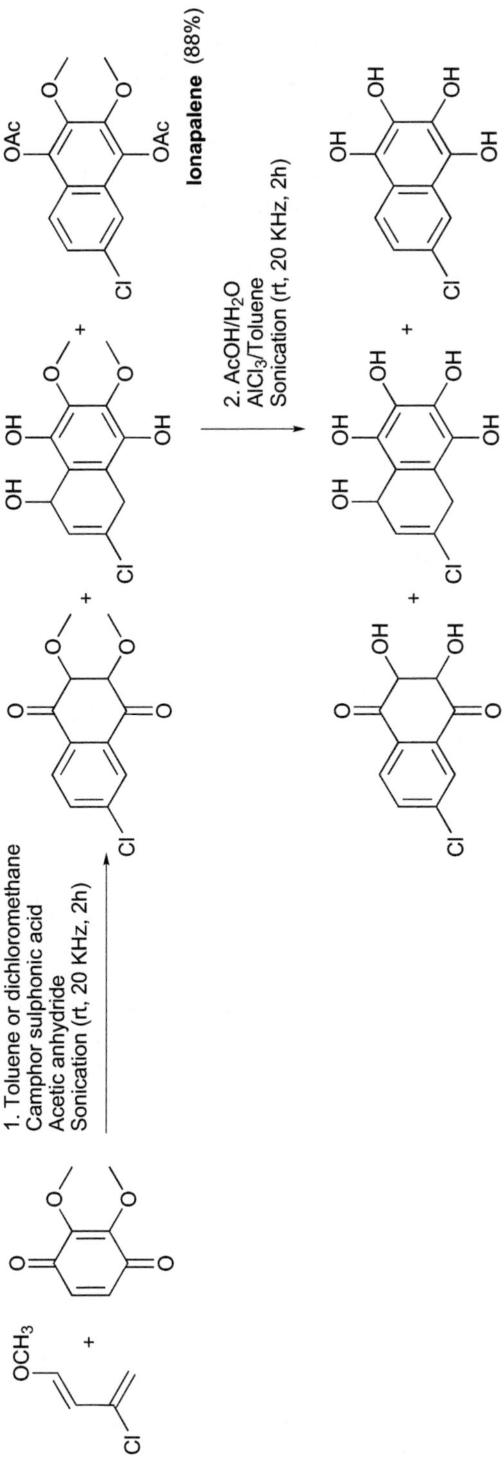

Scheme 3.9 Sonochemistry applied to the synthesis of lonapalene.[45,46]

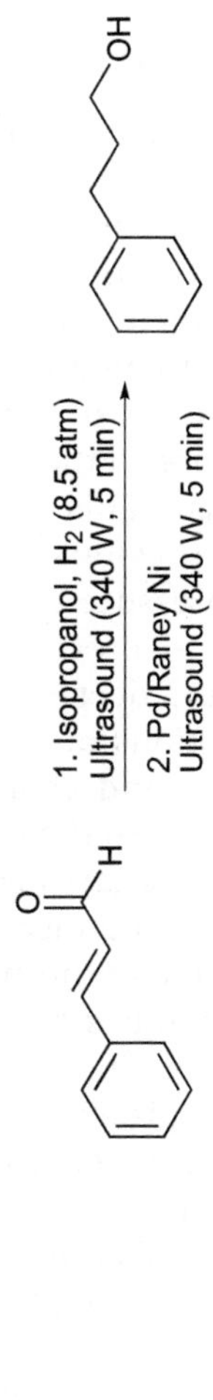

Scheme 3.10 Hydrogenation of cinnamaldehyde under ultrasound.[47]

Ultrasound-assisted three-component Mannich reaction catalysed by $Ga(OTf)_3$, a water-tolerant strong Lewis acid catalyst, of cycloketones with aromatic aldehydes and aromatic amines, in water, produced the corresponding β-amino cycloketones in good to excellent yields and good *anti* selectivities (Scheme 3.11).[48] This reaction has shown only trace or moderated yields using organic solvents such as THF, Et_2O or dichloromethane and moderated yields and *anti* selectivity using water and acetonitrile under conventional conditions with reaction times from 2 to 24 h. Therefore, the use of ultrasound decreases the reaction time and increases the yield and *anti* selectivity.

3.5 Mechanochemistry

Mechanochemistry comprises the chemical and physicochemical transformations of substances in all states of aggregation, produced by the effect of mechanical energy.[49] Although solvent-free reactions may efficiently evolve in homogenous conditions, *i.e.* solid–gas, solid–liquid and liquid–liquid mixtures, heterogeneous situations are more challenging to develop efficient chemical reactions. To carry out reactions in solid–solid state in the absence of solvents, it is necessary to provide the required energy to the reactants so that the molecules can suffer effective collisions and form the products. This can be achieved using some non-classical techniques, such as the previously discussed microwave and sonochemistry, but also under high-pressure conditions produced by mechanical action.

Several processes take place during the mechanical grinding of solids, such as a decrease of particle size, generation of new surfaces, formation of dislocations and point defects in the crystalline structure, phase transformations and chemical reactions. The occurrence of chemical reactions is attributed to the heat generated in the milling process, favoured by the large area of contact between the solids and the phase transformation. However, mechanochemical processes produce different reaction outcomes when compared to thermal processes, pointing out the importance of other effects. Many models have been developed to explain the influence of mechanical activation on chemical reactions. These include the hot-spot theory,[50,51] the magma-plasma model,[52] the hierarchic model,[53,54] or the theory of short-lived active centres.[55] The mechanical energy initially deforms and even melts the solids, forming hot spots where the molecules can reach very high vibrational excitation energies, leading to bond breaking. This period, called the plasma phase, is very short (10^{-7} s) and is followed by a slower post-plasma period ($>10^{-6}$ s), where relaxation processes dissipate the energy, reaching the equilibrium Maxwell–Boltzmann distribution and being responsible for many of the products formed. Finally, the energy accumulated in the defects of the crystalline structure can lead to slower chemical processes.[56–58]

Mechanochemistry is one of the newest techniques in organic synthesis and holds great promise in the development of faster and cleaner

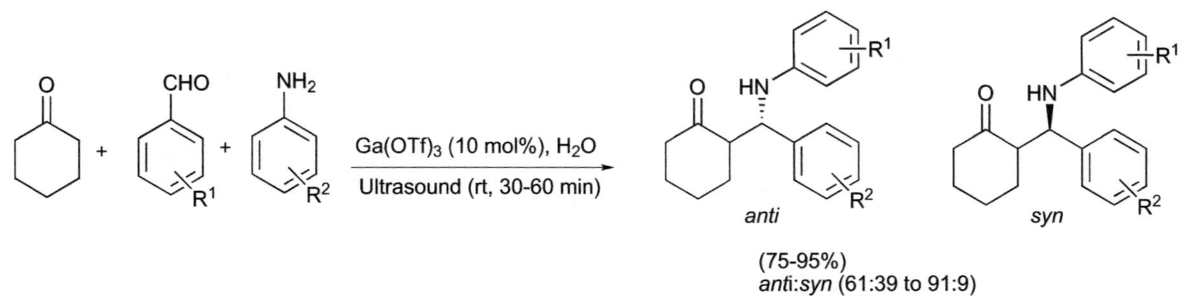

Scheme 3.11 Ultrasound-assisted three-component Mannich reaction.[48]

solvent-free synthetic methodologies of pharmaceuticals.[59,60] Below we provide some examples of the synthesis of pharmaceuticals through mechanical action.

The antiepileptic drug phenytoin was obtained through a one-pot two-step mechanical-assisted synthesis in 84% yield.[61] The hydrochloride salt of the initial amino acid was activated using trimethylsilylisocyanate and water under mechanical action for 8 h, followed by intramolecular cyclization catalysed by Cs_2CO_3 and mechanical activation for 3 h (Scheme 3.12).[61] The EcoScale and cEF values for this procedure, considering only the synthetic procedure and not the purification of the product, are 81 and 10 respectively, while for the classical synthesis of phenytoin, starting from benzyl and urea[62] we calculated an EcoScale value of 42 and a cEF of 28.3.

Chloropropamide and tolbutamide, first-generation antidiabetic sulfonylurea drugs, were synthetized under liquid-assisted mechanical activation,[63] using acetonitrile as the solvent, through a copper-catalysed reaction between the corresponding benzenesulfonamide and alkylisocyanates in very high yields, in 1 h. The copper catalyst was effectively removed by brief milling of the organic product with an aqueous solution of sodium ethylenediaminetetraacetate, followed by washing with water (Scheme 3.13).[63]

The amide bond of teriflunomide, the active metabolite of leflunomide Arava®, an immunomodulatory drug, was synthesized by ball-milling solvent-free treatment of 5-methyl-4-isoxazolecarboxylic acid with 1,1′-carbonyldiimidazole (CDI) for 20 min, followed by reaction with 4-(trifluoromethyl)aniline hydrochloride for 5 h, and finally 5 more min of ball milling after the addition of water.[64] Under classical stirring conditions the opening of the isoxazole ring with aqueous HCl afforded teriflunomide with 81% yield (Scheme 3.14).

Among other catalysed reactions relevant for the preparation of pharmaceuticals, metal-catalysed mechanochemical transformations such as metathesis of olefin catalysed by Grubbs' catalyst or C–H activation catalysed by rhodium, under mechanical activation, have also been reported.[65,66]

Ruthenium-catalysed olefin cross-metathesis and ring-closing metathesis using commercially available Grubb's catalyst gave high-yielding, rapid, room-temperature metathesis of solid or liquid olefins on a multigram scale, either in solvent-free conditions or using only a catalytic amount of solvent (Scheme 3.15).[65]

Rhodium(III)-catalysed aerobic *ortho*-C–H bond functionalization of acetanilides, under solventless conditions and in the presence of a catalytic amount of $Cu(OAc)_2$, led under mechanical action to a highly selective formation of *ortho*-olefin acetanilides derivatives at room temperature.[66] The authors pointed out that "the absence of an organic solvent, the avoidance of a high reaction temperature, the possibility of minimizing the amount of the metallic mediator, and the simplicity of the protocol result in a powerful and environmentally benign alternative to the common solution based standard protocol" (Scheme 3.16).[66]

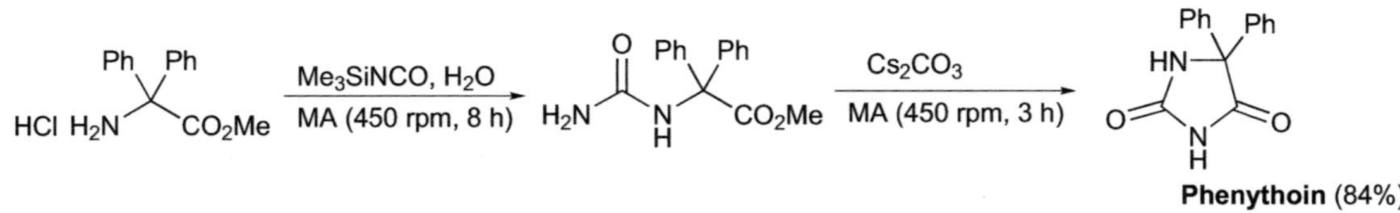

Scheme 3.12 Mechanochemistry applied to the synthesis of phenytoin.[61]

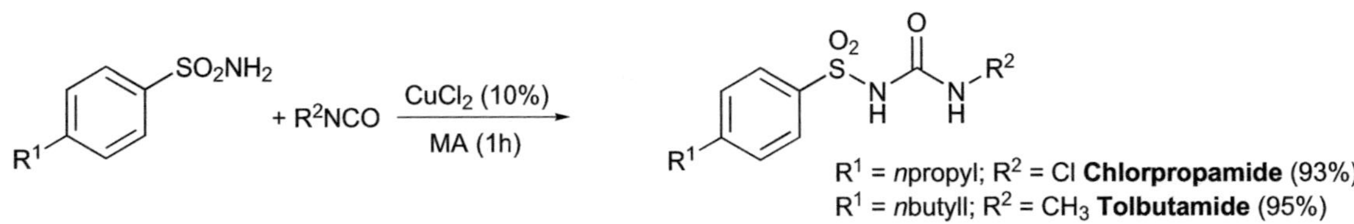

Scheme 3.13 Synthesis of chloropropamide and tolbutamide under mechanical activation.[63]

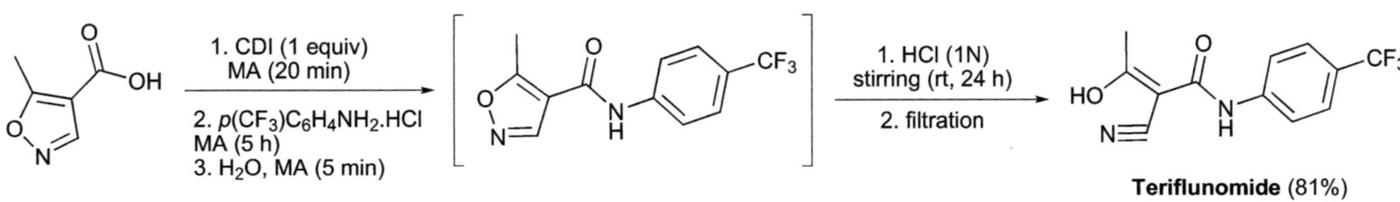

Scheme 3.14 Mechanochemical amide bond formation for the synthesis of teriflunomide.[64]

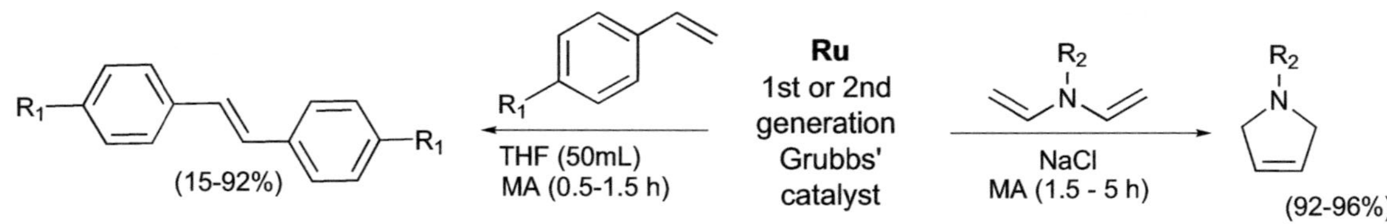

Scheme 3.15 Ruthenium-catalysed metathesis under mechanical action.[65]

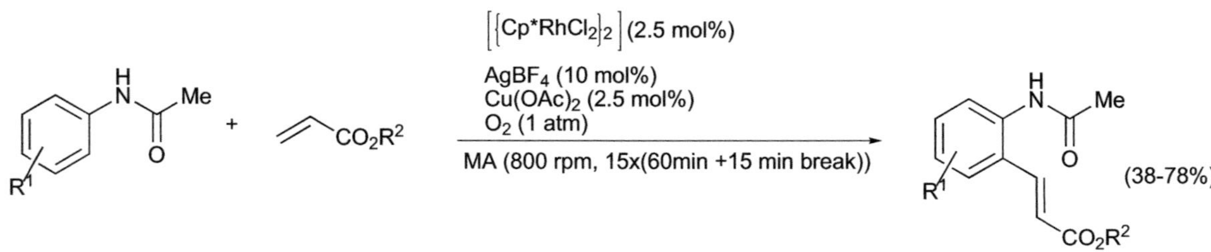

Scheme 3.16 Mechanochemical rhodium-catalysed *ortho* olefination of acetanilides.[66]

3.6 Conclusion

Microwave, ultrasound and mechanochemistry have been proven to be tools with capability to improve the sustainability and efficiency of synthetic processes, especially when combined with catalysis, alternative reaction media and solvent-free conditions. These technologies have been successfully integrated in the drug discovery and development process in the pharmaceutical industry, which is clearly corroborated by the high number of publications regarding the synthesis of drugs using these methodologies and herein illustrated by some interesting examples, in which several green chemistry metrics were used to compare methods and quantify the improvement in sustainability.

Acknowledgements

Mário J. F. Calvete thanks the FCT-Portugal (Portuguese Foundation for Science and Technology) for grant SFRH/BPD/99698/2014. The authors would like to thank Coimbra Chemistry Centre for nurturing chemical science with excellence.

References

1. United Nations, Report of the World Comission on Environmental and Development: Our Common Future, http://www.un-documents.net/wced-ocf.htm, Accessed December, 2016.
2. *Green Chemistry: Theory and Practice*, ed. P. T. Anastas and J. C. Warner, University Press, Oxford, 1998.
3. N. Winterton, *Green Chem.*, 2001, **3**, G73.
4. P. T. Anastas and J. B. Zimmerman, *Environ. Sci. Technol.*, 2003, **37**, 94A.
5. *Green Chemistry in the Pharmaceutical Industry*, ed. R. Sheldon, Wiley-VCH Verlag GmbH&Co. KGaA, Weinheim, 2010, p. 1.
6. US Environmental Protection Agency, Presidential Green Chemistry Challenge Award Winners, https://www.epa.gov/greenchemistry/information-about-presidential-green-chemistry-challenge.
7. American Chemical Society, http://www.acs.org/content/acs/en/greenchemistry/industry-business/pharmaceutical.html, Accessed December, 2016.
8. R. S. Varma, *Green Chem.*, 2008, **10**, 1129.
9. A. Bruckmann, A. Krebs and C. Bolm, *Green Chem.*, 2008, **10**, 1131.
10. R. A. Sheldon, *J. Chem. Technol. Biotechnol.*, 1997, **68**, 381.
11. P. T. Anastas, M. M. Kirchhoff and T. C. Williamson, *Appl. Catal., A*, 2001, **221**, 3.
12. R. Breslow, *Chemistry Today and Tomorrow: The Central, Useful, and Creative* Science, Jones & Bartlett Publishers, Sudbury, Massachussets, 1997.
13. *Green Chemistry Metrics: Measuring and Monitoring Sustainable Processes*, ed. A. Lapkin and D. Constable, Willey-Blackwell Publishing Ltd, United Kingdom, 2009.
14. D. J. C. Constable, A. D. Curzons and V. L. Cunningham, *Green Chem.*, 2002, **4**, 521.

15. R. A. Sheldon, *Chem. Ind.*, 1997, 12.
16. R. A. Sheldon, *Green Chem.*, 2017, **19**, 18–43.
17. K. Van Aken, L. Strekowski and L. Patiny, *Beilstein J. Org. Chem.*, 2006, **2**, 1.
18. G. Koller, U. Fischer and K. Hungerbühler, *Ind. Eng. Chem. Res.*, 2000, **37**, 960.
19. B. M. Trost, *Science*, 1991, **254**, 1471.
20. EPA, The presidential green chemistry challenge, https://www.epa.gov/greenchemistry/presidential-green-chemistry-challenge-winners, Accessed December, 2016.
21. The EcoScale, Fast and Transparent Evaluation of Organic Preparations, http://www.ecoscale.org, Accessed December, 2016.
22. R. Dach, J. J. Song, F. Roschangar, W. Samstag and C. H. Senanayake, *Org. Process Res. Dev.*, 2012, **16**, 1697.
23. M. G. T. C. Ribeiro, S. F. Yunes and A. S. C. Machado, *J. Chem. Educ.*, 2014, **91**, 1901.
24. M. G. T. C. Ribeiro, D. A. Costa and A. S. C. Machado, *Green Chem. Lett. Rev.*, 2010, **3**, 149.
25. M. G. T. C. Ribeiro and A. S. C. Machado, *J. Chem. Educ.*, 2013, **90**, 432.
26. EPA, Presidential Green Chemistry Challenge: 2002 Greener Synthetic Pathways Award, https://www.epa.gov/greenchemistry/presidential-green-chemistry-challenge-2002-greener-synthetic-pathways-award, Accessed December, 2016.
27. W. M. Welch, A. R. Kraska, R. Sarges and B. K. Koe, *J. Med. Chem.*, 1984, **27**, 1508.
28. G. P. Taber, D. M. Pfisterer and J. C. Colberg, *Org. Process Res. Dev.*, 2004, **8**, 385.
29. P. T. Anastas and R. L. Lankey, *Green Chem.*, 2000, **2**, 289.
30. *International Organization of Standardization, Environmental management-Life cycle Assessment-Requirements and Guidelines, ISO 14044*, Geneva, Switzerland, 2006.
31. A. S. Williams, *Life Cycle Analysis: A Step by Step Approach*, Illinois Sustainable Technology Center, Institute of Natural Resouce Sustainability, University of Illinois at Urbana-Champaign, 2009.
32. L. Perreux and A. Loupy, *Tetrahedron*, 2001, **57**, 9199.
33. R. Hoogenboom, T. F. A. Wilms, T. Erdmenger and U. S. Schubert, *Aust. J. Chem.*, 2009, **62**, 236.
34. C. O. Kappe and D. Dallinger, *Nat. Rev. Drug Discovery*, 2006, **5**, 51.
35. F. Leonetti, C. Capaldi and A. Carotti, *Tetrahedron Lett.*, 2007, **48**, 3455.
36. N. Elander, J. R. Jones, S. Y. Lu and S. Stone-Elander, *Chem. Soc. Rev.*, 2000, **29**, 239.
37. S. A. Stoneelander, N. Elander, J. O. Thorell, G. Solas and J. Svennebrink, *J. Labelled Compd. Radiopharm.*, 1994, **34**, 949.
38. C. Mugnaini, F. Manetti, J. A. Este, I. Clotet-Codina, G. Maga, R. Cancio, M. Botta and F. Corelli, *Bioorg. Med. Chem. Lett.*, 2006, **16**, 3541.
39. W. Q. Jiang, G. Allan, J. J. Fiordeliso, O. Linton, P. Tannenbaum, J. Xu, P. F. Zhu, J. Gunnet, K. Demarest, S. Lundeen and Z. Sui, *Bioorg. Med. Chem.*, 2006, **14**, 6726.

40. G. K. Jnaneshwara, A. V. Bedekar and V. H. Deshpande, *Synth. Commun.*, 1999, **29**, 3627.
41. *Handbook on Applications of Ultrasound: Sonochemistry for Sustainability*, ed. D. Chen, S. K. Sharma and A. Mudhoo, CRC Press, Boca Raton, 2012.
42. *Green Techniques for Organic Synthesis and Medicinal Chemistry*, ed. W. Zhang and B. Cue, John Wiley & Sons, Chichester, UK, 2012.
43. V. A. Chebanov, V. E. Saraev, S. M. Desenko, V. N. Chernenko, I. V. Knyazeva, U. Groth, T. N. Glasnov and C. O. Kappe, *J. Org. Chem.*, 2008, **73**, 5110.
44. T. Javed, T. J. Mason, S. S. Phull and N. R. Baker, *Ultrason. Sonochem.*, 1995, **2**, S3.
45. S. T. Perri and H. W. Moore, *Tetrahedron Lett.*, 1987, **28**, 4507.
46. R. S. Disselkamp, T. R. Hart, A. M. Williams, J. F. White and C. H. F. Peden, *Ultrason. Sonochem.*, 2005, **12**, 319.
47. G. L. Zhang, Z. H. Huang and J. P. Zou, *Chin. J. Chem.*, 2009, **27**, 1967.
48. C. Kajdas, in *Tribology in Engineering*, ed. H. Pihtili, InTech, Rijeka, 2013, p. 209.
49. F. P. Bowden and A. D. Yoffe, *Initiation and Growth of Explosion in Liquids and Solids*, Cambridge University Press, Cambridge, UK, 1985.
50. V. V. Boldyrev, *Bull. Acad. Sci. USSR, Div. Chem. Sci.*, 1990, **39**, 2029.
51. P. A. Thiessen, K. Meyer and G. Heinicke, *Grundlagen der Tribochemie*, Akademie Verlag, Berlin, 1967.
52. P. Balaz, *Mechanochemistry in Nanoscience and Minerals Engineering*, Springer Verlag, Berlin, Germany, 2008.
53. H. Heegn, *Sib. Chem. J.*, 1988, 3.
54. A. N. Strelets and P. Y. Butyagin, *Vysokomol. Soedin., Ser. A*, 1973, **15**, 654.
55. J. F. Fernandez-Bertran, *Pure Appl. Chem.*, 1999, **71**, 581.
56. S. L. James and T. Friscic, *Chem. Soc. Rev.*, 2013, **42**, 7494.
57. S. L. James and T. Friscic, *Chem. Commun.*, 2013, **49**, 5349.
58. D. Tan, L. Loots and T. Friscic, *Chem. Commun.*, 2016, **52**, 7760.
59. L. Konnert, B. Reneaud, R. M. de Figueiredo, J. M. Campagne, F. Lamaty, J. Martinez and E. Colacino, *J. Org. Chem.*, 2014, **79**, 10132.
60. A. Ashnagar, N. G. Naseri and M. Amini, *Asian J. Chem.*, 2009, **21**, 4976.
61. D. Tan, V. Strukil, C. Mottillo and T. Friscic, *Chem. Commun.*, 2014, **50**, 5248.
62. T. X. Metro, J. Bonnamour, T. Reidon, J. Sarpoulet, J. Martinez and F. Lamaty, *Chem. Commun.*, 2012, **48**, 11781.
63. J. L. Do, C. Mottillo, D. Tan, V. Strukil and T. Friscic, *J. Am. Chem. Soc.*, 2015, **137**, 2476.
64. G. N. Hermann, P. Becker and C. Bolm, *Angew. Chem., Int. Ed.*, 2015, **54**, 7414.
65. N. R. Pai and D. S. Dubhashi, *J. Liq. Chromatogr. Realt. Technol.*, 2010, **33**, 1359.
66. J.-L. Do, C. Mottillo, D. Tan, V. Štrukil and T. Friščić, *J. Am. Chem. Soc.*, 2015, **137**, 2476.
67. G. N. Hermann, P. Becker and C. Bolm, *Angew. Chem. Int. Ed.*, 2015, **54**, 7414.

CHAPTER 4

Carbonylation Reactions in the Synthesis of Pharmaceutically Active Compounds

R. SKODA-FÖLDES*[a] AND L. KOLLÁR[b]

[a] University of Pannonia, Department of Organic Chemistry, Egyetem. u. 10., Veszprém 8200, Hungary; [b] University of Pécs, Department of Inorganic Chemistry and MTA-PTE Research Group for Selective Chemical Syntheses, Ifjúság u. 6., 7624 Pécs, Hungary
*Email: skodane@almos.uni-pannon.hu

4.1 Introduction

Homogeneous catalytic reactions, ranging from hydrogenations to cross-coupling reactions, have become indispensable tools in synthesis.[1] Although key-compounds of organo–transition metal chemistry such as Zeise's salt (potassium-[trichloro-ethylene-platinate(II)]), the first transition metal–organic compound[2] and homoleptic carbonyl complexes [for instance Mond's tetracarbonyl-nickel(0)[3] and pentacarbonyl-iron(0)[4]] were discovered in the 19th century, their exploitation as synthetic tools is mainly attributed to the past half of the last century. The wide application of transition metal-catalysed homogeneous reactions in organic synthesis resulted in a real breakthrough of the chemistry in the last few decades, as demonstrated by the attribution of several Nobel prizes in this field.

The deeper understanding of the formation of the transition metal–carbon bond, the recognition of their properties, elementary reactions and the mechanistic investigation of the most widely used catalytic reactions

Green Chemistry Series No. 54
Sustainable Synthesis of Pharmaceuticals: Using Transition Metal Complexes as Catalysts
Edited by Mariette M. Pereira and Mário J. F. Calvete

Published by the Royal Society of Chemistry, www.rsc.org

(including rationalisation of the catalytic cycles) have rendered many of the transition metal-catalysed reactions as a solution for practical syntheses. Furthermore, the enhanced chemo-, regio- and enantioselectivities, in some cases close to those of enzymatic reactions, made transition metal-catalysed reactions important players in green chemistry. The applicability of homogeneous catalysts under mild conditions producing low amounts of waste, for instance *via* simplification of conventional multistep reaction sequences, made these reactions attractive in environmentally benign procedures. Most of the homogeneous catalytic reactions tolerate additional functional groups, *i.e.*, the desired functionalisation can be carried out without protection/deprotection reactions decreasing total yields.

Although carbon monoxide should definitely not be assigned as a green reactant due to its facile activation and transformation by various transition metal complexes, it still remains one of the key C1 building blocks, when used with appropriate safety precautions. It has to be noted that the first homogeneous catalytic reaction of practical importance, hydroformylation (oxo-reaction), discovered by Roelen,[5] is also based on the application of carbon monoxide as a reactant. The wide application of carbon monoxide has led to many modern carbonylation methods.[6–9]

Recently, some efforts have been made to introduce improved methodology into carbonylation reactions that meet novel environmental requirements. The use of green solvents, the application of recyclable catalysts instead of disposable conventional ones used under homogeneous conditions or the replacement of toxic CO with safer CO equivalents can be mentioned as examples.

The above facts, *i.e.*, the increasing selectivities and the potential green chemistry issues, have prompted us to summarise the most important recent achievements in the field of carbonylation reactions regarding their use in the sustainable synthesis of compounds of pharmaceutical interest.

The carbon monoxide-based carbonylation reactions, as in many textbooks and treatises, can be classified according to the substrates as shown in Scheme 4.1.

(a) Alkenes and alkynes undergo homogeneous catalytic carbonylations in the presence of a HX reagent. Although no clear nomenclature is used in general, these reactions can be named as hydrocarbonylation reactions: hydroformylation (X = H), hydroalkoxycarbonylation [X = OR (alkoxy)], hydroaryloxycarbonylation [X = OR (aryloxy)], hydroaminocarbonylation (X = NRR′) and hydrocarboxylation (X = OH). In this way, saturated or α,β-unsaturated aldehydes, esters, amides and carboxylic acids can be synthesised [Scheme 4.1, eqn (4.1) and (4.2)].

(b) Aryl or alkenylhalides/sulfonates (typically iodides, bromides or triflates) undergo carbonylations in the presence of various nucleophiles (HX). As close analogues to the above reactions, these carbonylations are alkoxycarbonylation [(X = OR (alkoxy)], aryloxycarbonylation [X = OR (aryloxy)], aminocarbonylation [X = NRR′] and carboxylation

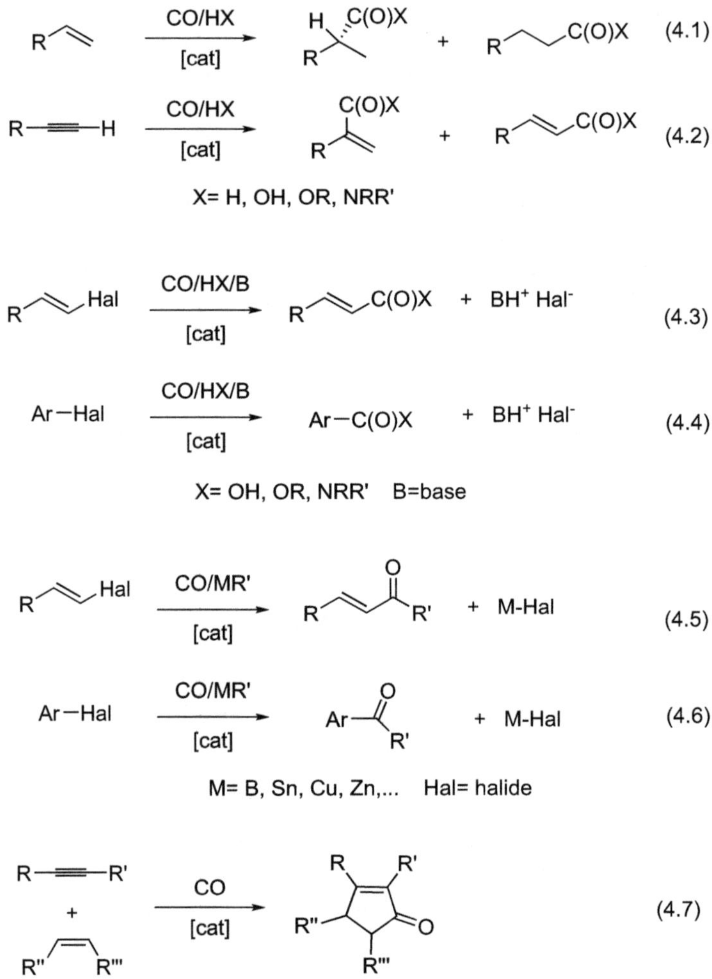

Scheme 4.1 The most important carbon monoxide-based carbonylation reactions.

(X = OH). In this way, aromatic or α,β-unsaturated esters, amides and carboxylic acids can be synthesised [Scheme 4.1, eqn (4.4) and (4.4)].

(c) Most of the transition metal-catalysed carbon–carbon bond forming reactions, involving aryl or alkenyl halides/sulfonates and an organometallic compound, can be accompanied by carbon monoxide insertion. In this way, carbonylative cross-coupling reactions (*e.g.* carbonylative Suzuki, Stille, Sonogashira and Negishi reactions) take place resulting in the formation of ketones [Scheme 4.1, eqn (4.5) and (4.6)].

(d) Carbon monoxide might also be involved in transition metal–catalysed cycloaddition reactions resulting in cyclic ketones. As a typical example, the Pauson–Khand 2 + 2 + 1 (alkyne–alkene–carbon monoxide) cycloaddition reaction could be mentioned [Scheme 4.1, eqn (4.7)].

Owing to the great variety of carbonylations, this chapter is restricted to the practical applications of selected reactions outlined in Scheme 4.1, eqn (4.1)–(4.6). That is, the application of some of the hydrocarbonylation reactions such as the 'parent' hydroformylation in the synthesis of pharmaceuticals will be described in Chapter 5. The main features of the reactions are demonstrated by examples from the past 10 years. Some industrial applications covering a longer interval are presented to show the practical importance of the processes.

When applying homogeneous catalytic reactions in synthesis, the formation of the active catalyst is sometimes overlooked. Often the isolated transition metal complex or salt, mainly commercially available, are considered as the catalyst. In fact, from the point of view of synthetic applications, it is convenient to start with an isolated precursor that is transferred into a kinetically labile, active catalyst. Owing to the complexity of the transition metal catalysts, *i.e.*, the simultaneous presence of the pre-catalyst, forming the real active 'starting' complex of the catalytic cycle, and the set of active species ('catalytic intermediates'), usually the term 'catalytic system' is used. The transformation of a catalytic precursor into pre-catalyst may involve the coordination (association) or dissociation of ligand(s) and the reduction of the transition metal.

4.2 Hydroalkoxycarbonylation of Alkenes

The application of transition metal-catalysed hydroalkoxycarbonylation reactions in the synthesis of compounds with pharmaceutical interest is still mainly restricted to the transformation of vinyl aromatics. Understandably, the palladium-catalysed asymmetric hydroalkoxycarbonylation of 4-isobutylstyrene and 2-methoxy-6-vinylnaphthalene is one of the most elegant solutions of the synthesis of the corresponding 2-arylpropionic acids, ibuprofen (**1**, Scheme 4.2) and naproxen (**2**), respectively, belonging to the widely known non-steroidal anti-inflammatory (NSAI) agents using (*S*)-2-dicyclopentylphosphino-2′-methoxy-1,1′-binaphthyl as chiral ligand L*.[10]

In a recent review, the hydroalkoxycarbonylation of various alkenes, such as norbornene, isobutene and various vinyl aromatics, was reported.[11] The recent advances of Pd-catalysed asymmetric mono- and bis-hydroalkoxycarbonylation of vinylarenes were reviewed by Claver *et al.*[12]

The hydroalkoxycarbonylation of 2-methoxy-6-vinylnaphthalene in the presence of an *in situ* formed catalyst, $PdCl_2$-(*S*)-2-dicyclopentylphosphino-2′-methoxy-1,1′-binaphthyl, was carried out and the corresponding branched methyl ester with pharmaceutical importance was obtained with 53% ee.[13]

The similar reaction with styrene resulted in optical yields of practical importance. The use of the *phanephos* ligand with 3,5-bis(trifluoromethyl)phenyl substituents in a dinuclear palladium complex enabled the synthesis of the desired branched ester in 79% ee with practically complete regioselectivity. For the first time, the simultaneous control of regioselectivity and enantioselectivity in the hydroxycarbonylation and hydroalkoxycarbonylation

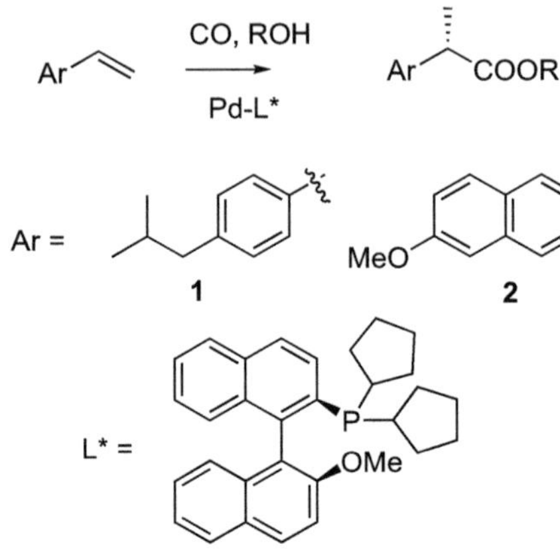

Scheme 4.2 Palladium-catalysed asymmetric hydroalkoxycarbonylation of 4-isobutylstyrene and 2-methoxy-6-vinylnaphthalene.

of styrene derivatives has been realised. These catalysts, characterised by x-ray crystallography, were derived from large cone angle ligands and gave high regioselectivity in the alkoxycarbonylation of styrene using various alcohols as nucleophiles.[14–16]

The use of pre-catalysts of the type PdCl(allyl)(monophosphine) [where monophosphine = tri-oxo-adamantyl cage phosphine (1,3,5,7-tetramethyl-6-phenyl-2,4,8-trioxo-6-phospha-adamantane)] resulted in very high reactivity in the regioselective hydroxycarbonylation and hydroalkoxycarbonylation (hydroesterification) of styrene even in the case of *tert*-butanol as the *O*-nucleophile.[17]

Phenoxy and phenylthio esters were synthesised in the regioselective alkoxycarbonylation of the corresponding allyl phenyl ethers. High regioselectivities towards normal esters were obtained by Pd/*dppb* catalysts under CO/H_2 (syngas) conditions [*dppb* = 1,4-bis(diphenylphosphino)butane].[18]

A diastereoselective synthesis of tetrahydropyran derivatives was achieved using palladium(II)-catalysed intramolecular hydroxycarbonylation of hexenols (Scheme 4.3). The Pd-cyclisation/carbonylation/hydroxylation of 1-phenylhex-5-en-1,3-diol was used as a key step in the total synthesis of (+)-2-[(2*S*,6*S*)-(6-methyltetrahydro-2*H*-pyran-2-yl)] acetic acid (**3**). The cyclisation/carbonylative phenylation (ketonylation) of the similar substrate was carried in a domino reaction leading to diospongin A (**4**).[19]

4.3 Carbonylation of Aryl/Alkenyl Halides

Reaction of aryl/alkenyl halides or halide equivalents (OTf, ONf, *etc.*) with *N*- or *O*-nucleophiles (Heck carbonylation) or organometallic reagents (carbonylative cross-coupling) and carbon monoxide produces carboxylic acid derivatives or carbonyl compounds, respectively.[20,21] One of the most

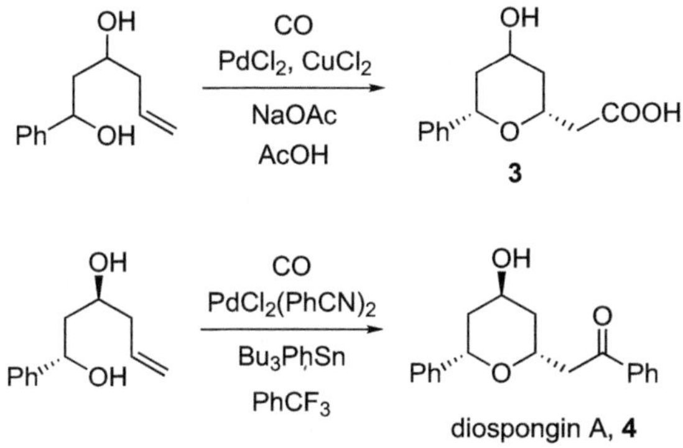

Scheme 4.3 Synthesis of tetrahydropyran derivatives by intramolecular hydroxycarbonylation and ketonylation reactions.

important issues influencing the success of the reaction is the selection of the substrate. Reactivity of halides decreases from iodides to chlorides ($I > Br \gg Cl$), unfortunately in reverse order to their price. Bromides can be converted into the products often only under forced conditions and/or using special catalysts. Catalytic activity and, in carbonylations with *N*- or *O*-nucleophiles, selectivity towards mono- or di-carbonylated products depend on the careful choice of the catalytic system (palladium precursor and ligands) and on the proper selection of reaction conditions, such as solvent, base, reaction temperature and CO pressure.

4.3.1 Aminocarbonylation Reactions

In accordance with the relevance of the carboxamide functionality in biologically active compounds, carbonylation in the presence of amine nucleophiles[22] is the most frequently applied version of carbonylation of unsaturated halides. Aryl, heteroaryl and alkenyl halides can be converted easily to amide products in the presence of simple amines, amino acid esters or ammonia equivalents. Moreover, carbonylation has been proved to be a convenient tool for the synthesis of radiolabelled compounds.

A series of biologically active amides, such as the antidepressant moclobemide (Figure 4.1, **5**), CX-546 (**6**) used for the treatment of schizophrenia, the stimulant agent nikethamide (**7**), and the antiarrhythmic drug procainamide (**8**), were obtained by the carbonylation of aryl bromides and the corresponding amines.[23] Palladium nanoparticles, deposited on a zeolitic imidazole framework (ZIF-8) that showed high activity in the conversion of aryl iodides, were found to be efficient only in the presence of phosphine ligands in carbonylation of the less reactive bromides. Furthermore, a moderate CO pressure (4 bar) had to be used to achieve optimal results. At the same time, moclobemide (**5**) could be produced at atmospheric pressure starting from 4-chloroiodobenzene.[24] The selective formation of the latter

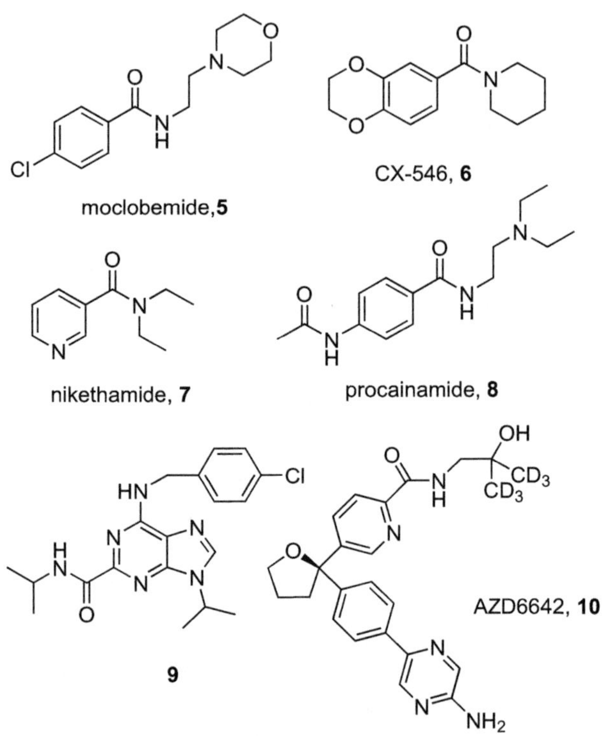

Figure 4.1 Pharmacologically active compounds obtained *via* aminocarbonylation of aryl halides.

compound is due to the inertness of the chloroarene moiety under the reaction conditions mentioned above.

Nikethamide (**7**) analogues were also obtained from 2-iodopyridine under homogeneous conditions in the presence of a catalyst composed of $Pd(OAc)_2$ and PPh_3.[25] While in this reaction selective monocarbonylation was observed even at high CO pressures (up to 40 bar), interestingly, a mixture of carboxamides and 2-ketocarboxamides was obtained in the whole pressure range (1–40 bar) from 3-iodopyridine. In the latter case, the 2-keto-carboxamides were obtained in a considerable amount even at atmospheric pressure while, upon using 40 bar, the reaction was selective towards the formation of double carbonylated products in the presence of aliphatic amines or amino acid esters as nucleophiles. This shows the decisive effect of not only the leaving group, but also the structure of the substrate on the outcome of the reaction.

Other heteroaryl halides were also converted successfully to the desired products. 2-Carboxamidopurines (Figure 4.1, **9**) with moderate inhibitory effect on cyclin-dependent kinases were obtained from 2-iodopurines in the presence of the preformed $Pd(PPh_3)_4$ catalyst, under atmospheric conditions.[26] Aminocarbonylation of a 2-bromopyridine functionality was a key reaction in the seven-step synthesis of a deuterium-labelled derivative of the 5-lipoxygenase activating protein inhibitor AZD6642 (**10**). The carbonylated intermediate was obtained in 80% yield using a moderate pressure (5 bar) and a catalyst with an activating ligand (Pd(*dppf*)Cl_2; *dppf*: 1,1′-bis(diphenylphosphino)ferrocene).[27]

Carboxamide derivatives with skeletons present in biologically active natural and synthetic compounds can be obtained under mild conditions from iodoalkenes. The latter moiety can be introduced in a two-step reaction sequence starting from readily available ketones *via* the transformation of the keto derivative to the corresponding hydrazone, followed by a treatment with iodine in the presence of a base.[28]

The efficiency of palladium-catalysed reactions of steroidal alkenyl iodides[29] can be well demonstrated by aminocarbonylations. The amido group can be introduced easily into different positions of a steroid molecule (Figure 4.2). 17-Carboxamides, analogues of known 5α-reductase inhibitors finasteride or dutasteride,[30] such as **11**, were produced from 17-iodo-16-enes.[31–36] Besides functionalisation of steroids with the natural 13β configuration, 17- and 16-carboxamido derivatives of unnatural 13α-androstanes[37–39] and -estranes[40] could be prepared in good to excellent yields. Among them, the 13α-18-nor-16-carboxamide **12** was shown to inhibit activation of the pain receptor TRPV1.[39] Functionalisation of D-homo-androstanes and -estranes[41] or that of the sterically more hindered 11- (*e.g.* in **13**)[42] and 12-positions (*e.g.* in **14**)[43] of the C-ring could easily be affected. Additionally, the homogeneous carbonylation reactions tolerate various functional groups, such as hydroxy substituents in positions C3[43] and C6,[33] as well as a spiroacetal moiety.[43]

The selective protection of keto groups *via* ethylene ketals during the introduction of the iodoalkene functionality enabled the synthesis of carboxamides with additional keto functionalities,[44] or made the introduction of two different carboxamido functionalities to the different positions (*e.g.* C3 and C17) of the steroidal skeleton possible.[45]

In these reactions, a great variety of *N*-nucleophiles could be used as reaction partners, including primary (*e.g.* in **12**)[39] and secondary amines (*e.g.* in **13**),[42] amino acid esters (*e.g.* in **14**),[43] and amino β-lactams (*e.g.* in **11**),[33] resulting in the formation of carboxamides in good yields even at atmospheric conditions. The application of diamines led to novel androstene-based dimers (such as **15**) linked through the A- and D-ring with 3-[36] and 17-dicarboxamide[34,36] spacers, respectively. Although in the latter reactions more severe conditions (30 bar CO) were needed to achieve high conversions, facile isolation of the target dicarboxamides could be achieved. Even primary amides could be synthesised easily using ammonium carbamate as an ammonia synthon under 1–6 bar CO pressure, depending on the structure of the substrate.[31]

Positron emission tomography (PET) is an imaging technique that allows for non-invasive *in vivo* quantification of receptor expression or enzyme and transporter activity. The method requires the preparation of radiolabelled

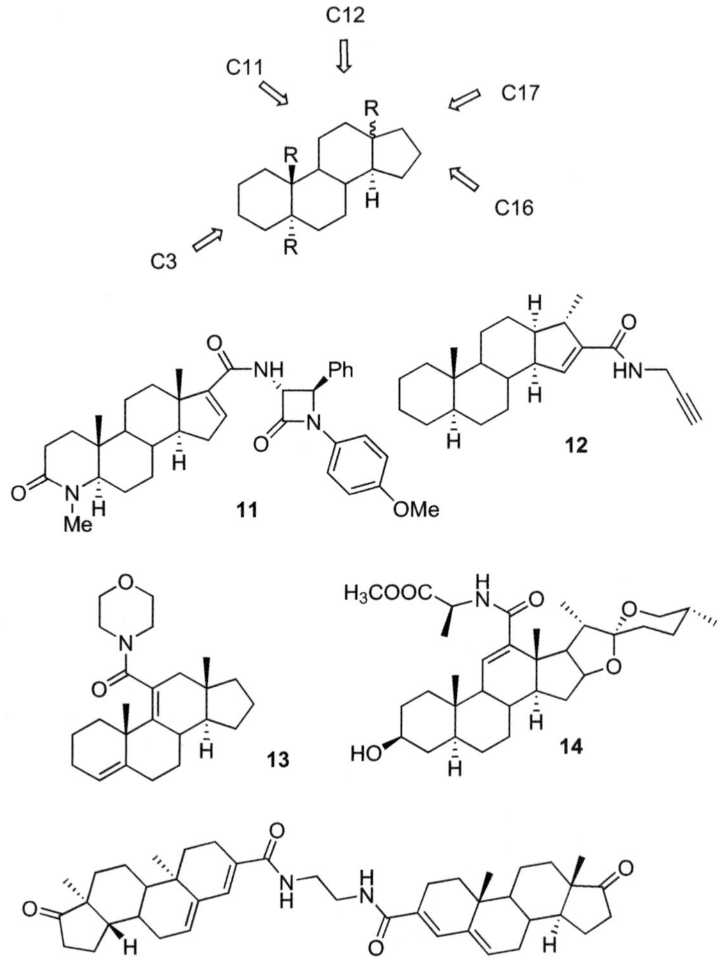

Figure 4.2 Functionalisation of the steroidal core by palladium-catalysed aminocarbonylation.

ligands. Aminocarbonylation using ^{11}CO and a stoichiometric amount of a palladium complex was proved to be an efficient way for the introduction of ^{11}C isotopes into a radioligand (**16**, Figure 4.3) for imaging brain metabotropic glutamate subtype receptor 1 (mGluR1).[46] as well as into irreversible TG2 inhibitors (*e.g.* **17**).[47]

The use of bifunctional substrates makes it possible to construct elaborate heterocyclic structures, present in natural products, *via* intramolecular or multicomponent reactions. Aminocarbonylation of 2-iodobenzylamine[48] or its *N*-Me derivative[23] resulted in the formation of 1-isoindolinones **18** and **19**, bearing the skeleton of an atypical antipsychotic agent. The products were obtained in good yields using a homogeneous and a heterogeneous palladium

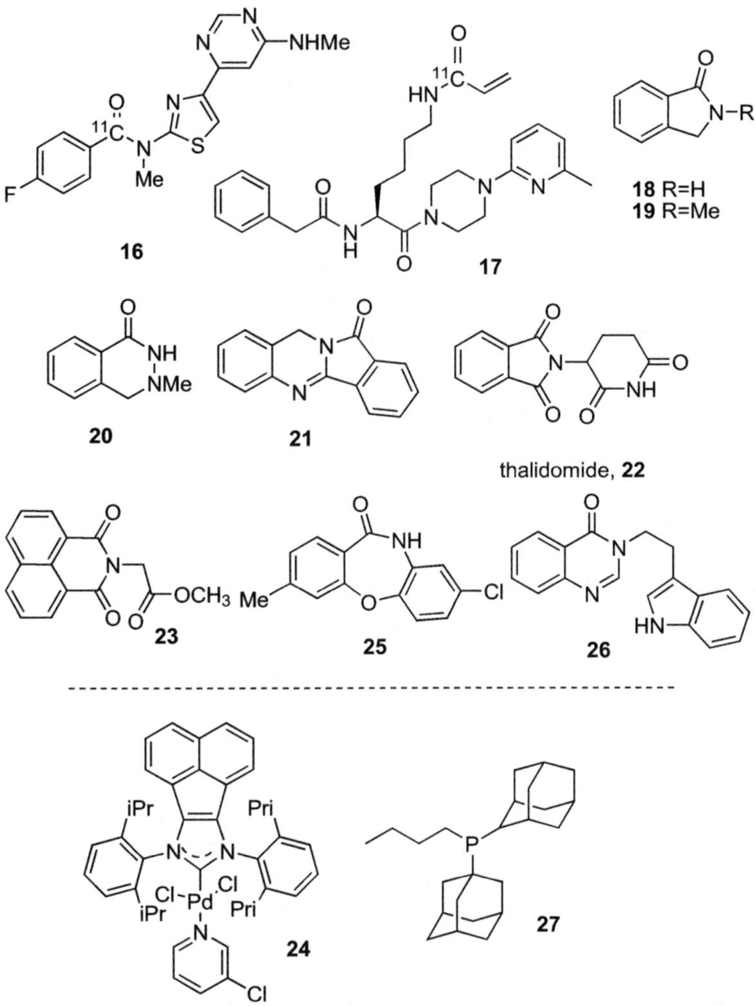

Figure 4.3 Various carboxamides (**16–23**, **25**, **26**) obtained in aminocarbonylation (above the dotted line), as well as a catalyst (**24**) and a ligand (**27**) used in some of these transformations (below the dotted line).

catalyst, respectively. Hydrazinocarbonylation of 2-iodobenzyl bromide led to tetrahydrophthalazinone **20** (Figure 4.3)[49] *via* a nucleophilic substitution–intramolecular aminocarbonylation reaction sequence. Incorporation of two molecules of carbon monoxide resulted in the synthesis of analogues of the anticancer agent batracylin, *e.g.* **21** from 1,2-dibromobenzenes and 2-aminobenzyl amine using $Pd(OAc)_2$ as the catalyst.[50]

Thalidomide **22** and the methyl ester (**23**) of the aldose reductase inhibitor alrestatin were produced in excellent yields in the presence of a Pd–NHC complex (**24**) starting from *o*-diiodobenzene and 1,8-diiodonaphthalene, respectively.[51] Dibenzo[b,e][1,4]oxazepin-11(5*H*)-ones **25** were obtained *via* a one-pot palladium-catalysed aminocarbonylation/aromatic nucleophilic substitution sequence, with 2-aminophenols and 2-bromofluorobenzenes as substrates.[52] The four-component reaction of 2-bromoaniline, trimethyl orthoformate and tryptamine under 10 bar of CO provided 3-[2-(3-indolyl)ethyl]-4(3*H*)-quinazolinone (**26**), the starting material for the synthesis of dihydrorutecarpin, a quinazolino carboline alkaloid.[53] In order to convert the less reactive bromoarenes, a catalyst system composed of $Pd(OAc)_2$ and a bulky phosphine ligand, $BuPAd_2$ (CataCXium A, **27**), was used in the latter two procedures. Alternatively, the four-component reaction leading to **26** could be affected by the use of the simple heterogeneous catalyst Pd/C starting from 2-iodoaniline instead of the bromo derivative.[54]

4.3.2 Alkoxy- and Hydroxycarbonylations

In contrast to the aminocarbonylation reaction, where the products can be obtained from a great variety of nucleophilic reaction partners, alkoxycarbonylation is mainly restricted to the use of methanol to synthesise methyl esters from the corresponding halides. As an example, while the methyl ester was obtained in acceptable yield (72%) during alkoxycarbonylation of a 13α-estrane with a 17-iodo-16-ene functionality, only traces of products could be detected under the same conditions with other primary alcohols (isopropyl- or benzyl alcohol) as reaction partners.[40]

Moreover, for less reactive substrates, considerably higher CO pressure had to be used to obtain acceptable yields. While the more reactive 13β epimer of 17a-iodo-D-homoestra-1,3,5(10),17-tetraene could be completely converted to the methoxycarbonylated compound at atmospheric CO pressure, only 6% conversion was achieved with the 13α isomer under the same reaction conditions. The conversion of the latter was efficiently increased to 88% by rising the carbon monoxide pressure to 40 bar.[41]

Methoxycarbonylation of a functionalised nitroresorcinol iodide allowed the regioselective introduction of the carboxyl group at 3 bar CO pressure, leading to the platensimycin aminoresorcylic acid core during the synthesis of sulfonamide analogues (*e.g.* **28**, Figure 4.4). The application of the $Pd(OAc)_2$/ligand **29** catalytic system proved to be critical in this reaction, especially when the conversion of the nitroaryl bromide was attempted. The latter could be converted to the desired product with much lower yield than the iodo compound (50% instead of 95%) even in the presence of this catalyst.[55] A palladium-catalysed carbonylation reaction of a Boc-protected aminonaphthyl bromide and 2-trimethylsilyl ethanol was a key step in the synthesis the pharmacophoric unit (**30**) of a novel carbamate pro-drug containing a pentagastrin moiety. Probably due to the relatively high electron density of the substrate, this synthetic step could be performed with only 56% yield using 1 bar CO and $Mo(CO)_6$ as the additional CO source, in the presence of a catalyst composed of $Pd(PPh_3)_2Br_2$ and *dppf* as the ligand.[56]

Efficient methoxycarbonylation reactions could be performed starting from sulfonic acid esters of enolates. Potential antagonists (*e.g.* **31**) of LRH-1,

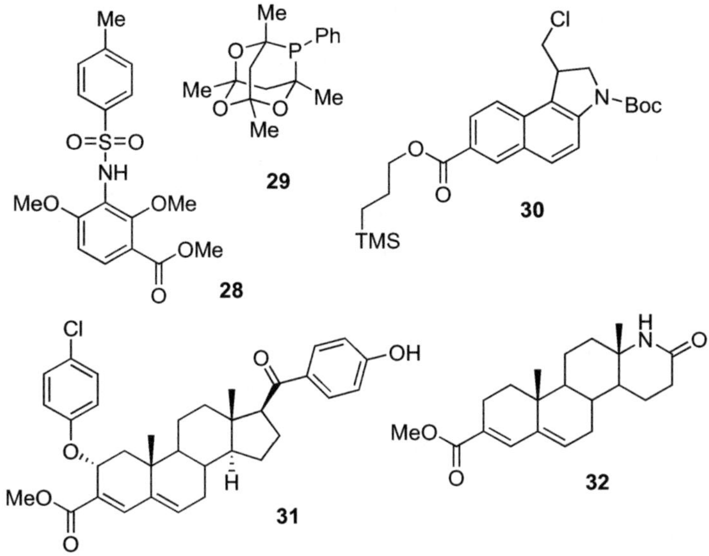

Figure 4.4 Products obtained in alkoxycarbonylation reactions.

a key regulator of estrogen receptor expression in breast cancer cells, were prepared employing a 15-step sequence from pregnenolone involving chemoselective carbonylation of an enol triflate under mild conditions [1 bar CO, 25 °C, catalyst: $Pd(PPh_3)_2(OAc)_2$]. As it could be expected, the chloroaryl moiety of C2 remained intact during the carbonylation step.[57] The methoxycarbonyl functionality was introduced to the core of a new hybrid (**32**) of known 5α-reductase inhibitors, finasteride and epristeride, starting from the C3 octafluoro-3-oxapentane-sulfonate of androst-4-en-3,17-dione.[58] Both C3 triflate and nonaflate derivatives of 20-protected pregnenes were converted to the methyl esters during the synthesis of potential *N*-methyl-D-aspartate receptor modulators. Although the nonaflates could be obtained using the air-stable and less expensive *N*-phenyltrifluoromethanesulfonimide reagent, their use resulted in somewhat lower yields (40% instead of 52% and 28% instead of 66%, in the 5α- and 5β-series, respectively).[59]

There are even less examples for hydroxycarbonylation of organic halides. Although hydroxycarbonylation may lead to the formation of unwanted side products in amino- or alkoxycarbonylation in the presence of less reactive nucleophiles due to traces of water present in the reaction mixtures,[37] carboxylic acid derivatives were mainly produced by the hydrolysis of methyl esters obtained by the methods mentioned above.[55,57–59] The insolubility of carboxylic acid derivatives in common solvents may make the purification of the reaction mixture cumbersome.[60] As an example of a successful direct hydroxycarbonylation, reaction of 3β-hydroxy-17-iodoandrosta-5,16-diene led to the formation of the 17-carboxy derivative in 65% yield in the presence of water as a reagent at atmospheric pressure.[37]

4.3.3 Carbonylative Coupling Reactions

Among carbonylation reactions involving organic halides and *C*-nucleophiles,[61] that of the cross-coupling of organoboron compounds, the carbonylative version of Suzuki–Miyaura coupling, can be considered as the most versatile tool in the synthesis of diaryl or α,β-unsaturated ketones.[62] Its success is due to the availability of the starting materials, the ease of handling and removal of the nontoxic boron-containing by-products, the relatively mild reaction conditions required and the tolerance of a broad range of functional groups. Owing to the latter feature, it is often applied as a key step during the synthesis of molecules with elaborate structures.

The biaryl ketone functionality of dronedarone (**33**, Figure 4.5), an antiarrhythmic agent, was formed at the last stage of a multistep synthesis by the carbonylative Suzuki–Miyaura cross-coupling of a 3-iodobenzo[b] furan derivative with an arylboronic acid.[63] In contrast to the original linear process involving a Friedel–Crafts acylation as one of the first steps, this method offers the advantage of the possibility to introduce a variety of substituents at the 2-position on the benzo[b] furan at a later stage.

The 2-aroylindole framework of 2-aroyltrimethoxyindoles, *e.g.* **34**, designed to investigate the effects of the replacement of the trimethoxyphenyl ring of phenstatin with a trimethoxyindole skeleton, was constructed by a domino reaction involving a carbonylative Suzuki coupling and an intramolecular nucleophilic substitution in the presence of $Pd(PPh_3)_4$ catalyst under 12 bar CO pressure.[64]

An elegant synthesis of the NSAI agents suprofen and ketoprofen was reported by Beller *et al.* involving the combination of carbonylative Suzuki coupling and palladium-catalysed hydroxycarbonylation. The two-step procedure was carried out in one pot, using the $Pd(OAc)_2$/Cata*CX*ium®A (**27**, Figure 4.3) system as the catalyst for both steps, starting from the corresponding aryl bromides.[65] A heterogeneous catalyst, obtained from palladium nanoparticles (Pd-NPs) supported on fibrous nanosilica, was efficiently used in the synthesis of (4-methoxyphenyl)(3,4,5-trimethoxyphenyl)methanone, a known cytotoxic compound inhibiting tubulin polymerization and inducing cell apoptosis. Due to the high activity of Pd-NPs in this reaction, a catalyst concentration as low as 0.1 wt% could be used and a

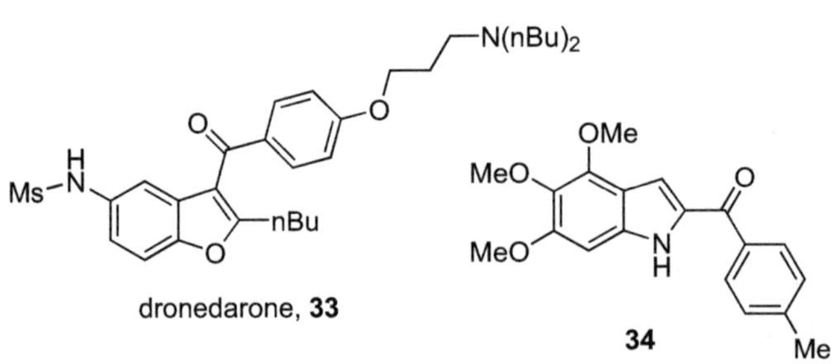

dronedarone, **33**

34

Figure 4.5 Diaryl ketones obtained *via* carbonylative Suzuki coupling.

TON (turnover number) of 1335 could be achieved, at 2 bar CO pressure, in the transformation of 3,4,5-trimethoxy-iodobenzene.[66]

Another possibility for the synthesis of unsymmetrical ketones involves the carbonylative version of Stille coupling using organostannanes as reaction partners. Due to its functional group tolerance and versatility it still offers an alternative to other carbonylative coupling procedures in spite of the toxicity of the tin compounds.[20] This method was applied during the synthesis of nakiterpiosin, a marine sponge metabolite exhibiting potent cytotoxicity against a leukemia cell line.[67] The Me_3Sn group of the organostannane component was introduced *via* palladium-catalysed stannylation of the corresponding bromide after the construction of the stereogenic centres of **35** (Scheme 4.4). The carbonylative coupling was achieved with a modified Stille's protocol using $Pd(PPh_3)_4$/CuCl in DMSO under 1 bar CO. The use of the CuCl additive and DMSO solvent were crucial to the success of this reaction.

The introduction of the C≡C triple bond during carbonylative Sonogashira coupling of organic halides and terminal alkynes offers the possibility of construction of heterocycles *via* a subsequent intramolecular addition step. This methodology was used in a convergent approach towards the quinolone subunit (**36**, Figure 4.6) of the hepatitis C virus (HCV) NS3 protease inhibitor, BILN 2061.[68] The desired quinolone was obtained in 70% isolated yield starting from 2-iodo-5-methoxyaniline and thiazolylacetylene in the presence of $PdCl_2$(*dppf*) as catalyst in Et_2NH at 120 °C under 17 bar of CO. Interestingly, under these conditions only 2.9% formation of amide **37** was detected in spite of the use of nucleophilic Et_2NH as solvent.

The permethylated kinase inhibitor BE-23372M (**38**) was obtained *via* an unusual double carbonylation process.[69] The alkynone product formed in the carbonylative Sonogashira reaction is supposed to undergo a

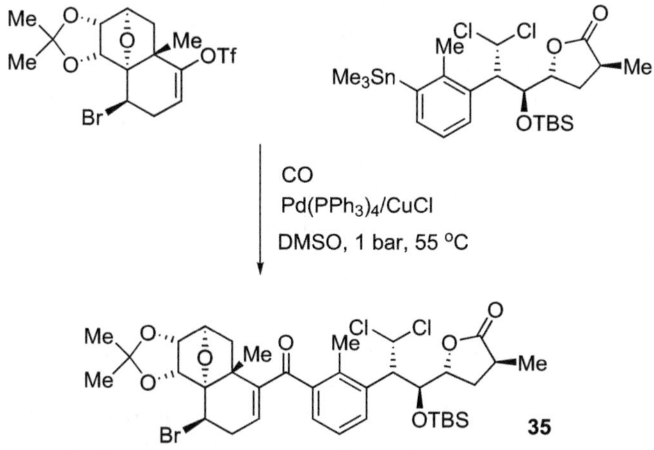

35

Scheme 4.4 Carbonylative Stille coupling, a key step in the total synthesis of nakiterpiosin.

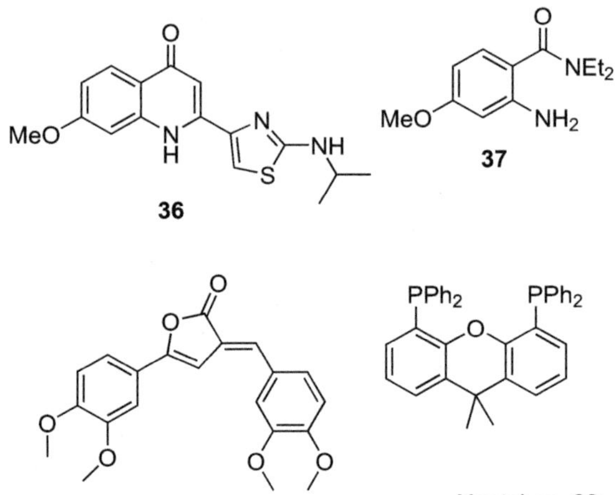

36 37 38

Figure 4.6 Heterocycles obtained *via* carbonylative Sonogashira–ring closure reaction sequences.

hydropalladation–CO insertion–ring closure reaction sequence in the presence of a catalyst system composed of [{(cinnamyl)PdCl}$_2$] and XantPhos (**39**).

4.3.4 The Use of CO Equivalents

To avoid the use of toxic CO, as well as the application of high-pressure equipment, a number of methods have been developed to use alternative sources for the introduction of carbon monoxide. The antitubercular agent pyrazinamide was synthesised from commercially available 2-bromopyrazine on a 10 mmolar scale in 76% yield using solid $Co_2(CO)_8$ and NH_4Cl as sources for CO and NH_3.[70] Carbonylation of aryl halides in the presence of $Mo(CO)_6$ was effected under microwave irradiation in the presence of a fluorous, oxime-based palladacycle.[71]

Skrydstrup developed a versatile methodology for the *ex situ* generation of carbon monoxide using a two-chamber system.[72] In one of the chambers, CO was derived by palladium-catalysed decarbonylation of a solid, stable and easy to handle tertiary acid chloride using a catalyst composed of $Pd(dba)_2$ and $P(tBu)_3$ and was used in carbonylation reactions taking place in the other chamber. This methodology was efficiently used for the synthesis of a number of biologically relevant structures with an amide or ester functionality including CX-546 (**6**, Figure 4.1), the gastroprokinetic drugs metaclopramide and bromopride, as well as a local anesthetic, butoxycaine. The products were obtained in good yields starting from aryl iodides in the presence of the $Pd(dba)_2/PPh_3$ catalyst system and near stoichiometric quantities of CO, at 80 °C, but the conversion of aryl bromides necessitated the use of more elaborate ligands, such as XantPhos (**39**) (4,5-bis(diphenylphosphino)-9,9-dimethylxanthene) or Cata*CX*ium® A.

Interestingly, the protocol made it possible to synthesise α-ketoamides, precursors for substituted phenethylamines, carrying out the carbonylation at room temperature and changing the catalyst to $Pd(dba)_2/P(tBu)_3$.[73] Among other compounds, fenofibrate, a triglyceride and cholesterol regulator could be obtained *via* a carbonylative Suzuki coupling of boronic acids starting from the corresponding aryl iodides in the presence of $PdCl_2$[74] or alternatively from the cheaper bromides using the $Pd(acac)_2$/CataCXium® A system.[75] The methodology also offers the possibility to introduce ^{11}C,[76] ^{13}C[60] and ^{14}C labels[77] into carbonyl compounds or carboxylic acid derivatives by the use of an acid chloride incorporating the corresponding isotope. Some important examples include the dopamine D2/D3 receptor antagonist Raclopride,[76] an intermediate of nordazepam, used in the treatment of anxiety,[74] as well as the PARP (poly (ADP-ribose) polymerase) inhibitor olaparib.[77]

4.3.5 Industrial Applications

Carbonylation reactions can be used effectively for the synthesis of carbonyl compounds/carboxylic acid derivatives on a laboratory scale. In spite of this, industrial applications are rare and most of these procedures are carried out only in pilot plants.

Lotrafiban is a 1,4-benzodiazepine with a chiral acetic acid moiety at C-2. Only the (*S*)-enantiomer is pharmacologically active as a GPIIb/IIIa antagonist. The key step is an aminocarbonylation of aryl iodide **40** (Scheme 4.5) in the presence of 4-pyridylpiperidine as the nucleophilic partner that was carried out on 100 kg scale in a pilot plant. Using 4,4′-bipiperidine directly in the carbonylation reaction without prior protection led to the formation of the corresponding α-ketoamide as well as a bis-carbonylated dimer. Instead, carbonylation was carried out with 4-pyridylpiperidine in the presence of $Pd(PPh_3)_2Cl_2$ catalyst in anisole at 100 °C and atmospheric pressure. Interestingly, the application of dicyclohexylamine (DCHA) as the base instead of DIPEA (*N*,*N*-diisopropyl-ethylamine) increased the yield of the amide product from 65 to 90%. However, it should be mentioned that this bulky secondary

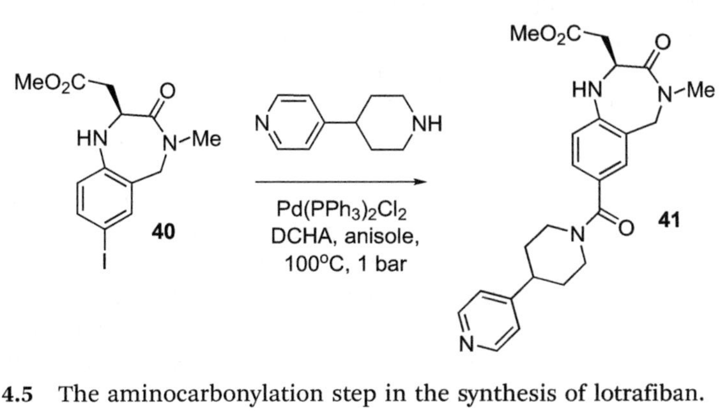

Scheme 4.5 The aminocarbonylation step in the synthesis of lotrafiban.

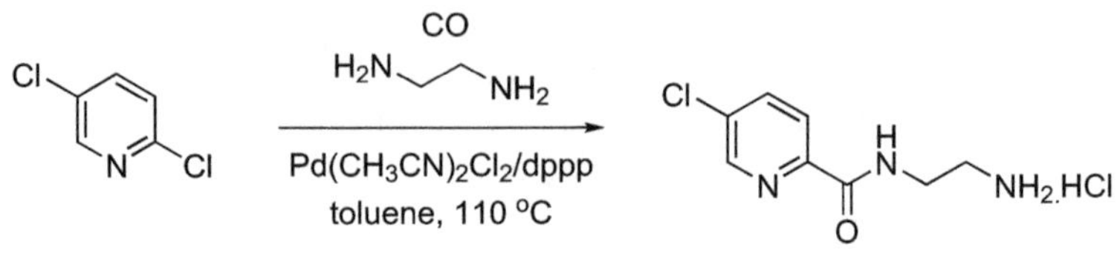

Scheme 4.6 Synthesis of lazabemide *via* aminocarbonylation.

amine was chosen carefully to avoid competition with pyridylpiperidine. Moreover, the insolubility of the DCHA–HI salt by-product was also an important feature. Lotrafiban was obtained by the Pd/C-catalysed hydrogenation of the pyridine ring of **41** as the last synthetic step.[78]

The synthesis of monoamine oxidase B inhibitor lazabemide (**42**, Scheme 4.6) was patented by Hoffmann-LaRoche[79] This one-step synthesis involving the aminocarbonylation of 2,5-dichloropyridine afforded the HCl salt of the desired product in 65% yield with 100% atom efficiency and replaced the original eight-step synthesis with 8% overall yield. The reaction was carried out with an excess of ethylenediamine in the presence of the $Pd(CH_3CN)_2Cl_2$/*dppp* catalyst system [*dppp*: 1,3-bis(diphenylphosphino)propane] under 10 bar CO pressure. Due to the high activity of the catalyst (TON = 3000), it could be applied in a small amount and traces of palladium could be removed from the product by an appropriate workup. The reactivity of the chloroarene substrate and complete selectivity for aminocarbonylation at C2 is due to the activating effect of the nitrogen of the heteroaryl moiety on the C–Cl bond of this position.

4.4 Oxidative Carbonylation Reactions

One of the drawbacks of the procedures described above involves the formation of stoichiometric amounts of wastes due to the formation of HX salts in Heck carbonylation or organometallic by-products in carbonylative cross-coupling reactions. Instead, direct functionalisation of C–H bonds may provide an ideal alternative pathway. Although oxidative carbonylation reactions was mainly investigated using simple terminal and internal alkenes with the aim of producing (chiral) building blocks, some examples for the synthesis of compounds of pharmaceutical importance are also known. The mechanistic details, the structure–reactivity features of palladium-catalysed oxidative carbonylation reactions have been reviewed recently.[80,81]

The oxidative carbonylation of 4-yn-1-ones and propargylic esters was studied in detail (Scheme 4.7). The asymmetric version of this reaction was applied for the synthesis of (-)-AL-2 (**43**), which inhibits PDA-induced tumour promotion. A 3,4-dioxygenated-9-hydroxy-1-nonyn-5-one derivative, obtained from diethyl tartrate, underwent intramolecular acetalization and stereoselective construction of *E*-methoxycarbonylmethylidene functionality in palladium-catalysed methoxycarbonylation.[82]

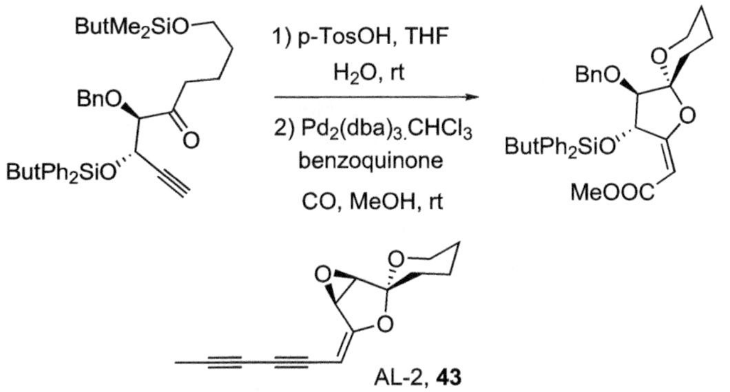

Scheme 4.7 Oxidative carbonylation of 4-yn-1-ones.

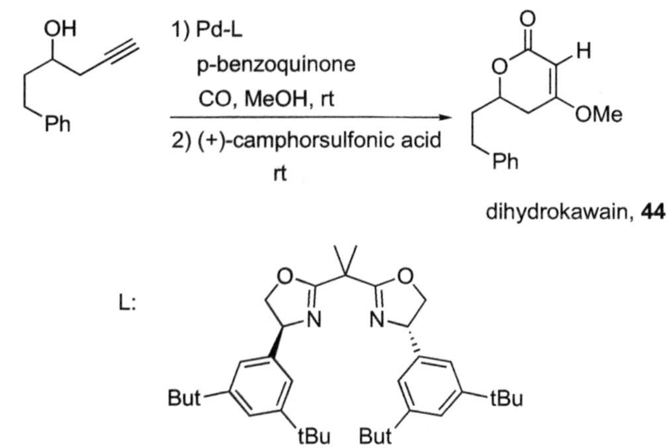

Scheme 4.8 Oxidative methoxycarbonylation of terminal alkynes.

The oxidative methoxycarbonylation of terminal alkynes by Pd-bis(oxazoline)-type complexes led to methoxyacrylates (Scheme 4.8). Homopropargyl alcohols were transferred to (±)-dihydrokawain (**44**) (a six-membered lactone) and the corresponding β-hydroxyacrylate.[83]

The application of the same methodology carried out with propargyl alcohols (Scheme 4.9) led to the total synthesis of annularin G (**45**) and annularin H (**46**).[84]

Functionalised alcohols such as menthol (**47**) and dehydroepiandrosterone (**48**) reacted towards the corresponding oxidative carbonylation products, the desired unsaturated esters (Scheme 4.10).[85]

The oxidative iodination and subsequent alkoxycarbonylation of indol derivatives resulted in indole-3-carboxylates (Scheme 4.11). Using tropine as secondary alcohol the biologically active tropisetron, a 5-HT_3 antagonist was synthesised.[86]

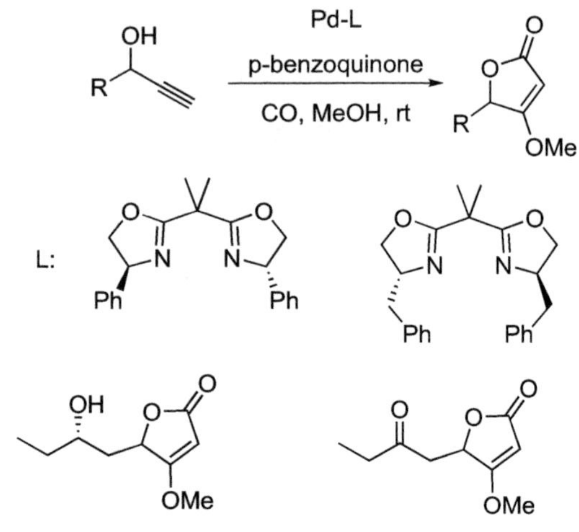

Scheme 4.9 Oxidative methoxycarbonylation of propargyl alcohols.

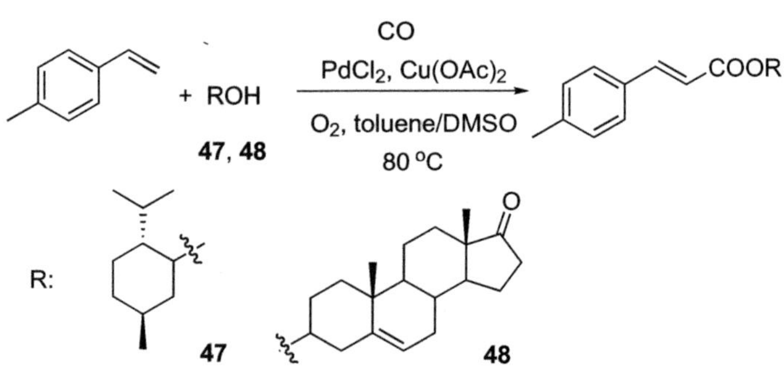

Scheme 4.10 Oxidative carbonylation of menthol (**47**) and dehydroepiandrosterone (**48**).

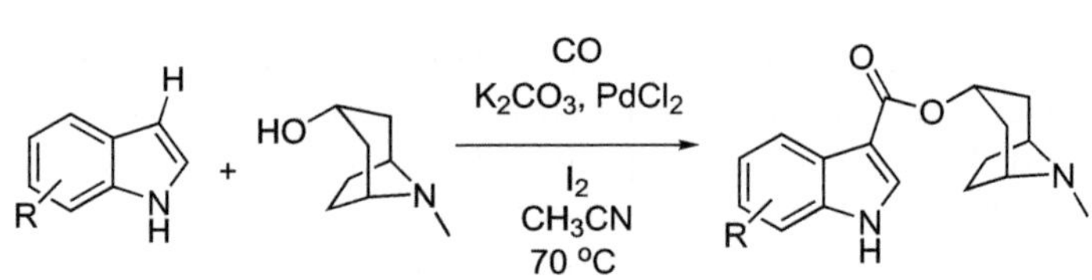

Scheme 4.11 Oxidative iodination and alkoxycarbonylation of indol derivatives.

The intramolecular version of palladium-catalysed oxidative alkoxycarbonylation (lactonization) (Scheme 4.12) was used in the stereoselective total synthesis of crisamicin A (**49**). The annulation with desired stereochemistry required the application of thioureas such as *N,N,N′,N′*-tetramethyl thiourea.[87]

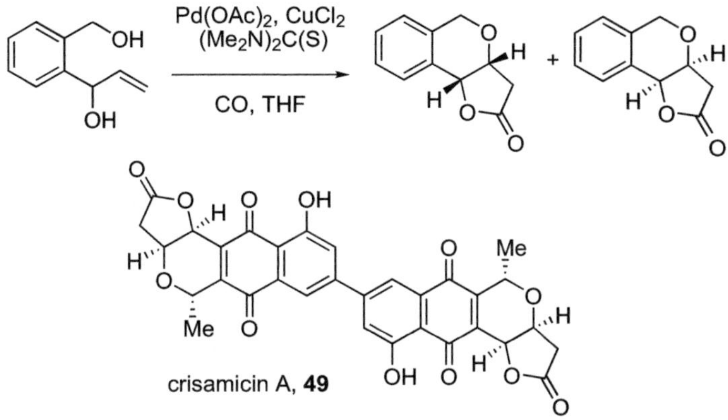

Scheme 4.12 Intramolecular palladium-catalysed oxidative alkoxycarbonylation used in the total synthesis of crisamicin A (**49**).

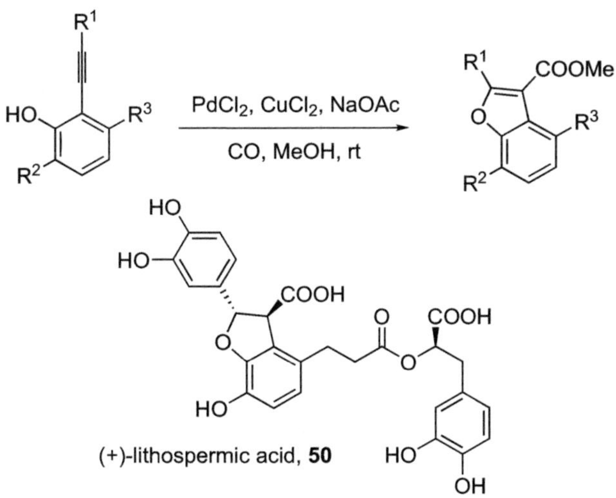

Scheme 4.13 Palladium-catalysed carbonylative annulation as a key step in the total synthesis of (+)-lithospermic acid (**50**).

Palladium-catalysed carbonylative annulation was published as a key step in the formal total synthesis of (+)-lithospermic acid (**50,** Scheme 4.13). An *ortho*-hydroxydiarylalkyne was subjected to undergo cyclisation resulting in the corresponding tetrasubstituted benzofuran possessing ester functionality.[88]

A sterically congested furanolactone motif, found in Caribbean sponges of the genus *Plakortis*, was assessed by palladium-catalysed hydroxycyclization–carbonylation–lactonization cascade (Scheme 4.14). The total synthesis of various plakortones was achieved using the appropriate ene-1,3-diol.[89]

Quinoline-2(1*H*)-ones, with wide applications in pharmaceutical chemistry, were synthesised in iridium-catalysed carbonylative C–H activation of

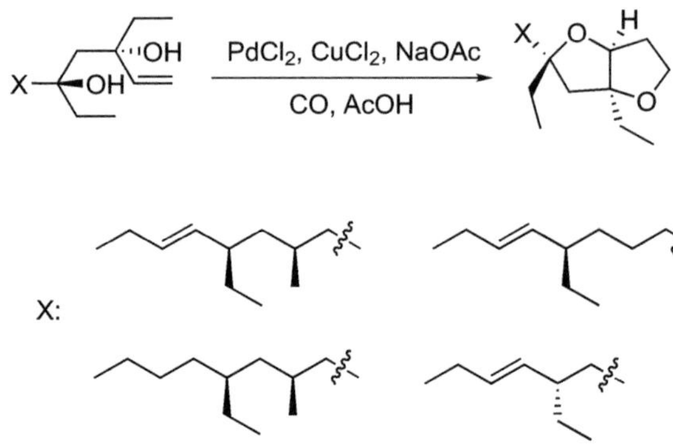

Scheme 4.14 Palladium-catalysed hydroxycyclization–carbonylation–lactonization cascade leading to plakortones.

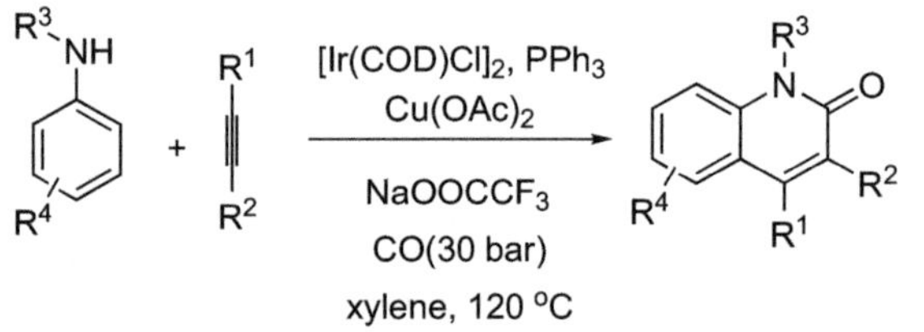

Scheme 4.15 Iridium-catalysed carbonylative C–H activation of anilines.

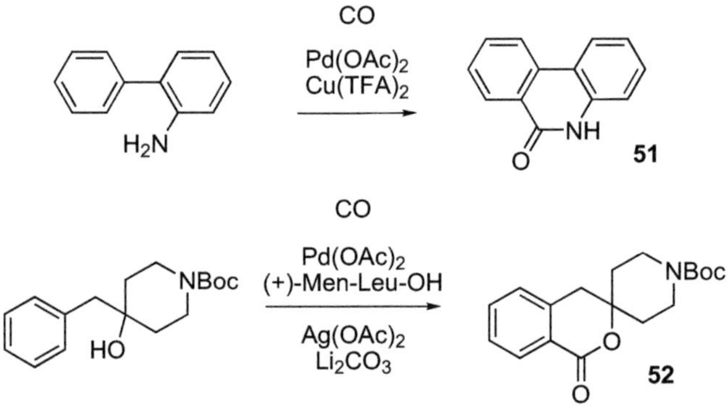

Scheme 4.16 Synthesis of lactams and lactones *via* oxidative carbonylation of arenes.

anilines (Scheme 4.15). The carbonylative annulation reaction tolerates both the halide substituents of the aniline and the substituents of the alkyne, enabling further functionalisation of the products.[90]

The combination of C–H carbonylation of arenes with a subsequent intramolecular cyclisation serves as an efficient and sustainable tool for the construction of different heterocycles, especially lactams and lactones. Some recent examples include phenanthridinone **51** (Scheme 4.16),[91] benzolactams, key intermediates for the synthesis of staurosporinone, an indolocarbazole alkaloid[92] and the histamine release inhibitor **52**.[93]

4.5 Conclusion and Outlook

As the above reactions demonstrate, a great variety of building blocks of pharmaceuticals as well as elaborate heterocyclic structures can be produced *via* carbonylation reactions. However, industrial applications of these procedures are still rare. To achieve widespread application, more efficient catalysts should be developed that ensure higher turnover numbers and make it possible to carry out carbonylations at atmospheric conditions as well as to replace the starting material aryl iodides with cheaper bromides or, still preferably, with chlorides. Although there are already some examples of the use of recyclable catalysts that minimise contamination of products, up to now these catalysts have mainly been used in simple model reactions.

Acknowledgements

The support of the National Research, Development and Innovation Office (OTKA 120014 and K113177) is acknowledged.

References

1. *Transition Metals for Organic Synthesis*, ed. M. Beller and C. Bolm, Wiley-VCH, Weinheim, 1998, vol. I–II.
2. W. C. Zeise, *Ann. Phys.*, 1827, **9**, 932.
3. L. Mond, *J. Chem. Soc.*, 1890, **57**, 749.
4. L. Mond and C. Langler, *J. Chem. Soc.*, 1891, **59**, 1090.
5. O. Roelen, *Angew. Chem.*, 1948, **60**, 62.
6. L. Ojima, C. Y. Tsai, M. Tzamarioudaki and D. Bonafoux, in *The Hydroformylation Reaction in Organic Reactions*, ed. L. Overman, J. Wiley & Sons,Weinheim, 2000, p. 1.
7. *Rhodium Catalysed Hydroformylation*, ed. C. Claver and P. W. N. M. van Leeuwen, Kluwer Academic Publishers, Dordrecht, 2000.
8. *Modern Carbonylation Methods*, ed. L. Kollar, Wiley-VCH, Weinheim, 2008.
9. R. Franke, D. Selent and A. Boerner, *Chem. Rev.*, 2012, **112**, 5675.
10. G. P. Stahly and R. M. Starrett, Production methods for chiral non-steroidal antiinflammatory Profen drugs, in *Chirality in Industry II*, ed. A. N. Collins, G. N. Sheldrake and J. Crosby, John Wiley and Sons, 1997, p. 19.
11. Ph. Kalck and M. Urrutigoïty, *Inorg. Chim. Acta*, 2015, **431**, 110.
12. C. Godard, B. K. Munoz, A. Ruiz and C. Claver, *Dalton Trans.*, 2008, 853.
13. Y. Kawashima, K. Okano, K. Nozaki and T. Hiyama, *Bull. Chem. Soc. Jpn.*, 2004, **77**, 347.

14. T. M. Konrad, J. T. Durrani, C. J. Cobley and M. L. Clarke, *Chem. Commun.*, 2013, **49**, 3306.
15. A. Grabulosa, J. J. R. Frew, J. A. Fuentes, A. M. Z. Slavin and M. L. Clarke, *J. Mol. Catal. A*, 2010, **330**, 18.
16. T. M. Konrad, J. A. Fuentes, A. M. Z. Slawin and M. L. Clarke, *Angew. Chem., Int. Ed.*, 2010, **49**, 9197.
17. J. A. Fuentes, A. M. Z. Slawin and M. L. Clarke, *Catal. Sci. Technol.*, 2012, **2**, 715.
18. M. Amezquita-Valencia and H. Alper, *J. Org. Chem.*, 2016, **81**, 3860.
19. O. Karlubikova, M. Babjak and T. Gracza, *Tetrahedron*, 2011, **67**, 4980.
20. A. Brennführer, H. Neumann and M. Beller, *Angew. Chem., Int. Ed.*, 2009, **48**, 4114.
21. R. Grigg and S. P. Mutton, *Tetrahedron*, 2010, **66**, 5515.
22. M. Kiss, A. Takács and L. Kollár, *Curr. Green Chem.*, 2015, **2**, 319.
23. T. T. Dang, Y. Zhu, J. S. Y. Ngiam, S. C. Ghosh, A. Chen and A. M. Seayad, *ACS Catal.*, 2013, **3**, 1406.
24. F. Tinnis, O. Verho, K. P. J. Gustafson, C. W. Tai, J. E. Bäckvall and H. Adolfsson, *Chem. - Eur. J.*, 2014, **20**, 5885.
25. A. Takács, B. Jakab, A. Petz and L. Kollár, *Tetrahedron*, 2007, **63**, 10372.
26. L. Vandromme, M. Legraverend, S. Kreimerman, O. Lozach, L. Meijer and D. S. Grierson, *Bioorg. Med. Chem.*, 2007, **15**, 130.
27. Å. Lindelöf, C. Ericsson, R. Simonsson, G. Nilsson, G. Grönberg and C. S. Elmore, *J. Labelled Compd. Radiopharm.*, 2016, **59**, 340.
28. D. H. R. Barton, G. Bashiardes and J. L. Fourrey, *Tetrahedron Lett.*, 1983, **24**, 1605.
29. D. Czajkowska-Szczykowska, J. W. Morzycki and A. Wojtkielewicz, *Steroids*, 2015, **97**, 13.
30. S. Aggarwal, S. Thareja, A. Verma, T. R. Bhardwaj and M. Kumar, *Steroids*, 2010, **75**, 109.
31. J. Balogh, S. Mahó, V. Háda, L. Kollár and R. Skoda-Földes, *Synthesis*, 2008, 3040.
32. E. Szánti-Pintér, J. Balogh, Z. Csók, L. Kollár, Á. Gömöry and R. Skoda-Földes, *Steroids*, 2011, **76**, 1377.
33. J. Balogh, R. Skoda-Földes, K. Vazdar and I. Habuš, *J. Organomet. Chem.*, 2012, **703**, 51.
34. R. M. B. Carrilho, M. M. Pereira, M. J. S. M. Moreno, A. Takács and L. Kollár, *Tetrahedron Lett.*, 2013, **5**, 2763.
35. G. Mikle, B. Boros and L. Kollár, *Tetrahedron: Asymmetry*, 2014, **25**, 1527.
36. M. Kiss, S. Mahó, K. Böddi, B. Boros and L. Kollár, *Monatsh. Chem.*, 2015, **146**, 357.
37. P. Ács, A. Takács, A. Szilágyi, J. Wölfling, G. Schneider and L. Kollár, *Steroids*, 2009, **74**, 419.
38. E. Szánti-Pintér, Z. Csók, Z. Berente, L. Kollár and R. Skoda-Földes, *Steroids*, 2013, **78**, 1177.

39. E. Szánti-Pintér, J. Wouters, Á. Gömöry, É. Sághy, É. Szőke, Z. Helyes, L. Kollár and R. Skoda-Földes, *Steroids*, 2015, **104**, 284.
40. P. Ács, A. Takács, A. Szilágyi, J. Wölfling, G. Schneider and L. Kollár, *Steroids*, 2008, **73**, 669.
41. A. Takács, P. Ács, J. Wölfling, G. Schneider and L. Kollár, *Steroids*, 2010, **75**, 1075.
42. P. Ács, B. Jakab, A. Takács and L. Kollár, *Steroids*, 2007, **72**, 627.
43. P. Ács, E. Müller, G. Czira, S. Mahó, M. Pereira and L. Kollár, *Steroids*, 2006, **71**, 875.
44. P. Ács, A. Takács, M. Kiss, N. Pálinkás, S. Mahó and L. Kollár, *Steroids*, 2011, **76**, 280.
45. M. Kiss, N. Pálinkás, A. Takács, S. Mahó and L. Kollár, *Steroids*, 2013, **78**, 693.
46. J. Hong, S. Lu, R. Xu, J. S. Liow, A. E. Woock, K. J. Jenko, R. L. Gladding, S. S. Zoghbi, R. B. Innis and V. W. Pike, *Nucl. Med. Biol.*, 2015, **42**, 967.
47. B. van der Wildt, M. M. M. Wilhelmus, J. Bijkerk, L. Y. F. Haveman, E. J. M. Kooijman, R. C. Schuit, J. G. J. M. Bol, C. A. M. Jongenelen, A. A. Lammertsma, B. Drukarch and A. D. Windhorst, *Nucl. Med. Biol.*, 2016, **43**, 232.
48. D. Marosvölgyi-Haskó, A. Takács, Z. Riedl and L. Kollár, *Tetrahedron*, 2011, **67**, 1036.
49. D. Marosvölgyi-Haskó, A. Petz, A. Takács and L. Kollár, *Tetrahedron*, 2011, **67**, 9122.
50. J. Chen, H. Neumann, M. Beller and X. F. Wu, *Org. Biomol. Chem.*, 2014, **12**, 5835.
51. S. Liu, Q. Deng, W. Fang, J. F. Gong, M. P. Song, M. Xu and T. Tu, *Org. Chem. Front.*, 2014, **1**, 1261.
52. C. Shen, H. Neumann and X. F. Wu, *Green Chem.*, 2015, **17**, 2994.
53. L. He, H. Li, H. Neumann, M. Beller and X. F. Wu, *Angew. Chem., Int. Ed.*, 2014, **53**, 1420.
54. K. Natte, H. Neumann and X. F. Wu, *Catal. Sci. Technol.*, 2015, **5**, 4474.
55. J. McNulty, J. J. Nair and A. Capretta, *Tetrahedron Lett.*, 2009, **50**, 4087.
56. L. F. Tietze, O. Panknin, B. Krewer, F. Major and I. Schuberth, *Int. J. Mol. Sci.*, 2008, **9**, 821.
57. J. Rey, T. J. C. O'Riordan, H. Hu, J. P. Snyder, A. J. P. White and A. G. M. Barrett, *Eur. J. Org. Chem.*, 2012, 3781.
58. Z. Yao, Y. Xu, M. Zhang, S. Jiang, M. C. Nicklaus and C. Liao, *Bioorg. Med. Chem. Lett.*, 2011, **21**, 475.
59. E. Štastná, H. Chodounská, V. Pouzar, J. Borovská and L. Vyklicky Jr, *Collect. Czech. Chem. Commun.*, 2011, **76**, 1141.
60. M. Burhardt, R. Taaning, N. C. Nielsen and T. Skrydstrup, *J. Org. Chem.*, 2012, **77**, 5357.
61. X. F. Wu, H. Neumann and M. Beller, *Chem. Soc. Rev.*, 2011, **40**, 4986.

62. M. Blangetti, H. Rosso, C. Prandi, A. Deagostino and P. Venturello, *Molecules*, 2013, **18**, 1188.
63. T. Okitsu, M. Ogasahara and A. Wada, *Chem. Pharm. Bull.*, 2016, **64**, 1149.
64. M. Arthuis, R. Pontikis, G. G. Chabot, L. Quentin, D. Scherman and J. C. Florent, *Eur. J. Med. Chem.*, 2011, **46**, 95.
65. H. Neumann, A. Brennführer and M. Beller, *Adv. Synth. Catal.*, 2008, **350**, 2437.
66. P. Gautam, M. Dhiman, V. Polshettiwar and B. M. Bhanage, *Green Chem.*, 2016, **18**, 5890.
67. S. Gao, Q. Wang and C. Chen, *J. Am. Chem. Soc.*, 2009, **131**, 1410.
68. N. Haddad, J. Tan and V. Farina, *J. Org. Chem.*, 2006, **71**, 5031.
69. X. F. Wu, B. Sundararaju, P. Anbarasan, H. Neumann, P. H. Dixneuf and M. Beller, *Chem. – Eur. J.*, 2011, **17**, 8014.
70. A. S. Suresh, P. Baburajan and M. Ahmed, *Tetrahedron Lett.*, 2015, **56**, 4864.
71. W. J. Ang, L. C. Lo and Y. Lam, *Tetrahedron*, 2014, **70**, 8545.
72. P. Hermange, A. T. Lindhardt, R. H. Taaning, K. Bjerglund, D. Lupp and T. Skrydstrup, *J. Am. Chem. Soc.*, 2011, **133**, 6061.
73. D. U. Nielsen, K. Neumann, R. H. Taaning, A. T. Lindhardt, A. Modvig and T. Skrydstrup, *J. Org. Chem.*, 2012, **77**, 6155.
74. A. Ahlburg, A. T. Lindhardt, R. H. Taaning, A. E. Modvig and T. Skrydstrup, *J. Org. Chem.*, 2013, **78**, 10310.
75. K. M. Bjerglund, T. Skrydstrup and G. A. Molander, *Org. Lett.*, 2014, **16**, 1888.
76. T. L. Andersen, S. D. Friis, H. Audrain, P. Nordeman, G. Antoni and T. Skrydstrup, *J. Am. Chem. Soc.*, 2015, **137**, 1548.
77. A. T. Lindhardt, R. Simmonsen, R. H. Taaning, T. M. Gøgsig, G. N. Nilsson, G. Stenhagen, C. S. Elmore and T. Skrydstrup, *J. Labelled Compd. Radiopharm.*, 2012, **55**, 411.
78. R. J. Atkins, A. Banks, R. K. Bellingham, G. F. Breen, J. S. Carey, S. K. Etridge, J. F. Hayes, N. Hussain, D. O. Morgan, P. Oxley, S. C. Passey, T. C. Walsgrove and A. S. Wells, *Org. Process Res. Dev.*, 2003, **7**, 663.
79. M. Scalone and P. Vogt (Hoffmann-LaRoche), EP385210, 1990.
80. X. F. Wu, H. Neumann and M. Beller, *ChemSusChem*, 2013, **6**, 229.
81. Q. Liu, H. Zhang and A. Lei, *Angew. Chem., Int. Ed.*, 2011, **50**, 10788.
82. N. Miyakoshi and C. Mukai, *Org. Lett.*, 2003, **5**, 2335.
83. K. Kato, S. Motodate, T. Mochida, T. Kobayashi and H. Akita, *Angew. Chem., Int. Ed.*, 2009, **48**, 3326.
84. S. Motodate, T. Kobayashi, M. Fujii, T. Mochida, T. Kusakabe, S. Katoh, H. Akita and K. Kato, *Chem. – Asian J.*, 2010, **5**, 2221.
85. L. Wang, Y. Wang, C. Liu and A. Lei, *Angew. Chem., Int. Ed.*, 2014, **53**, 5657.
86. R. Lang, L. Shi, D. Li, C. Xia and F. Li, *Org. Lett.*, 2012, **14**, 4130.

87. Z. Li, Y. Gao, Y. Tang, M. Dai, G. Wang, Z. Wang and Z. Yang, *Org. Lett.*, 2008, **10**, 3017.
88. J. Fischer, G. P. Savauge and M. J. Coster, *Org. Lett.*, 2011, **13**, 3376.
89. P. Y. Hayes, S. Chow, F. Rahm, P. V. Bernhardt, J. J. De Voss and W. Kitching, *J. Org. Chem.*, 2010, **75**, 6489.
90. F. Zhu, Y. Li, Z. Wang and X. F. Wu, *Adv. Synth. Catal.*, 2016, **358**, 3350.
91. Z. Liang, J. Zhang, Z. Liu, K. Wang and Y. Zhang, *Tetrahedron*, 2013, **69**, 6519.
92. Y. Wada, H. Nagasaki, M. Tokuda and K. Orito, *J. Org. Chem.*, 2007, **72**, 2008.
93. Y. Lu, D. Leow, X. Wang, K. M. Engle and J. Q. Yu, *Chem. Sci.*, 2011, **2**, 967.

CHAPTER 5

Applications of Catalytic Hydroformylation in the Synthesis of Biologically Relevant Synthons and Drugs

M. M. PEREIRA

University of Coimbra, Department of Chemistry, Rua Larga, Coimbra, Portugal
Email: mmpereira@qui.uc.pt

5.1 Introduction

The catalytic hydroformylation reaction, discovered by Otto Roelen[1,2] in 1935, consists of the *cis* addition of hydrogen and carbon monoxide (*syngas*) to a C=C double bond, allowing the one-pot synthesis of aldehydes, enclosing one more carbon atom than the starting olefin (Figure 5.1). As a pure addition reaction, in which all raw materials are incorporated in the products, the catalytic hydroformylation meets all requirements of a process with total atomic economy. Nowadays, catalytic hydroformylation is one of the largest scale industrial processes of homogeneous catalysis, widely used in both the bulk and fine chemicals industries, with a production capacity of more than 9 million tons of aldehydes, or the corresponding alcohols, per year.[3,4]

In addition, the application of hydroformylation as the central reaction of sequential processes to transform aldehydes into other functional groups for

Green Chemistry Series No. 54
Sustainable Synthesis of Pharmaceuticals: Using Transition Metal Complexes as Catalysts
Edited by Mariette M. Pereira and Mário J. F. Calvete

Published by the Royal Society of Chemistry, www.rsc.org

Figure 5.1 Otto Roelen, the "father" of the catalytic hydroformylation reaction, when he worked for Ruhrchemie SA.
Photograph reprinted from *Journal of Organometallic Chemistry*, **754**, G. D. Frey, 75 Years of oxo synthesis – The success story of a discovery at the OXEA Site Ruhrchemie, 5–7, copyright 2014, with permission from Elsevier.

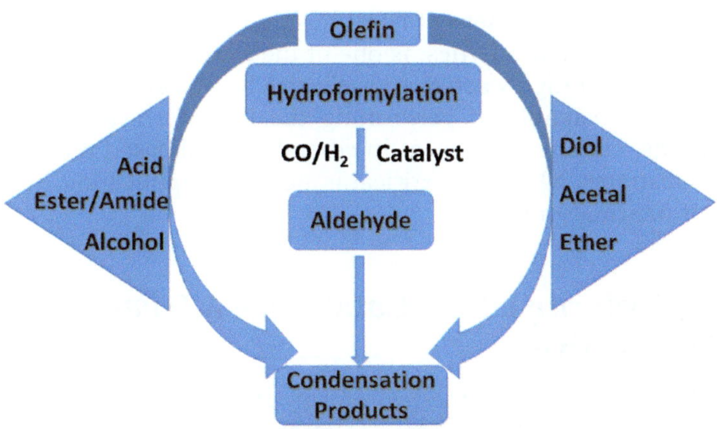

Figure 5.2 Products obtained in one-pot processes using hydroformylation as the central reaction.

the production of added-value products (Figure 5.2), with a high atom and eco-scale economy, have turned it into a paradigmatic example of the implementation of green chemistry principles at the industrial level.[5–9]

Originally, the hydroformylation reaction was only useful for the synthesis of poorly functionalized products, but the development of new, efficient and selective catalysts has allowed its application to largely functionalized substrates, turning hydroformylation into a relevant process with a variety of applications in fine chemistry, including pharmaceuticals.[10] Despite these advances, the application of catalytic hydroformylation as a real tool for drug

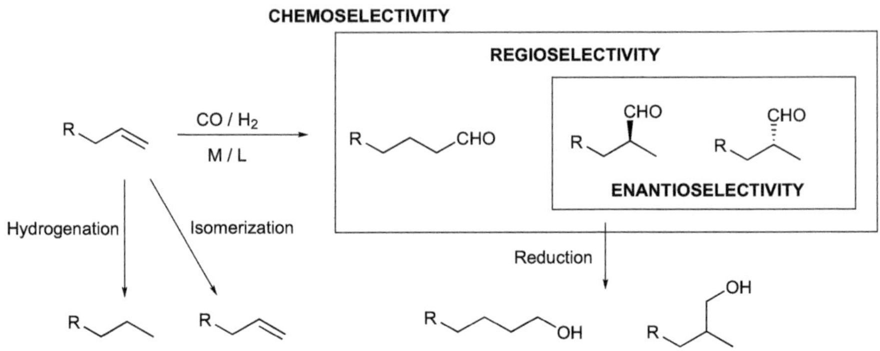

Scheme 5.1 Selectivity is an important issue when designing a catalytic hydroformylation reaction.

synthesis, under the green chemistry principles, is still a challenge and several aspects require improvements, namely, the optimization of selectivity,[3–5,11] (chemo-, regio-, enantio- and diastereo-selectivity; Scheme 5.1), the recovering and reutilization of the catalysts,[7] and the replacement of expensive and toxic metals (such as rhodium) by less polluting and cheaper metals.

Therefore, the development of new phosphorus ligands, with appropriate stereo and electronic properties, is one of the greatest challenges regarding the development of highly active and selective hydroformylation catalysts, in particular those capable of performing the hydroformylation of internal double bonds, commonly present in several natural-based and biologically relevant substrates, which is a critical issue for its application in the development of pharmaceuticals.

5.2 Hydroformylation Catalysts—A Historical Perspective

Initially, Roelen considered the hydroformylation reaction a heterogeneous process catalyzed by $Co(OAc)_2$ (Ac = acetate) (**1**) but, approximately 30 years later, Heck and Breslow identified the species $[CoH(CO)_4]$ (**2**, Scheme 5.2) as being the homogeneous catalytically active species of cobalt-catalyzed hydroformylation reactions.[12] However, this first hydroformylation catalyst (**2**) was shown to be active only under very drastic reaction conditions (*syngas* pressures higher than 20 MPa and temperatures in the range of 200 °C).

The performance of cobalt carbonyl type catalysts was clearly enhanced by the discovery of $CoH(CO)_n(PAlkyl_3)_n$, (**3**, Scheme 5.2), by Shell Corporation, which introduced it in the so-called Shell High Olefin Process (SHOP),[13] which is capable of promoting a sustainable tandem process where an internal olefin suffers isomerization into the terminal position, followed by hydroformylation and finally hydrogenation of the aldehyde to the corresponding alcohol in one pot (Scheme 5.3). However, the low

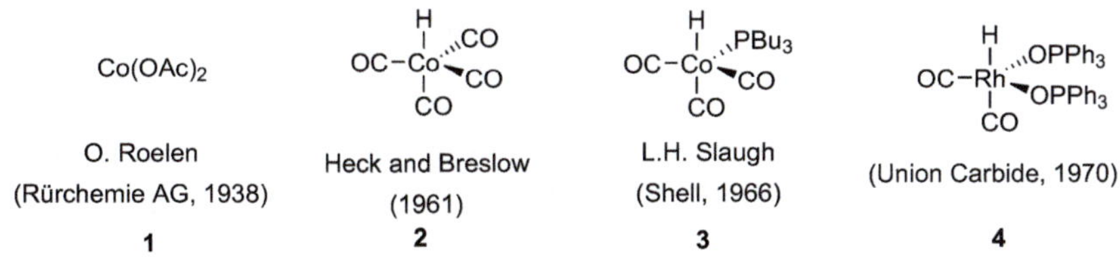

Scheme 5.2 The first hydroformylation catalysts.

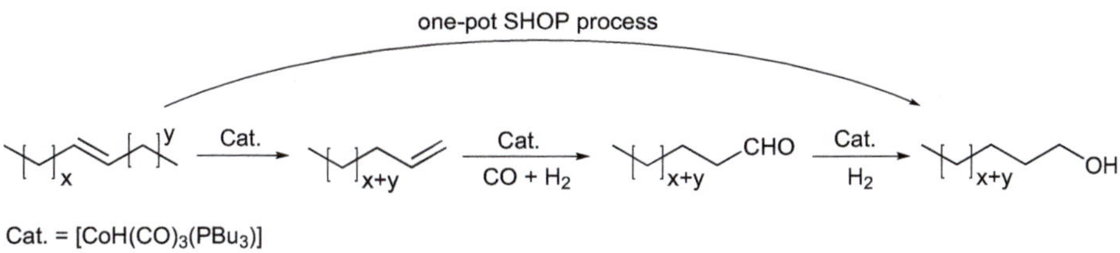

Scheme 5.3 Shell High Olefin Process (SHOP).

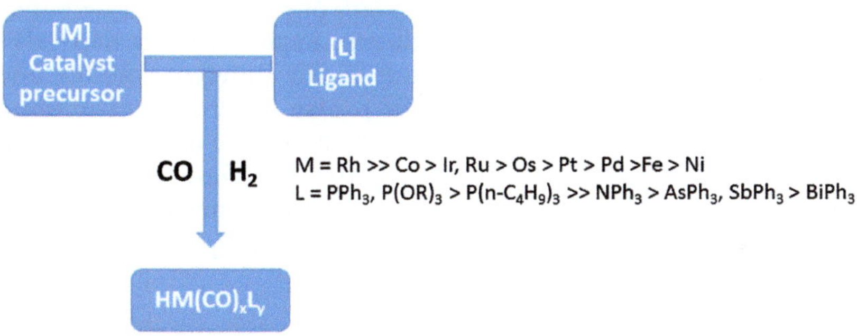

Figure 5.3 Generation of hydroformylation catalytic species.

chemoselectivity for aldehydes, poor regioselectivity and low activity for hydroformylation of internal olefins obtained with cobalt catalysts led to a continuous search for the development of new, more active and selective catalysts.

Nowadays, the generally accepted catalytic activity[5] of transition metal–ligand ([M]–L) pairs is depicted in Figure 5.3, with rhodium still being the most active metal, by far, for catalytic hydroformylation reactions.

The substitution of cobalt carbonyl catalysts **2** and **3** by Rh/phosphorus-modified catalyst **4** (Scheme 5.2), patented by Union Carbide Corporation in 1970,[14] almost solved the problem of chemoselectivity for aldehydes of terminal olefins and, since then, the structural modulation and synthesis of phosphorus ligands[15–22] is a parallel area of research, which provides a significant contribution to the industrial application of the hydroformylation reaction to fine chemical synthesis. Among them, phosphites, due to their easier structural modulation and higher π-acceptor properties, have contributed to great enhancements in activity and also selectivity of hydroformylation catalysts. Van Leeuwen[23,24] described the synthesis of a bulky tris(*o*-^{t}butylphenyl)phosphite (**5**), which is a landmark in the hydroformylation of 1,2- and 2,2-disubstituted olefins. This surprising high reactivity towards the hydroformylation of substituted olefins was explained by the large cone angle of the phosphite ligand (170°) that just allows the coordination of one ligand molecule (regardless of the ligand concentration) and consequently this catalytic species exhibits less steric hindrance than the corresponding Rh/triphenylphosphine, where two or more phosphine molecules can be coordinated (Scheme 5.4). More recently, Bayon and Pereira[25] described the synthesis of a set of *tris*-binaphthyl-based chiral monophosphites (**6**, Scheme 5.4) with large cone angles (240–270°), whose rhodium complexes were also very active and selective in the hydroformylation of internal double bonds.[26,27] These Rh/phosphite catalysts opened the way for application of this reaction in the synthesis of fine chemicals with particular relevance for the pharmaceutical industry, including the hydroformylation of Δ4-steroidal double bonds.[27,28]

The control of the regioselectivity of the hydroformylation reaction, catalyzed by rhodium-phosphorus ligands, is still one of the major demands for academics and the pharmaceutical industry. Regioselectivity depends first and foremost on substrate structure, due to the different stability and reactivity of hydrometalated products, but it also depends on reaction parameters (pressure and temperature) and, greatly, on the electronic and steric parameters of the phosphorus ligands coordinated to the rhodium. A brief description of the reaction mechanism of rhodium catalysts of type $RhH(PPh_3)_n(CO)_n$ (**A**) is presented in Scheme 5.5, as proposed by Lazzaroni.[29]

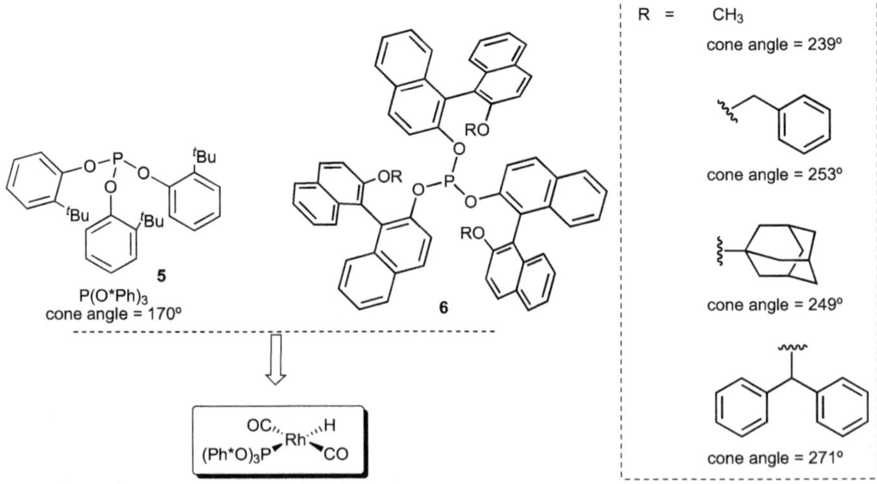

Scheme 5.4 Examples of bulky aryl phosphite ligands.

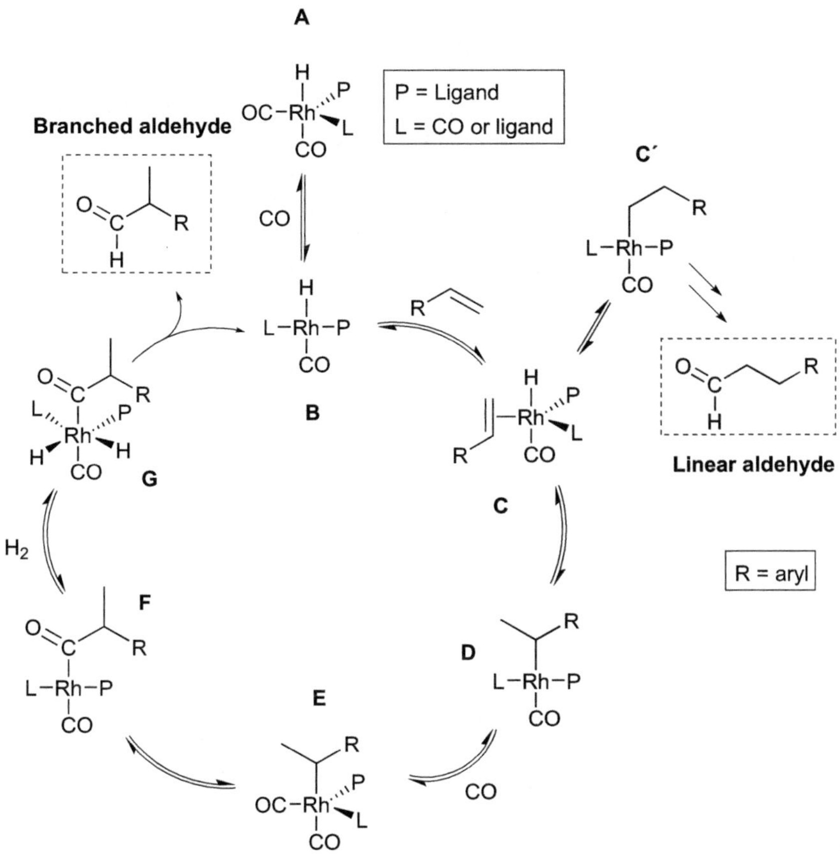

Scheme 5.5 Mechanism of Rh-catalyzed hydroformylation reaction.

The developments in IR[30] and NMR[31] spectroscopies have given an important contribution to the identification of the hydroformylation catalytically active species, but it is out of the scope of this chapter and additional information can be read in a number of excellent books and review papers.[4,5,32]

It is well accepted that the active catalytic species $RhH(PPh_3)_n(CO)_{4-n}$ (**A**) is prone to dissociate a CO or phosphorus molecule to form the 16-electron unsaturated complex **B** (dissociative mechanism), which is then able to coordinate the olefin to form species **C** (Scheme 5.5). The Rh–H insertions can give rise to the branched **D**, or linear square plane alkyl complexes. Complex **C′** can undergo β-hydride elimination, leading to the original olefin or its isomeric form, or it can react with CO to form the bipyramidal complex **E**. This complex can undergo the migratory insertion of the alkyl group into CO, generating the square planar acyl complex **F**. The oxidative addition of H_2 to complex **F** gives the catalytic species **G**. Then, complex **G** can go through reductive elimination to form the aldehyde and generate the initial catalytic species **B.** Thus, at a temperature of <70 °C and CO pressure of >1 MPa the insertion reaction is usually irreversible, and the regioselectivity is basically determined by this linear : branched ratio. However, the control of regioselectivity *via* reaction conditions (pressure and temperature) is also possible : High CO pressure, where less isomerization is expected, combined with low temperatures, which diminishes β-elimination, clearly contributes to the increase of the branched aldehyde, while low CO pressures and higher temperatures contribute to an increase of β-elimination and consequently of the amount of linear aldehyde (Scheme 5.5).

It is also well established that the number of phosphines and the stereochemistry of the Rh complex is a key factor in regioselectivity, for the same substrate and the same reaction conditions (T and P), but it is not our purpose to discuss in detail the mechanisms and kinetics of these processes, which are already described in the literature.[4]

The development of bidentate ligands, with designed backbones and bite angles,[33] also allowed significant improvements to the regioselectivity of rhodium-catalyzed hydroformylation of terminal olefins. The ones with bite angles around 90°, like (diphenylphosphino)ethane (**7**, DPPE), form preferentially less hindered axial-equatorial active species (Scheme 5.6a), with moderate regioselectivity for the linear aldehyde. On the other hand, bidentate ligands, like XantPhos (**8**)[34] and BiPhePhos (**9**),[35] with bite angles around 120°, form more sterically constrained bis-equatorial complexes, allowing the preferential formation of the terminal σ-alkyl complex, with a consequent increase of the linear aldehyde[4] (Scheme 5.6b).

The introduction of bidentate phosphorus ligands not only contributed to the improvement of the regioselectivity of the Rh-catalyzed hydroformylation

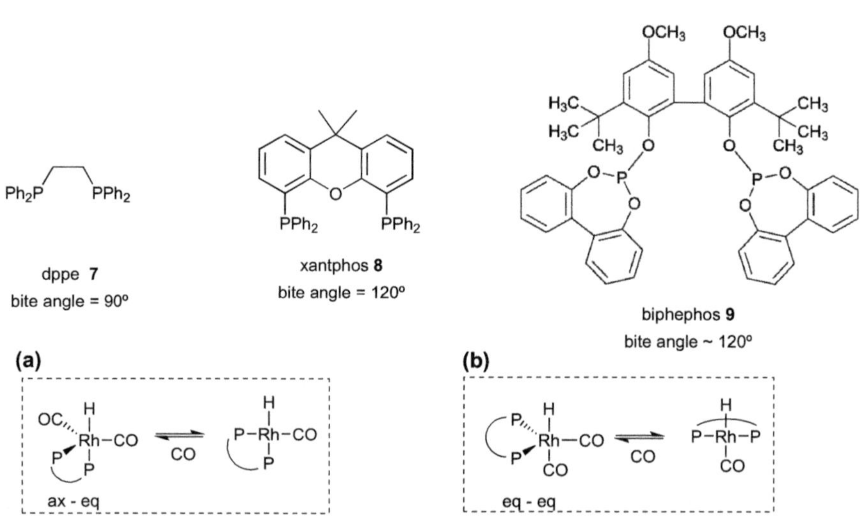

Scheme 5.6 Examples of bidentate phosphorus ligands used in catalytic hydroformylation and preferred geometry of their rhodium complexes.

but was also the driving force for the progress achieved in asymmetric hydroformylation reactions.[36,37] An important advance was achieved by the introduction of bulky diphosphites derived from homochiral (2*R*,4*R*)-pentane-2,4-diol, reported by Babin and Whiteker at Union Carbide Corporation.[38] The best results were achieved with ligand **10** (Scheme 5.7), which led to 90% ee in the Rh-catalyzed hydroformylation of styrene. The excellent enantioselectivity obtained with this type of diphosphite ligand was attributed to the formation of eight-membered chelates with rhodium(I), through a bis-equatorial coordination mode. Another relevant class of

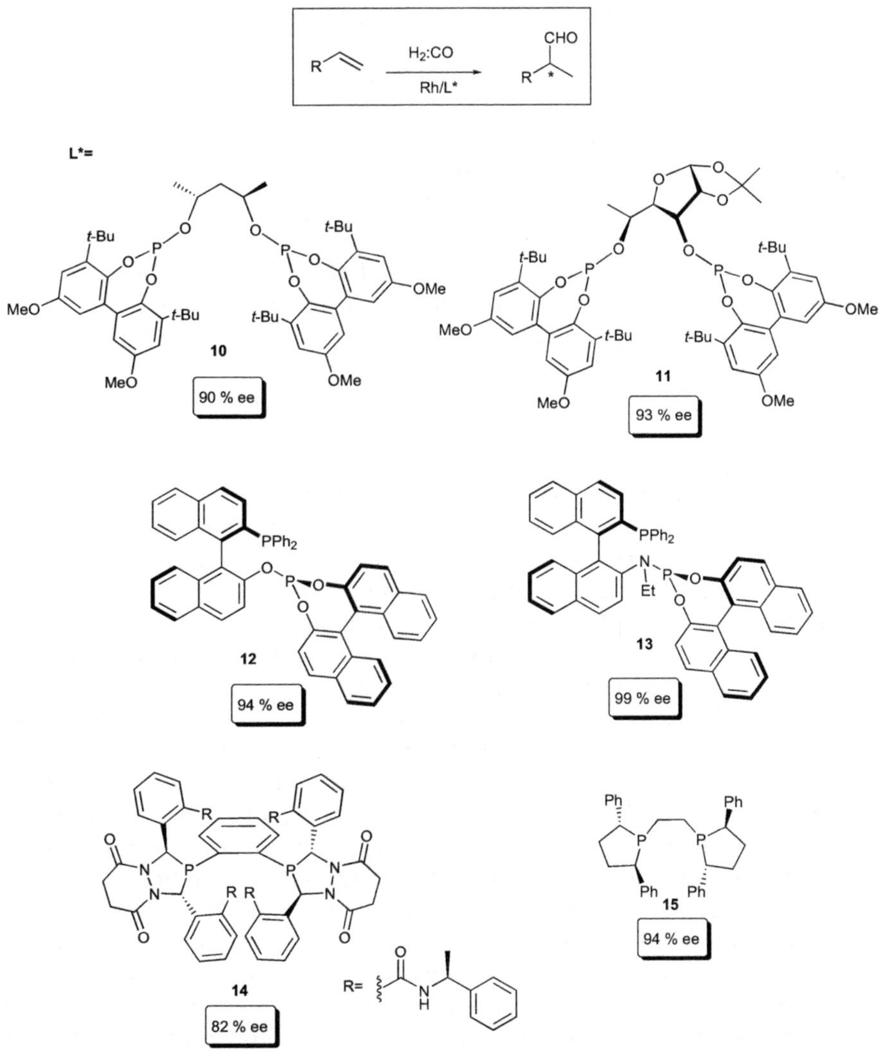

Scheme 5.7 Emblematic bidentate phosphorus ligands used in asymmetric Rh-catalyzed hydroformylation of olefins.

diphosphite ligands are those based on carbohydrate backbones, developed by Claver.[39] The rhodium(I) complex of **11**, containing a three-carbon bridge, achieved 93% ee in the hydroformylation of styrene. However, the major breakthrough was the discovery of mixed phosphite/phosphine ligands that form stable eight-membered chelate rings like BINAPHOS (**12**), synthesized by Takaya and Nozaki,[40,41] leading to high enantioselectivities for a variety of olefins (Scheme 5.7). Up to 95% ee together with 91% regioselectivity towards the branched product was obtained in the hydroformylation of styrene, and more than 90% ee with a range of other substrates, such as functionalized and internal alkenes. A unique dissymmetric environment in the Rh(I)-generated catalytically active species was ascribed to be the main factor for the high enantioselectivity obtained, where **12** coordinates to Rh(I) in equatorial–axial fashion, in which the more σ-donor phosphine P atom sits in the plane with the CO ligands, while the more π-accepting phosphite P atom binds apically to the hydride.[40]

Zhang and coworkers synthesized the mixed phosphine–phosphoramidite ligand **13**[42] from the chiral NOBIN (2-amino-2′-hydroxy-1,1′-binaphthyl). The crowded *N*-substituent was intended to supply a conformationally rigid structure able to provide a deeper and more closed chiral pocket than the corresponding complex formed from ligand **12**. Ligand **13** indeed produced a better performance in a number of substrates,[43] providing 99% ee along with 90% iso-regioselectivity, in the Rh-catalyzed hydroformylation of styrene (Scheme 5.7).[42,43]

Later, Landis[44] described the easy synthesis of chiral 3,4-diazaphospholanes **14** (Scheme 5.7) and its application to Rh-catalyzed enantioselective hydroformylation of styrene, allyl cyanide and vinyl acetate (relevant drug precursors) at a temperature of 60 °C and 3.5 MPa of $CO–H_2$ pressure. The best results using bidentate ligands that form stable five-membered chelate rings gave excellent enantioselectivities for all the relevant substrates studied and turnover frequencies up to 3000 h^{-1}. Furthermore, bisphospholane ligand **15** (Scheme 5.7), reported by the same authors,[45,46] was also efficiently used in the Rh-catalyzed hydroformylation of several olefins, achieving enantiomeric excesses of up to 94%.

Even though the asymmetric hydroformylation reaction has scarcely been used in industrial processes, the straightforward synthesis of phosphorus ligands and their stability cannot fail to open new prospects for its future application in asymmetric chemical synthesis. Furthermore, a detailed discussion of these aspects and the influence of the substrate structure on the chemo-, regio-, and enantioselectivity can be read in additional references.[4,10,36]

Regardless of the high activity and selectivity for aldehydes obtained with rhodium catalysts, their large-scale industrial application is still a problem, not only due to the continuously increasing rhodium prices but also due to lack of process sustainability caused by difficulties in rhodium availability. Thus, the search for alternative and less toxic metal catalysts for hydroformylation like iridium, ruthenium and even iron is currently an area of great interest.[47–51]

5.3 Hydroformylation with Alternative Catalytic Systems

In 2011, Beller opened the way for the potential application of iridium and ruthenium metal complexes as catalysts for tandem catalytic hydroformylation/hydrogenation of olefins.[52,53] Nozaki enlarged the scope and demonstrated that the reduction reaction of the aldehyde, generated in the hydroformylation step, to the corresponding alcohol could be conducted by the same ruthenium catalyst. To achieve this purpose, a Shvo's type catalyst was developed, using XantPhos (**8**) as the co-ligand, and only traces of aldehyde have been obtained, under optimized reaction conditions (Scheme 5.8).[54]

Beller[55] expanded this methodology to convert several olefins into the corresponding alcohols using a new ruthenium/imidazole phosphine **16** as the catalyst, the best results being obtained with linear terminal olefins, with chemoselectivities for alcohols reaching up 99% (Scheme 5.9).

The great interest of this reaction for industrial purposes led Behr to develop continuous mini-plants, at kilogram scale, for hydroformylation reactions using iridium[56] and ruthenium[57] metal complexes as catalysts. One of the most important features making this process more attractive in terms of cost and environment is the recycling mechanism involving an extraction with a non-polar solvent mixture of isooctane–DMF followed by distillation of the products and reutilization of the catalyst (Figure 5.4). Under these conditions, very low iridium and phosphorus leaching rates were observed ($<0.1\%\ h^{-1}$), which was the key to successful steady state operation.

Moreover, it should be mentioned that the great majority of hydroformylation processes using alternative metals as catalysts have been designed for bulk chemistry applications, where the preparation of alcohols is of utmost interest. However, in the pharmaceutical industry, preferential

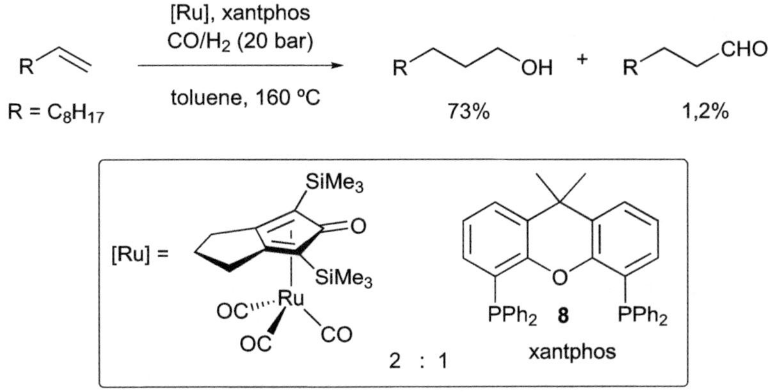

Scheme 5.8 Sequential hydroformylation/hydrogenation reaction of dec-1-ene using Ru/XantPhos catalysts.

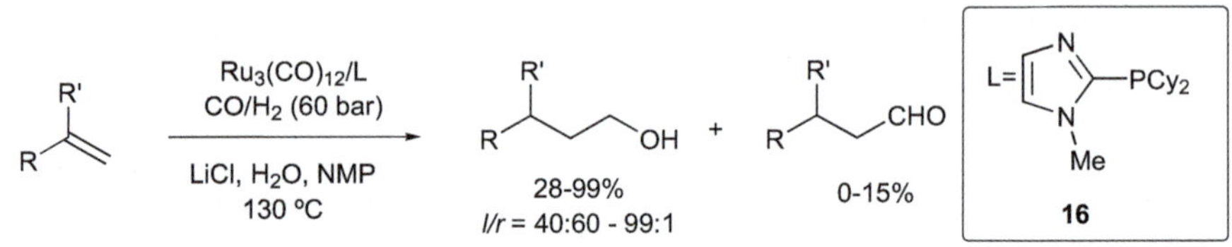

Scheme 5.9 Synthesis of alcohols from olefins catalyzed by ruthenium/phosphine-imidazole complexes.

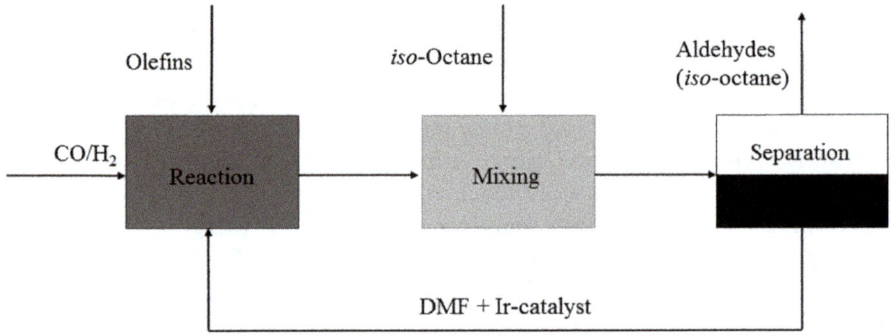

Figure 5.4 Continuous mini-plant schematics for hydroformylation reactions using iridium catalysts.[56]

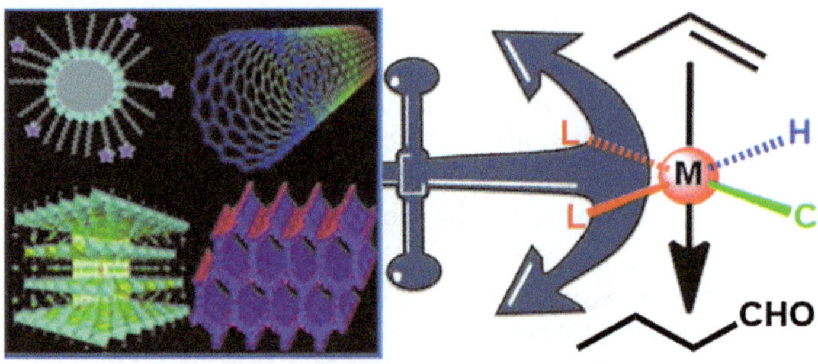

Figure 5.5 Generic principle for immobilization of hydroformylation catalysts.[58]

selectivity for aldehydes is still the major goal with rhodium still being the metal of choice. So, the challenge to develop less polluting catalytic processes using rhodium catalysts should be centered on processes where the catalyst can be recycled (Figure 5.5). Regarding this goal, several approaches have been reported in the literature: (i) immobilization of the homogeneous catalysts onto solid supports[58] like micro- and meso-porous silica materials,[59] clays,[60] organic polymers[61] and carbon nanotubes;[62] (ii) two-phase processes using alternative solvents;[63] and (iii) liquid phase-phase extraction.[64] These immobilized catalysts present similar catalytic performance to the homogeneous ones but with the advantage of being recycled and

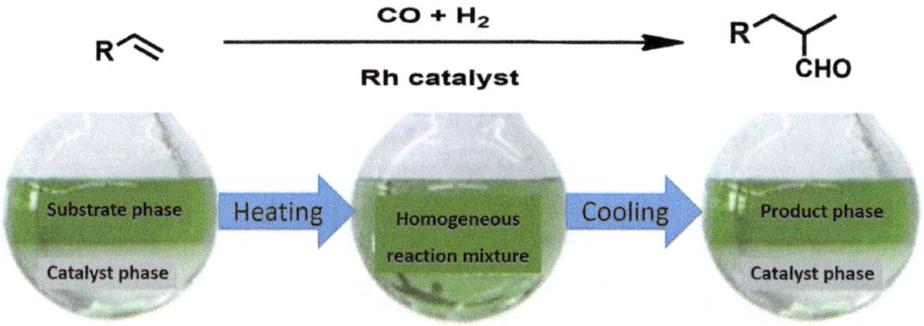

Figure 5.6 Biphasic alternative reaction system used for catalytic hydroformylation.

reused. This is an important perspective for future application at the industrial pharmaceutical level but it requires big reactors and the immobilization should be covalent, to avoid the leaching of the catalysts.[65]

Other attempts to reuse the catalyst were carried out using biphasic solvent systems based on ionic liquids,[66] fluorinated solvents,[67] water[68] and supercritical CO_2.[69] The major drawback of these alternative media methods, besides the lower activity and sometimes selectivity, is the requirement of specifically modified phosphines that require intricate synthetic processes in their preparation (Figure 5.6).

Nevertheless, several important examples have emerged in the last two decades, where different types of functionalized ligands have been applied in hydroformylation reactions under alternative reaction media (Scheme 5.10). For instance, for biphasic water–organic solvent systems, an emblematic catalytic system was developed by Ruhrchemie and Rhône-Poulenc based on ligand **17**, also known as TPPTS (triphenylphosphine-3,3′,3″-trisulfonic acid trisodium salt), a water-soluble phosphine used in the industrial hydroformylation of propene with an annual production over 400 000 tons.[70] Hoechst AG also developed a water-soluble binaphthyl-based diphosphine ligand (**18**), which was used in propene hydroformylation.[71] Other ligands based on ionic liquids have also been developed for hydroformylation catalysis in a liquid–liquid biphasic reaction environment, namely ligands **19** and **20**,[72,73] while fluorinated ligands **21** and **22** have been used in biphasic fluorous–organic solvent mixtures or in supercritical CO_2 as the reaction environment.[74,75]

5.4 Catalytic Hydroformylation in the Synthesis of Biologically Active Molecules: Selected Examples

As previously mentioned, rhodium-catalyzed hydroformylation of olefins has become an important tool for the multi-kilogram scale manufacturing of drugs or simple building blocks for active pharmaceutical ingredients.[3–5,76,77] One of the first examples of its application in the synthesis of pharmaceuticals on a commercial scale is the synthesis of aldehyde **24**, a key

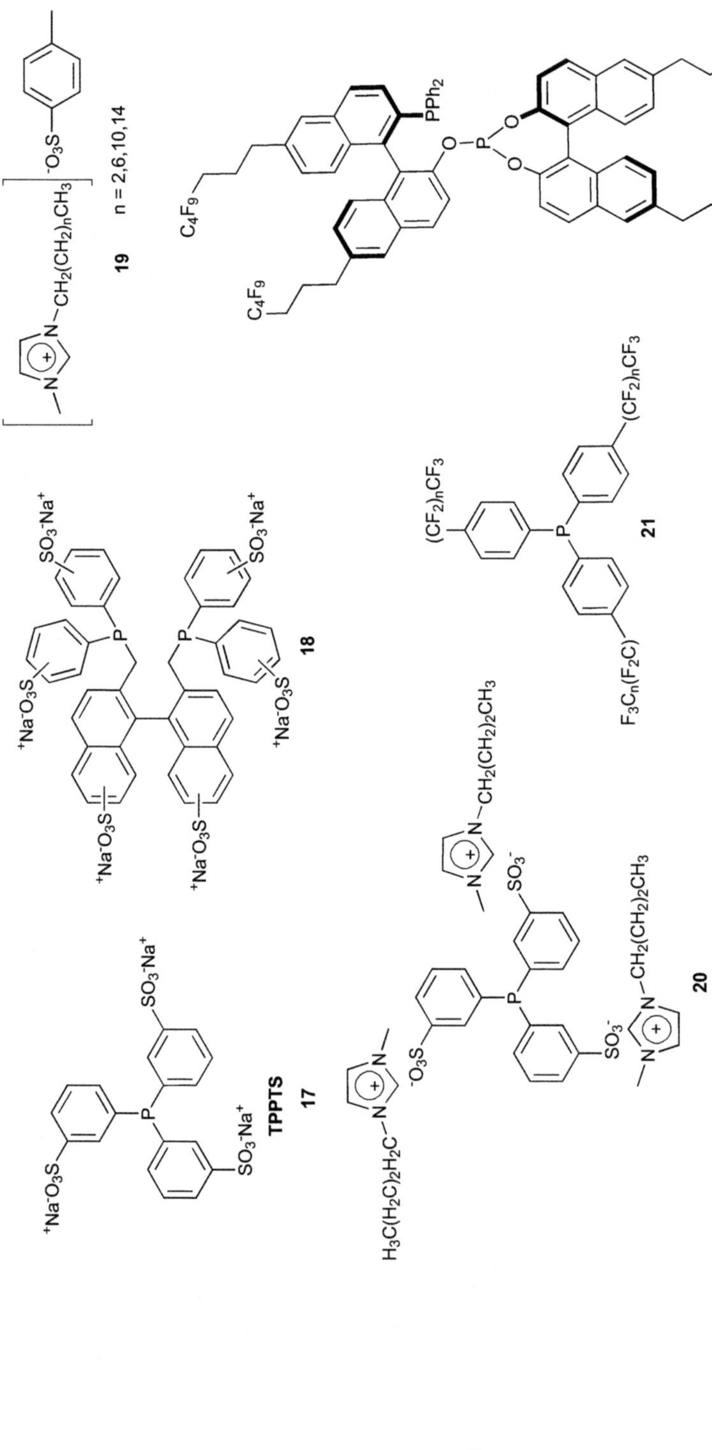

Scheme 5.10 Catalytic hydroformylation ligands applied in alternative reaction media.

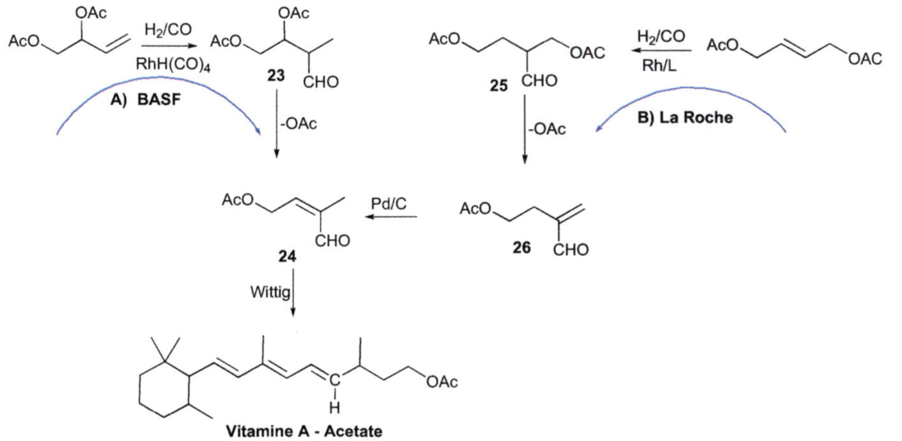

Scheme 5.11 Hydroformylation reactions in the industrial synthesis of vitamin A, developed by BASF and Hoffmann-La Roche.

precursor in the manufacturing process of vitamin A (Scheme 5.11).[78] The BASF process[79] starts with the hydroformylation of 1,2-diacetoxy-3-butene,[80] using $RhH(CO)_4$ as the catalyst, at $T=70$ °C and a CO–H_2 pressure of 60 MPa, yielding the branched aldehyde, 2-methyl-3,4-diacetoxybutanal (**23**) with 80% regioselectivity. After elimination of acetic acid, the α,β-unsaturated aldehyde **24** is obtained and further transformed into vitamin A through a Wittig reaction (Scheme 5.11A). On the other hand, the Hoffmann-La Roche process[81] is based on the hydroformylation of 1,4-diacetoxy-2-butene using $HRh(CO)PPh_3$ as the catalyst, at a temperature of 75 °C and a CO–H_2 pressure of 14 MPa. Under these reaction conditions, aldehyde **25** is obtained in 77% yield. This aldehyde can also eliminate acetic acid to give **26** with 90% yield, which can be isomerized to the desired aldehyde **24** in good yield, using Pd/C as a heterogeneous catalyst, and further transformed into vitamin A, also *via* a Wittig reaction (Scheme 5.11B).

In 2007, Chaudhari[82] described an alternative process for manufacturing the vitamin A intermediate **26**, where the hydroformylation and the deacetoxylation steps were carried out using a single tandem process with high conversion and selectivity (Scheme 5.12). To make this process economically viable for industry, a sustainable biphasic system was implemented using tri-(*m*-sulfophenyl)phosphine-TPPTS (**13**) as the rhodium ligand, which allowed the reutilization of the catalyst without loss of activity over several cycles. This is a paradigmatic example of the application of a green chemistry approach where a tandem hydroformylation-based protocol is an appealing alternative to the use of toxic phosgene.

Another relevant example of the application of the hydroformylation reaction in the pharmaceutical industry is the synthesis of (*S*)-allysine ethylene acetal, implemented by Chirotec.[83] The (*S*)-allysine ethylene acetal is an important intermediate in the manufacture of important angiotensin I-converting

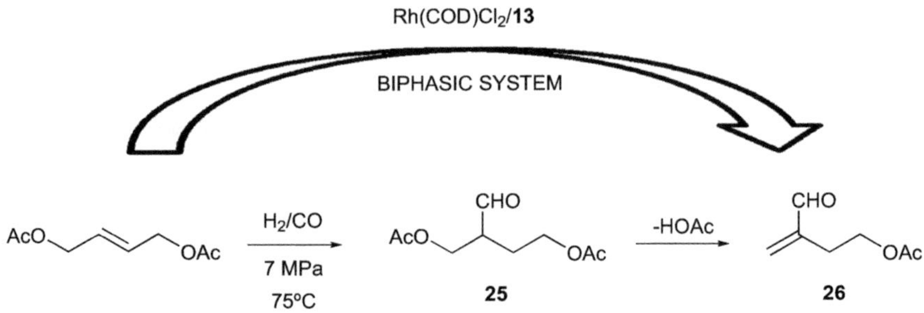

Scheme 5.12 Synthesis of vitamin A intermediate **26** using a tandem biphasic hydroformylation/deacetoxylation process.

enzyme and neutral endopeptidase inhibitors,[84] such as ilepatril and omapatrilat (Scheme 5.13). In this process, the hydroformylation of crotonaldehyde ethylene acetal **27** is performed using Rh/BiPhePhos (**9**) as the catalyst, under a CO–H_2 pressure of 0.3 MPa and a temperature of 80 °C, yielding aldehyde **28** in 90% (96% chemoselectivity, 94% *n*-regioselectivity). This rhodium catalyst, modified with BiPhePhos ligand (**9**), with a bite angle of 120°, is able to catalyze a tandem process involving isomerization of the olefin to the terminal position followed by regioselective hydroformylation. Then, **29** was obtained by a cyanoamination reaction, followed by protection of the amino group with a benzylic moiety affording the racemic *N*-protected amino acid **30**. This racemate is selectively resolved to the (*S*)-isomer *via* an enzymatic process with (*L*)-acylase. This is an interesting example where a catalytic process combined with enzymatic resolution allows the preparation of enantiopure drugs under sustainable conditions.

Storz,[85] at Pfizer, reported a three-step one-pot procedure for the high yielding diastereoselective synthesis of derivative **33**, a commonly used building block in the preparation of a variety of pharmacologically active molecules. The process was developed on a multi-kilogram scale *via* hydroformylation of norbornylene using $Rh(CO)_2(acac)$/DPPF (DPPF = 1,1′-bis(diphenylphosphino)ferrocene) as the catalyst, at 35 °C and 0.3 MPa of H_2–CO pressure. These reaction conditions allowed the preparation of *exo*-aldehyde **31** with 100% diastereoselectivity, which after oxidation with $NaClO_2$ in the presence of TEMPO catalyst gave the desired intermediate 2-*exo*-norbornyl carboxylic acid (**32**) and, finally, the respective sodium salt (**33**) in yields up to 90% (Scheme 5.14).

Another relevant example of the use of the hydroformylation reaction as a key tool for the synthesis of biologically active molecules was reported by Bredenkamp, who described the preparation of indoles.[86,87] These natural molecule analogues have a wide range of functions in living organisms and are considered, one of the most important structural motifs in drug discovery with a demand of about 20 000 t per year. The classic Fischer's indole synthesis, discovered in 1883, based on the reaction of a substituted phenylhydrazine with an aldehyde or a ketone (Scheme 5.15) is still the

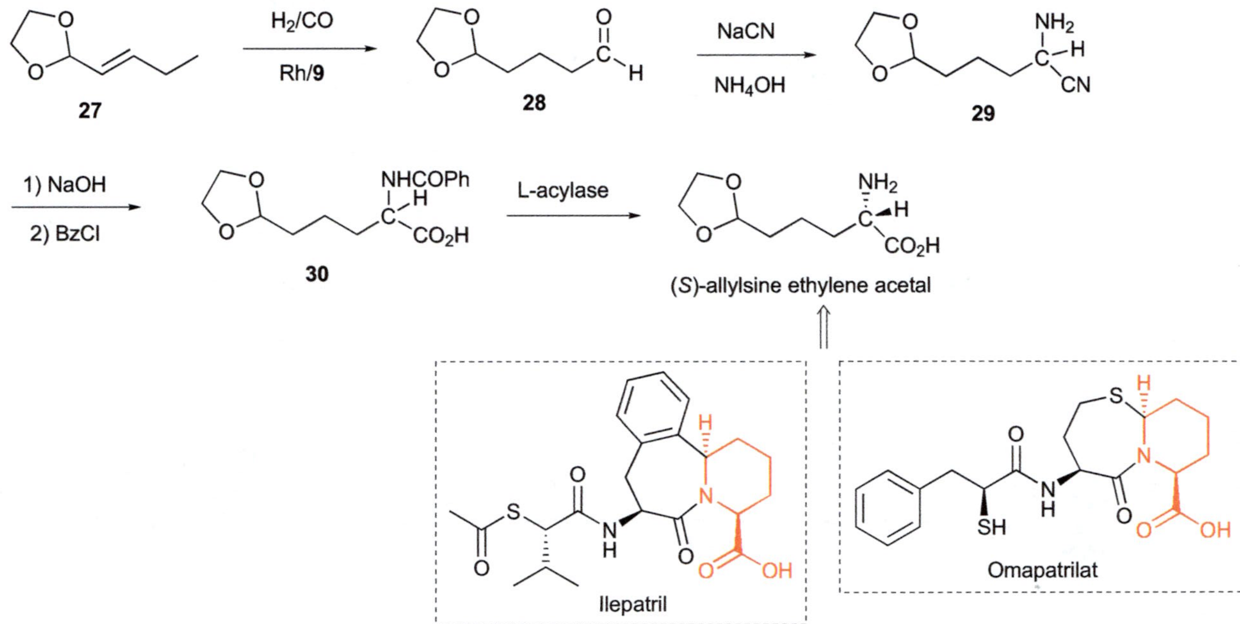

Scheme 5.13 Synthesis of (*S*)-allysine ethylene acetal, an important intermediate in the synthesis of ilepatril and omepatrilat.

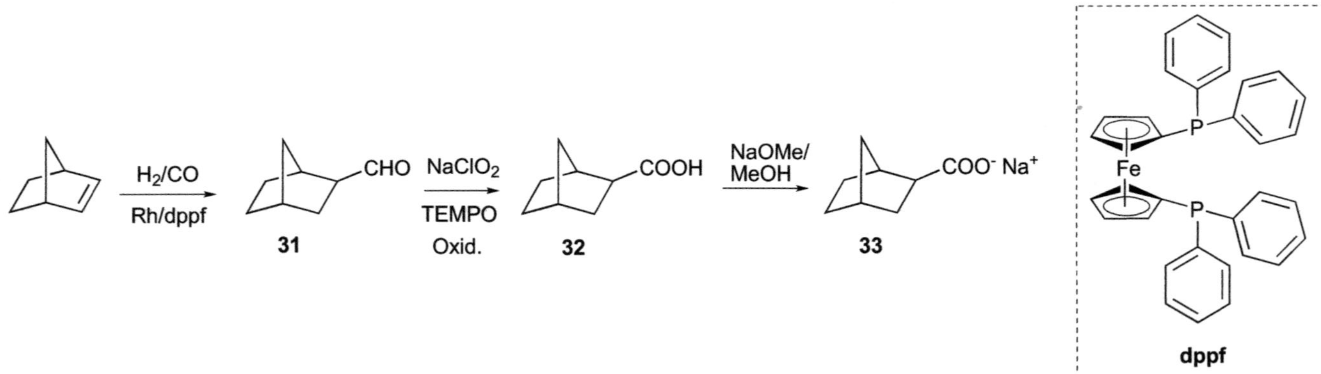

Scheme 5.14 Synthesis of 2-*exo*-norbornyl carboxylic acid *via* Rh/DPPF-catalyzed hydroformylation of norbornylene.

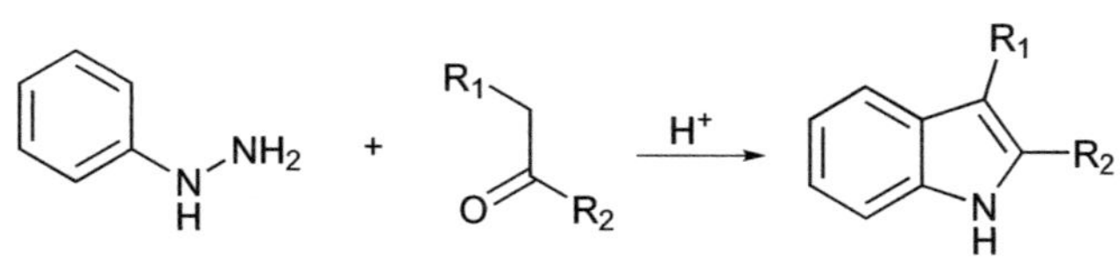

Scheme 5.15 Classic Fischer's indole synthesis.

most commonly used process for the preparation of 2,3-indole-substituted molecules.[88] However, these reactions usually require the use of toxic Brønsted or Lewis acid catalysts (HCl, H_2SO_4, boron trifluoride, zinc chloride, *etc.*). So, the search for sustainable alternative methodologies for the preparation of indoles is a great challenge. The Bredenkamp[86] approach (Scheme 5.16) is based on the hydroformylation of 2-alkynylanilines (**34**), which were previously prepared *via* palladium-catalyzed Sonogashira coupling. Subsequently, a tandem hydroformylation/intramolecular cyclization reaction was implemented, using Rh/DPPE as the catalyst at 100 °C and a CO–H_2 (1 : 1) pressure of 2.5 MPa, which afforded the desired 3-substituted indoles (**35**) with high selectivities and isolated yields of up to 92% (Scheme 5.16).

In 2008, Dow Global Technologies Inc.[89] patented the hydroformylation of *N*-allyl-phthalimide (**36**) at a 200 g scale using Rh/BiPhePhos (**9**) as the catalyst at 80 °C and 0.3 MPa CO–H_2, having obtained the linear *N*-acylamino aldehyde **37** with 92% regioselectivity (Scheme 5.17). The subsequent derivatization into the corresponding acetal and cleavage of the phthalimide functionality afforded the desired 4,4-dimethoxybutan-1-amine (**38**), which is an important bifunctional synthetic intermediate for the preparation of fine chemicals, such as several pharmaceuticals and fragrances.

Chiou[90] described a sustainable approach for the synthesis of pyrrolidinoindolines *via* one-pot domino Rh-catalyzed hydroformylation/double cyclization of amino cinnamyl derivatives, which can be used as intermediates for the preparation of physostigmine and physovenine,[91] efficient drugs against Alzheimer's disease (Scheme 5.18). Through this methodology, the carbamate **39** was subjected to hydroformylation conditions using $Rh(CO)_2(acac)$/triphenylphosphite as the catalyst at 75 °C and a CO–H_2 pressure of 8 MPa. It should be noted that the crude aldehyde was not isolated, being used without further purification in the double cyclization step, affording the desired product **40** in 80% yield (Scheme 5.18).

Diederich[92] reported the hydroformylation of 7-azanorbornene (**41**) to prepare the aldehyde **42**, using the catalytic system Rh/**43** as the catalyst at 65 °C and a CO–H_2 pressure of 4 MPa. The resulting aldehyde **42** is a crucial synthon for the preparation of 7-azanorbornane-plasmepsins, which are potent inhibitors of malarial aspartic proteases (Scheme 5.19).

Weinreb[93] described a rhodium-catalyzed hydroformylation strategy as one of the intermediary steps in the total synthesis of lepadiformine (Scheme 5.20). This alkaloid compound is known to have cytotoxic activity against several tumor cell lines, as well as cardiovascular activity.[94,95] Through this synthetic approach, the key intermediate dimethyl acetal **45** was obtained in 81% yield from the hydroformylation of hydroxymethyl compound **44** using the Rh/triphenylphosphite catalyst under moderate reaction conditions (60 °C, 0.4 MPa CO–H_2), followed by *in situ* aldehyde protection with trimethyl orthoformate in methanol.

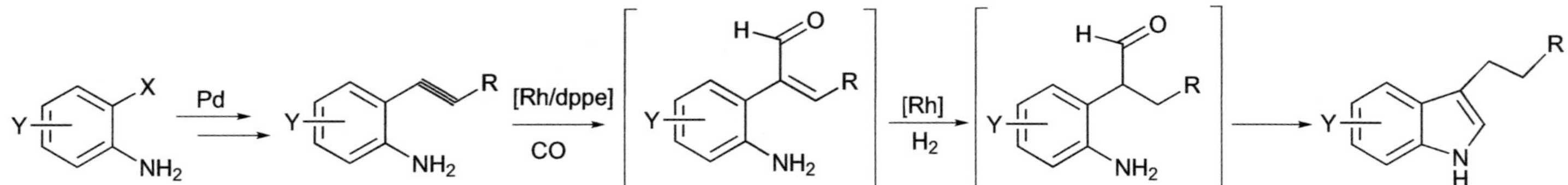

34 **35** up to 92%

Scheme 5.16 Synthesis of 3-substituted indoles *via* hydroformylation/cyclization.

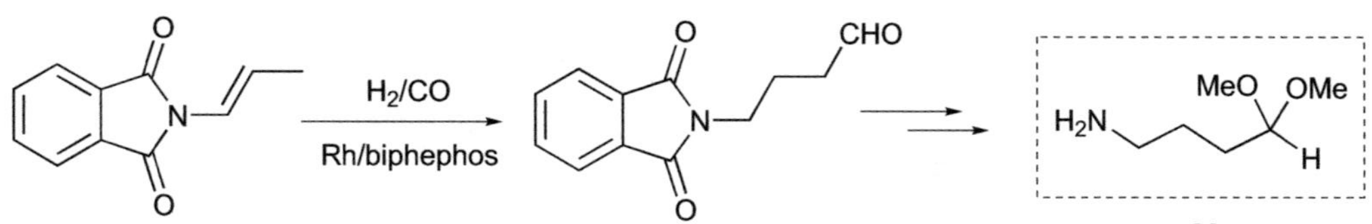

36 **37** **38**

4,4-dimethoxybutan-1-amine

Scheme 5.17 Hydroformylation of *N*-allyl-phthalimide as the key step in the synthesis of amino dialkoxy compounds.

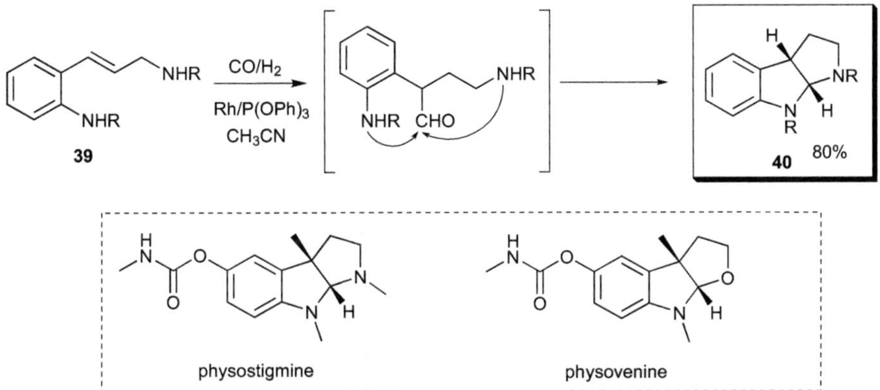

Scheme 5.18 Tandem hydroformylation/double cyclization reaction of amino cinnamyl derivatives for the synthesis of pyrrolidinoindolines.

5.4.1 Enantioselective and Diastereoselective Hydroformylation in Drug Synthesis

As stated above, the discovery and development of new phosphorus ligands, such as mono- and di-phosphines and/or phosphites, with modulated structures, and the advances of *in situ* spectroscopic techniques, made significant contributions to the implementation of the hydroformylation reaction in the pharmaceutical industry. Most of the catalysts consist of rhodium complexes, which have allowed sustainable methods for manufacturing aldehydes from olefins with 100% atomic economy and high chemo- and regio-selectivity. In addition, the development of active and enantioselective chiral catalytic systems represents a considerable challenge in synthetic chemistry, regarding the effective application of asymmetric hydroformylation to industrial pharmaceutical processes, in order to transform prochiral olefins into enantiomerically pure aldehydes.

The asymmetric hydroformylation reaction may be considered as one of the most powerful and sustainable processes for the one-pot transformation of a C=C bond into an aldehyde, with an additional carbon atom, and the generation of stereogenic centers resulting from the construction of new C–C and C–H bonds is of utmost relevance (Figure 5.7).

However, the difficulties in obtaining highly active, selective and sufficiently stable catalysts currently constitute one of the greatest issues regarding the transposition of this reaction to the pharmaceutical industry. Therefore, this is a major challenge that still needs to be addressed.

In this section, we present a set of selected examples of chiral catalysts and their application in the preparation of biologically relevant enantiopure drugs/synthons.

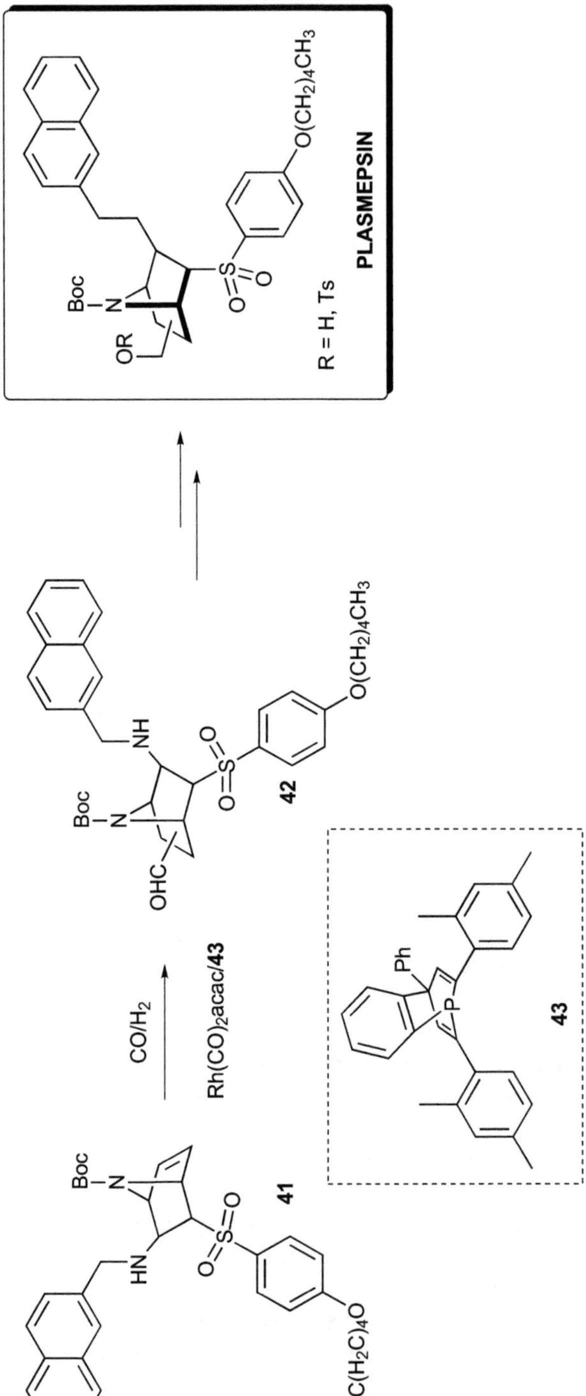

Scheme 5.19 Hydroformylation of 7-azanorbornene, leading to the synthesis of a 7-azanorbornane-plasmepsin intermediate.

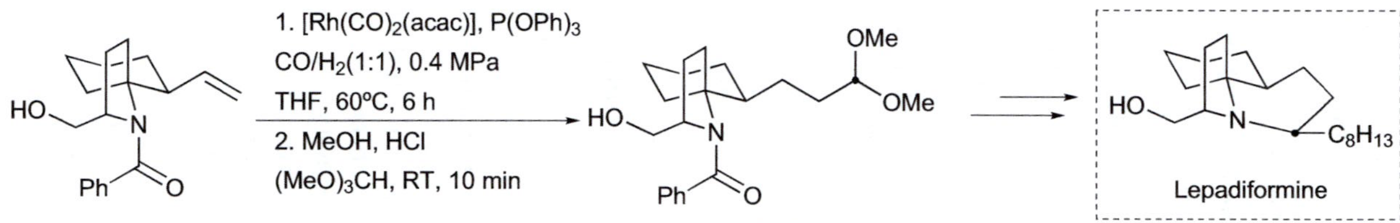

44 **45** 81%

Scheme 5.20 Rh-catalyzed hydroformylation leading to aldehyde **45**, an intermediate in the synthesis of lepadiformine.

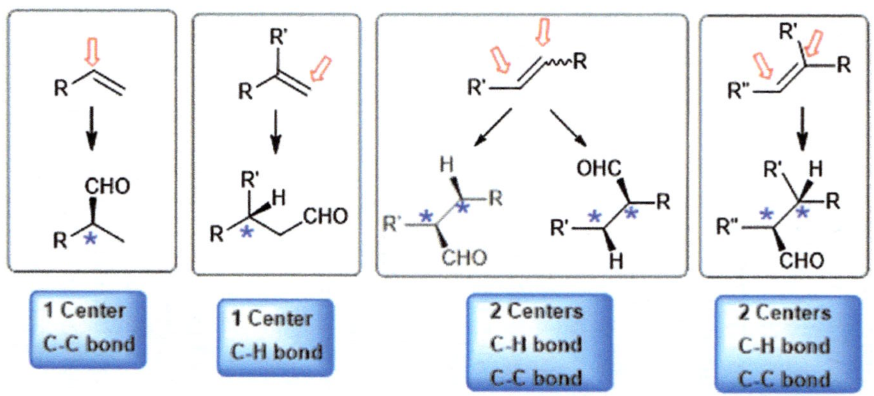

Figure 5.7 New C–C bonds and stereogenic centers, resulting from the asymmetric hydroformylation of C=C double bonds.

A great number of the academic studies regarding the implementation of asymmetric hydroformylation to the synthesis of pharmaceuticals have used vinylarenes (styrene and vinyl naphthalene derivatives) as model substrates. This may be attributed to the interest of the pharmaceutical industry in the search for sustainable approaches for the synthesis of α-aryl propionic acids, which constitute one of the most promising classes of molecules among nonsteroidal anti-inflammatory drugs (NSAIDs). Among them (*S*)-naproxen is a paradigmatic example that is sold in enantiomerically pure form. To the best of our knowledge, the only synthetic process reported so far for manufacturing (*S*)-naproxen at an industrial level is the one developed by Synthex SA,[96,97] which involves several sequential synthetic reactions and requires a chiral resolution step at the end *via* Pope–Peachey methodology (Scheme 5.21).

The first report on the use of catalytic hydroformylation for the synthesis of racemic naproxen was described by Brown[98] with the use of Rh/sugar-derived phosphine catalyst (**46**, Scheme 5.22). However, the first enantioselective approach for manufacturing (*S*)-naproxen was later described by Stille using a Pt/biphosphine catalyst (**47**), which achieved a high enantiomeric excess (81%) although with low regioselectivity for the branched aldehyde (b : l ratio = 0.7).[99]

An environmentally benign catalytic route for the synthesis of racemic naproxen has been recently reported by Rajurkar[100] *via* a two-step hydroformylation/oxidation. The development of this new strategy based on catalytic hydroformylation of 2-methoxy-6-vinylnaphthalene[97,101] followed by oxidation of the branched aldehyde is an interesting and "greener" synthetic alternative that involves fewer steps and generates smaller amounts of undesired by-products (Scheme 5.23).

More recently, Landis reported an efficient methodology for the in-flow enantioselective synthesis of (*S*)-naproxen based on asymmetric hydroformylation using Rh/**14** catalyst, followed by Pinnick oxidation of the chiral aldehyde **48** (Scheme 5.24).[102]

Despite these and other significant works in both academic and industrial labs, which led to efficient asymmetric hydroformylation catalysts of vinylarenes with both high enantioselectivity and regioselectivity, the high costs of the olefinic substrate and the rhodium catalysts, combined with the laborious synthesis of chiral phosphorus ligands, make the asymmetric hydroformylation economically unviable as an industrial route for manufacturing (*S*)-naproxen when compared with the efficient and economical enantiomeric resolution process used by Synthex.

Several other studies report the crucial implementation of the asymmetric hydroformylation reaction as a key step in the synthesis of biologically active compounds. For instance, Burke[103] described the asymmetric hydroformylation of enol ether **49** to obtain aldehyde **50**, which is a relevant intermediate for the total synthesis of the biologically active (+)-patulolide-C,[104] a naturally occurring 12-membered ring macrolide with antifungal

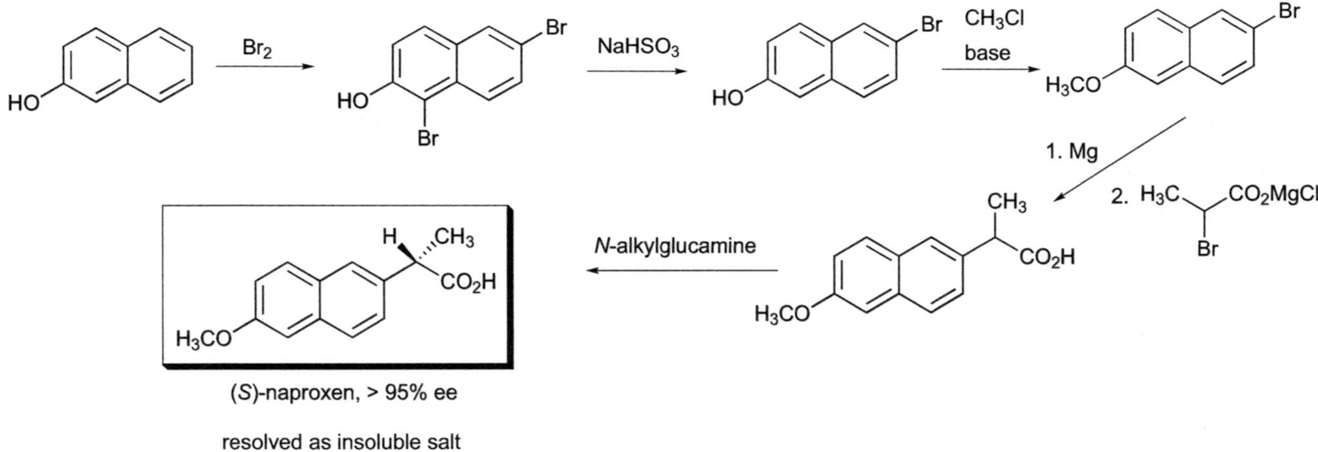

Scheme 5.21 Synthetic process for (*S*)-naproxen, developed by Synthex SA.

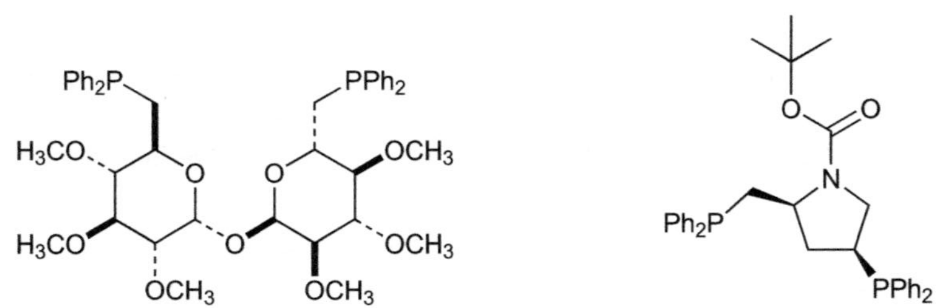

46 **47**

Scheme 5.22 Phosphorus ligands used in the first hydroformylation processes for the synthesis of naproxen.

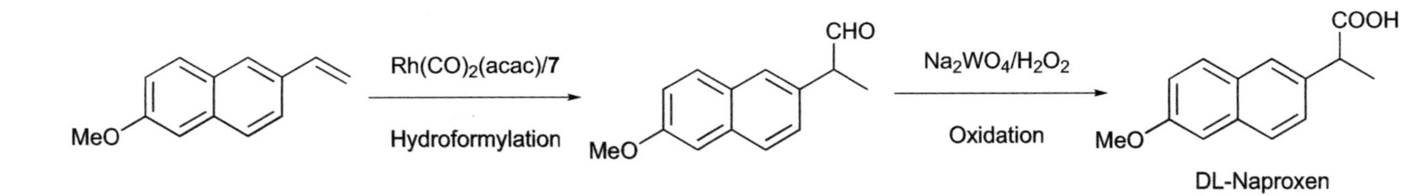

Scheme 5.23 Synthesis of racemic naproxen *via* hydroformylation-oxidation route.

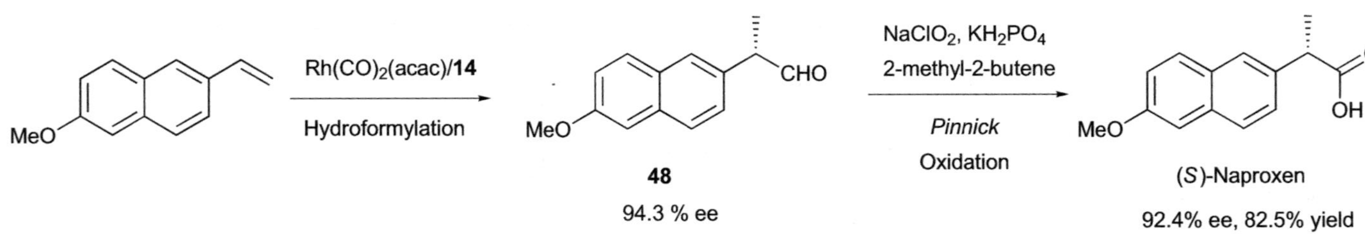

Scheme 5.24 Enantioselective in-flow synthesis of (*S*)-naproxen *via* hydroformylation.

and antibiotic activity.[105] The starting enol ether **49** was obtained *via* rhodium-catalyzed hydroacetoxylation. Then, aldehyde **50** was obtained with 100% conversion, full regioselectivity and high diastereoselectivity through rhodium-catalyzed hydroformylation under moderate reaction conditions (1 MPa H_2–CO at 50 °C) using the diazaphospholane **14** as the ligand. In the search for a sustainable process, the aldehyde intermediate was not isolated, with the crude product being subsequently submitted to an intramolecular Wittig reaction, in which the ylide was generated *in situ* by reaction of the terminal hydroxyl group with the ketene moiety, leading to cyclized product **51**. Finally, the desired molecule (+)-patulolide-C was obtained with 97% diastereomeric excess, after enzymatic hydrolysis with pseudomonas fluorescent lipases (IPL) (Scheme 5.25).

This methodology, based on a tandem hydroformylation/Wittig reaction, was applied by the same authors to the synthesis of other natural molecules with biological applications.[106] Such examples, despite making use of expensive rhodium catalysts, illustrate the efficiency of sustainable one-pot synthetic strategies centered on catalytic asymmetric hydroformylation for preparation of biologically relevant chiral products in high yields with chemo- and stereoselectivity, meeting the demands of green chemistry.

Other significant examples of the synthesis of biologically relevant compounds include the highly active, chemo-, regio- and stereo-selective rhodium-catalyzed hydroformylation of 1,3-dienes, using the chiral phosphine-phosphite ligand (*S,R*)-BINAPHOS (**12**).[40,41] For example, Jacobsen[107] described the application of this Rh catalytic system to promote the asymmetric hydroformylation of diene **52** under moderate reaction conditions (CO–H_2 pressure = 2 MPa, $T = 35$ °C) as a crucial step in the total synthesis of (+)-ambruticin (Scheme 5.26). Through this synthetic step, aldehyde **53** was obtained with 91% regioselectivity and a 96 : 4 diastereomeric ratio. (+)-Ambruticin is an antifungal agent, naturally isolated from fermentation extracts of the myxobacterium *Polyangium cellulosum*, that exhibits prominent activity against systemic medical pathogens such as *Coccidioides immitis*, *Histoplasma capsulatum*, and *Blastomyces dermatitidis*, and displays a potent inhibitory activity against the yeast strain *Hansenula anomala* (MIC = 0.03 μg mL^{-1}).[108,109]

Breit and Mann demonstrated the efficient application of rhodium-catalyzed hydroformylation of homoallylic azides for the syntheses of piperidinyl alkaloids and analogues with biological relevance.[110,111] The catalytic hydroformylation reactions were carried out using the catalytic system Rh/BiPhePhos (**9**) under moderate conditions (65 °C, 0.5 MPa CO–H_2) and applied to a diversity of homoallylic azides, reaching the corresponding aldehydes in moderate to good yields (51–92%) (Scheme 5.27a). This reaction was further applied as a key synthetic step in the total syntheses of several relevant alkaloids, such as (*S*)-anabasine (*via* one-pot hydroformylation/hydrogenation reaction) and (*S*)-nicotine (through a one-pot hydroformylation/Schmidt rearrangement process) (Scheme 5.27b).[110]

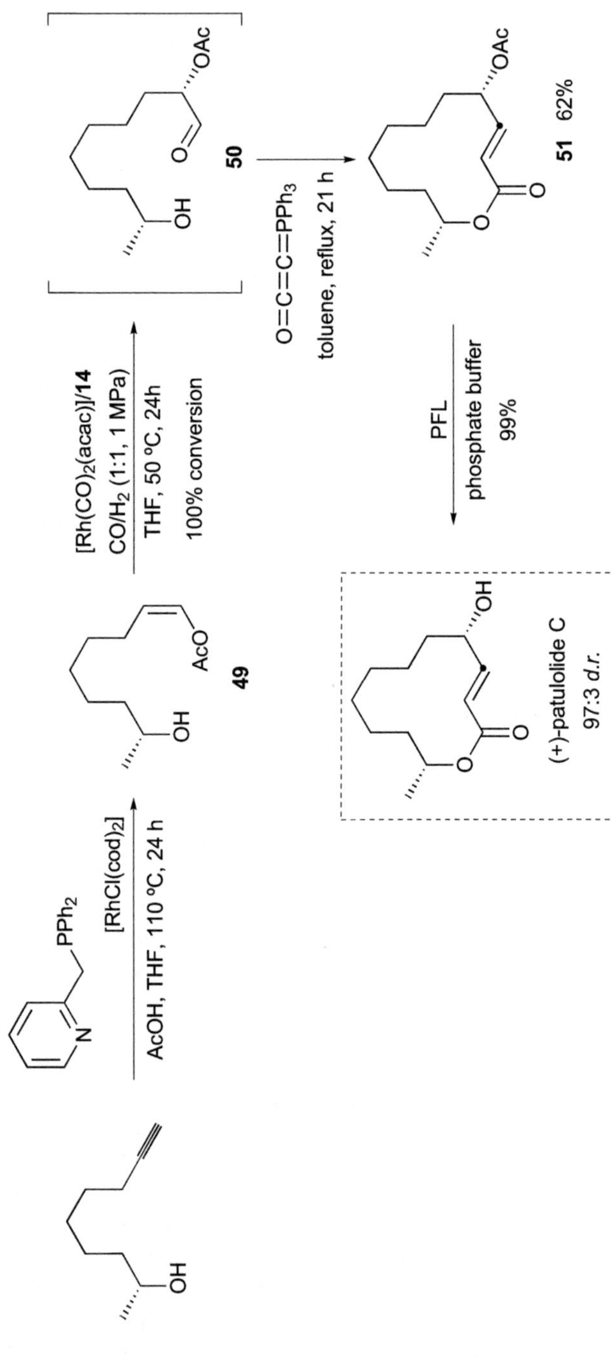

Scheme 5.25 Synthesis of (+)-patulolide C *via* one-pot asymmetric hydroformylation/Wittig reaction.

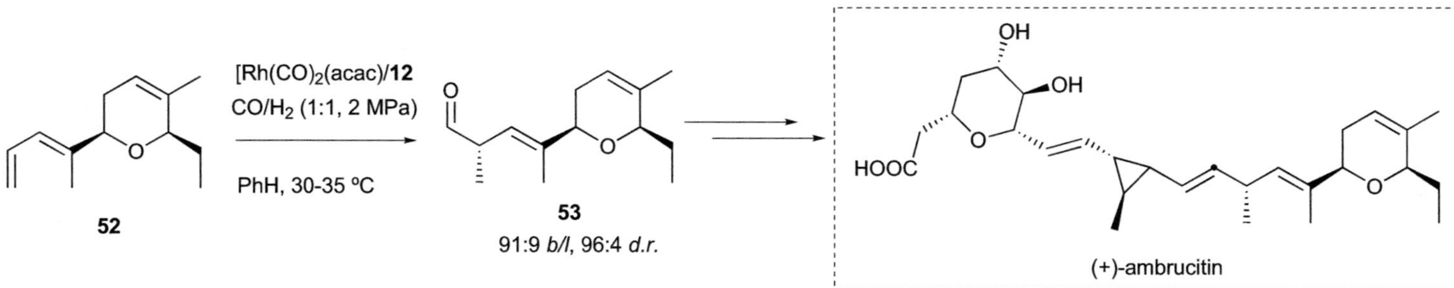

Scheme 5.26 Regio- and diastereo-selective hydroformylation of a 1,3-diene in the synthesis of (+)-ambruticin.

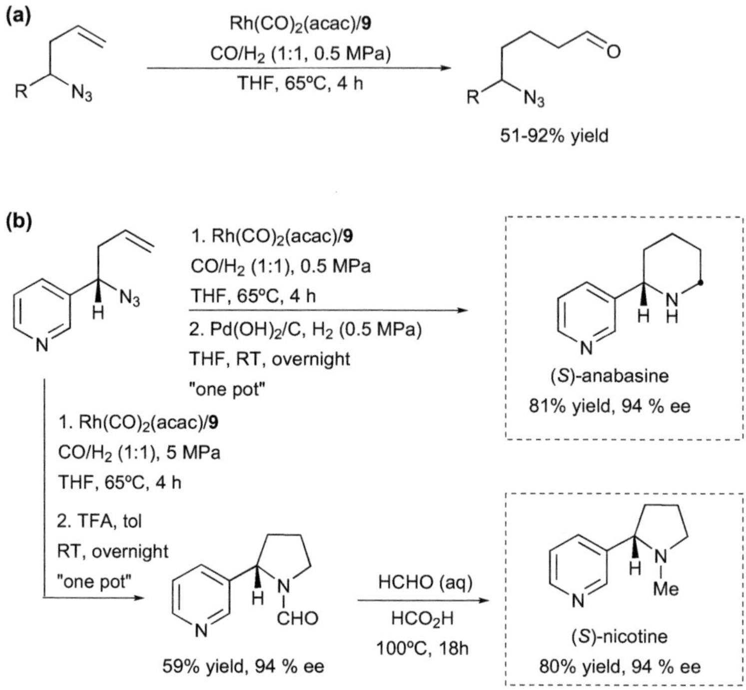

Scheme 5.27 Rh-catalyzed hydroformylation of homoallylic azides and its application to the synthesis of (*S*)-anabasine and (*S*)-nicotine.

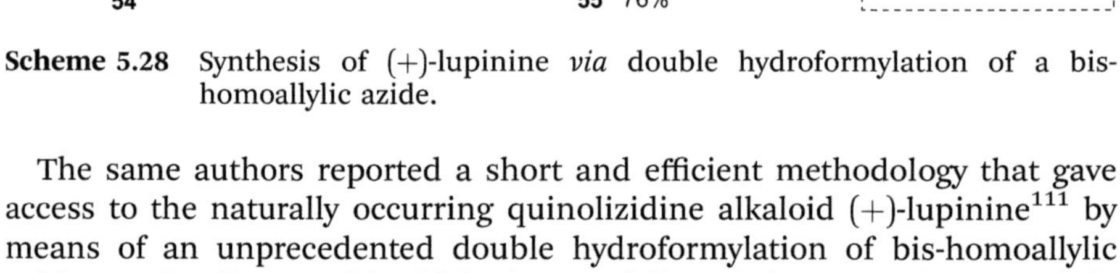

Scheme 5.28 Synthesis of (+)-lupinine *via* double hydroformylation of a bis-homoallylic azide.

The same authors reported a short and efficient methodology that gave access to the naturally occurring quinolizidine alkaloid (+)-lupinine[111] by means of an unprecedented double hydroformylation of bis-homoallylic azide **54**, leading to bis-aldehyde **55**, followed by a tandem catalytic hydrogenation/reductive bis-amination (Scheme 5.28).

In 2015, Krische[112] described the total synthesis of the emblematic polyketide ionophore antibiotic (+)-zincophorin methyl ester (Scheme 5.29), a potent agent with *in vivo* activity ($\leq$1 ppm) against Gram-positive bacteria.[113,114] One of its key steps consisted of the hydroformylation of **56**, using a XantPhos (**8**) modified rhodium catalyst,[33] providing the linear

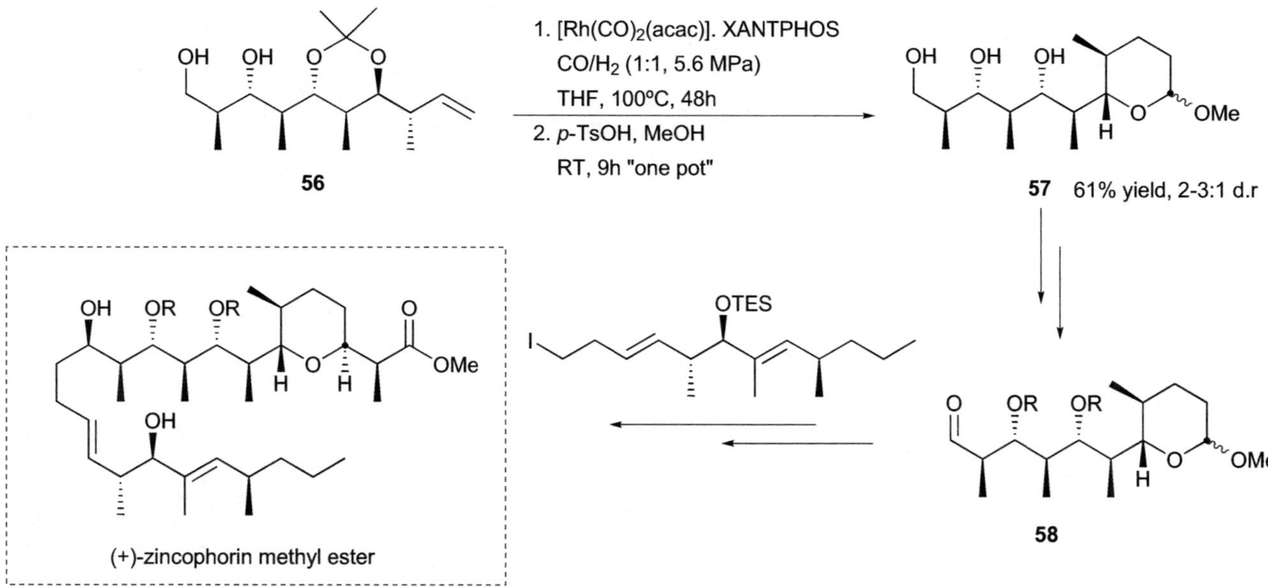

Scheme 5.29 Synthesis of (+)-zincophorin methyl ester, including a one-pot hydroformylation/acetalization step.

aldehyde **57**, which upon exposure to methanol, in the presence of *p*-toluenesulfonic acid, delivered the pyran **58** as a mixture of diastereomers at the anomeric position.

Breit[115] illustrated, for the first time, the concept of an RDG (reagent-directing group) controlled organic reaction to the enantioselective total synthesis of (*R*,*R*,*R*)-α-tocopherol, which is one of the most prominent and biologically active naturally occurring vitamin E compound, an essential food ingredient and the most important fat-soluble antioxidant. One of the key steps of this synthesis was the *o*-DPPB-directed (*o*-DPPB = *ortho*-diphenylphosphanylbenzoyl) diastereoselective hydroformylation of homometallic precursor **59**, using a Rh/triphenylphosphite catalyst, which afforded aldehyde **60** in 81% yield with a diastereomeric ratio of 91 : 9 (Scheme 5.30). The RDG (*o*-DPPB) served to control the stereoselectivity during the course of the Rh-catalyzed hydroformylation reaction for the construction of the C16 isoprenoid side chain, and the same *o*-DPPB group also acted as a reagent-directing leaving group during the course of a directed copper-mediated allylic substitution in **61**, which simultaneously served as the fragment-coupling step and led to the removal of the *o*-DPPB group, to obtain the intermediate **62**.

The same author[116] reported the total synthesis of (+)-clavolonine, as well as the first enantioselective synthesis of (−)-deacetylfawcettiine and (+)-acetylfawcettiine (Scheme 5.31), which are complex alkaloid compounds widely used in traditional Chinese medicine and homeopathic therapies.[117] At the essence of their syntheses is a new strategy for the construction of a central cyclohexane core, consisting of an (*o*-DPPB)-directed hydroformylation/carbonyl-ene reaction sequence, followed by a second (*o*-DPPB)-directed hydroformylation reaction, thus providing a straightforward increase of molecular complexity. Subjecting the *o*-DPPB ester **63** to the conditions of a directed hydroformylation allowed for a smooth chemo-, regio-, and diastereoselective reaction of the 1,1-disubstituted alkene function, in the presence of a trisubstituted alkene, providing *syn*-aldehyde **64** in excellent yield and diastereoselectivity (diastereomeric ratio = 97 : 3). Subsequently, the directed rhodium-catalyzed hydroformylation of the exocyclic double bond of **65**, using a Rh/phosphite catalyst, sets the stereocenter at C6 (diastereomeric ratio = 95 : 5), with concomitant formation of aldehyde **66**, which is further transformed into (+)-clavolonine (Scheme 5.31). This was the first application of such a reaction sequence in the context of complex alkaloid synthesis and highlights the potential usefulness of a strategy involving catalyst-directing group-controlled organic reactions in total synthesis.

5.5 Conclusion and Future Perspective

The hydroformylation reaction can be considered as one of the most paradigmatic and pioneering examples of the use of transition metal complexes for the synthesis of biologically relevant synthons and drugs, obeying several

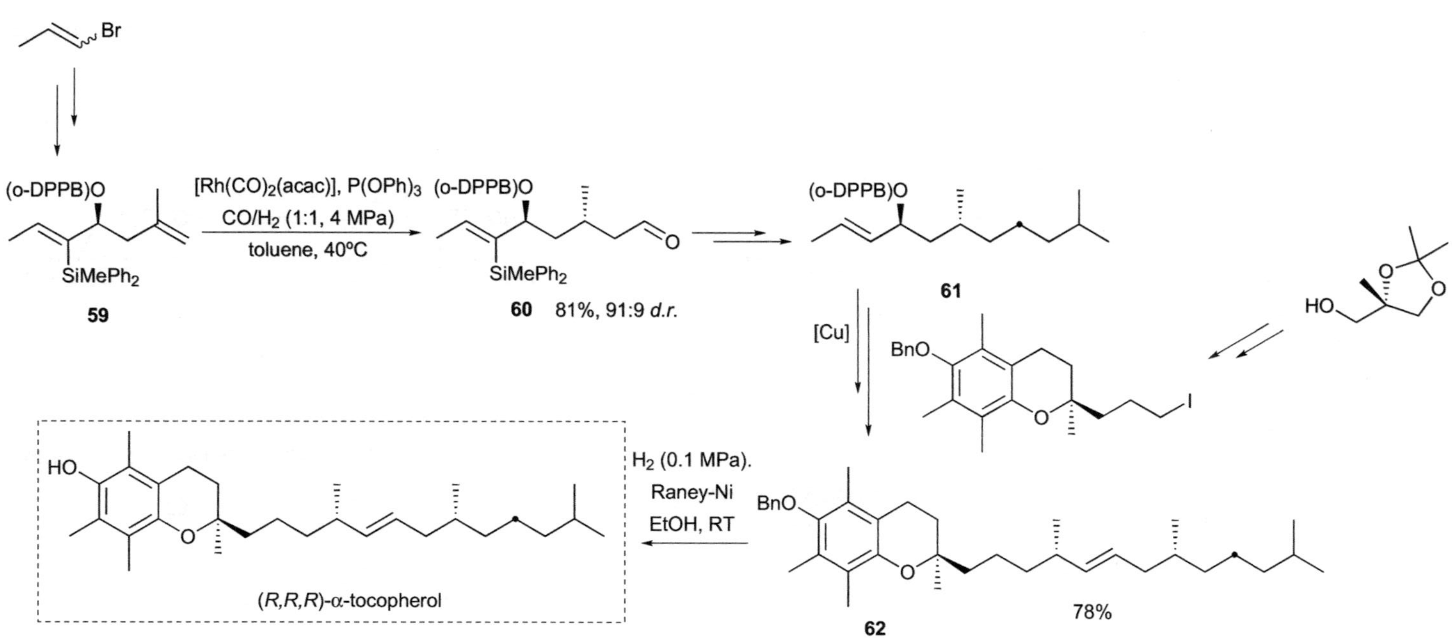

Scheme 5.30 Synthesis of (*R*,*R*,*R*)-α-tocopherol, involving diastereoselective RDG-controlled hydroformylation of a homometallic compound.

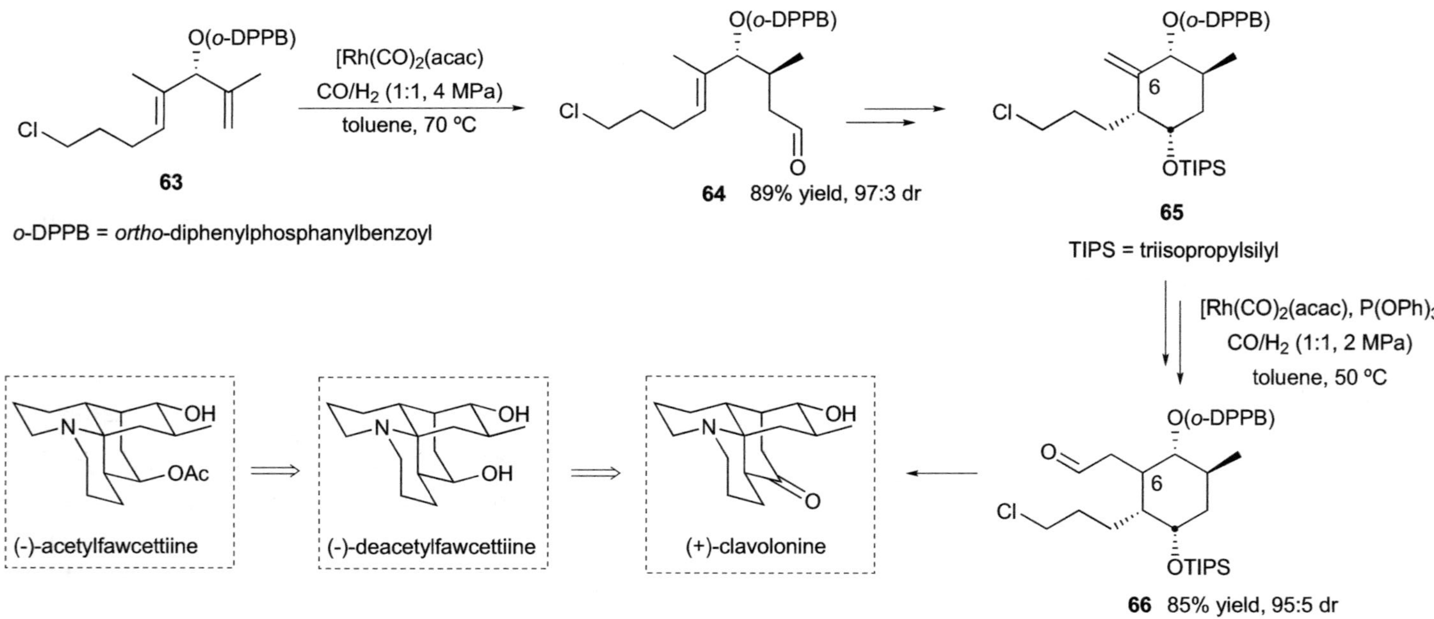

Scheme 5.31 Synthesis of (+)-clavolonine-derived alkaloids *via* catalyst-directing group (CDG) hydroformylation.

green chemistry principles: first, it is a catalytic reaction that allows the one-pot transformation of olefins into aldehydes in a 100% atom-economy process (second, eighth and ninth principles). In addition, the development of active and selective metal catalysts has allowed the use of milder reaction conditions (pressure and temperature), leading to a decrease in energy consumption (sixth principle), also enabling highly chemoselective processes with exclusive formation of aldehydes, reducing and preventing waste formation (first principle).

Although during the last few decades the hydroformylation reaction has been almost exclusively used in the bulk chemistry industry, there has recently been an increasing interest in its application to fine chemical processes, particularly in the pharmaceutical industry. In this context, the design and discovery of chiral ligands have led to the development of highly enantioselective synthetic processes, which permitted the syntheses of enantiomerically pure synthons and drugs, constituting a huge advance for medicinal chemistry. However, the high price associated with the development and use of metal catalysts and the often demanding syntheses of chiral phosphorus ligands require an urgent search for more economically viable and less polluting alternative solutions, namely through the replacement of rhodium catalysts by less toxic metals (such as ruthenium and iron), as well as through immobilization of homogeneous catalysts in solid supports and/or the development of biphasic systems in order to allow their recovery and reuse.

In summary, the application of the hydroformylation reaction in the pharmaceutical industry, according to the principles of green chemistry, is currently an important area with growing interest and significant promise in the near future.

Acknowledgements

The author would like to thank Coimbra Chemistry Centre for nurturing chemical science with excellence and also project "SunStorage—Harvesting and storage of solar energy", with reference POCI-01-0145-FEDER-016387, funded by the European Regional Development Fund (ERDF), through COMPETE 2020—Operational Programme for Competitiveness and Internationalisation (OPCI), and by national funds, through FCT—Fundação para a Ciência e a Tecnologia I.P.

References

1. O. Roelen, Chem. Verwertungsges. Oberhausen mbH, 1938, DE 849548.
2. B. Cornils, W. A. Herrmann and M. Rasch, *Angew. Chem. Int. Ed.*, 1994, **33**, 2144.
3. *Homogeneous Catalysis. Understanding the Art*, ed. P. W. N. M. van Leeuwen, Kluwer Academic Publ., Dordrecht, 2004.
4. *Rhodium Catalyzed Hydroformylation*, ed. P. W. N. M. van Leeuwen and C. Claver, Kluwer Academic Publ., Dordrecht, 2002.

5. R. Franke, D. Selent and A. Börner, *Chem. Rev.*, 2012, **112**, 5675.
6. B. P. Bondzic, *J. Mol. Catal. A: Chem.*, 2015, **408**, 310.
7. A. Galvan, F. J. Fananas and F. Rodriguez, *Eur. J. Inorg. Chem.*, 2016, 1306.
8. *Green and Sustainable Medicinal Chemistry: Methods, Tools and Strategies for the 21st Century Pharmaceutical Industry*, ed. L. Summerton, H. F. Sneddon, L. C. Jones. J. H. Clark, RSC Green Chemistry Book Series, RSC Publishing, Cambridge, 2016.
9. A. Behr, A. J. Vorholt, K. A. Ostrowski and T. Seidensticker, *Green Chem.*, 2014, **16**, 982.
10. M. Beller and K. Kumar, in *Transition Metals for Organic Synthesis. Building Blocks and Fine Chemicals*, 2nd edn, ed. M. Beller and C. Bolm, Wiley-VCH Verlag GmbH, Weinheim, 2004, p. 28.
11. A. F. Peixoto, M. M. Pereira and A. A. C. C. Pais, *J. Mol. Catal. A: Chem.*, 2007, **267**, 234.
12. R. F. Heck and D. S. Breslow, *J. Am. Chem. Soc.*, 1961, **83**, 4023.
13. L. H. Slaugh and R. D. Mullineaux, Shell, 1966, US 3239569 and 3239570.
14. R. L. Pruett and J. A. Smith, Union Carbide Corporation, 1970, US 3527809.
15. M. M. Pereira, M. J. F. Calvete, R. M. B. Carrilho and A. R. Abreu, *Chem. Soc. Rev.*, 2013, **42**, 6990.
16. M. J. F. Calvete, A. R. Abreu, R. M. B. Carrilho and M. M. Pereira, in *Advances in Chemistry Research*, ed. J. C. Taylor, Nova Science Publishers, New York, 2013, vol. 19, p. 111.
17. M. M. Pereira, R. M. B. Carrilho and M. J. F. Calvete, in *Organophosphorus Chemistry*, ed. D. W. Allen, D. Loakes and J. C. Tebby, Specialist Periodical Reports Series, RSC Publishing, Cambridge, 2017, **46**, 52.
18. Y. Jiao, M. S. Torne, J. Gracia, J. W. H. Niemantsverdriet and P. W. N. M. van Leeuwen, *Catal. Sci. Technol.*, 2017, **7**, 1404.
19. S. Allmendinger, H. Kinuta and B. Breit, *Adv. Synth. Catal.*, 2015, **357**, 41.
20. R. Tan, X. Zheng, B. Qu, C. A. Sader, K. R. Fandrick, C. H. Senanayake and X. Zhang, *Org. Lett.*, 2016, **18**, 3346.
21. X. Zheng, K. Xu and X. Zhang, *Tetrahedron Lett.*, 2015, **56**, 1149.
22. A. R. Abreu, A. F. Peixoto, A. R. Almeida, M. A. O. Lourenco, A. C. B. Neves, J. C. Bayón and M. M. Pereira, *Chem. Lett.*, 2013, **42**, 37.
23. P. W. N. M. Van Leeuwen and C. F. Roobeek (to Shell), 1980, US 4467116.
24. P. W. N. M. Van Leeuwen and C. F. Roobeek, *J. Organomet. Chem.*, 1983, **258**, 343.
25. R. M. B. Carrilho, A. R. Abreu, G. Petöcz, J. C. Bayón, M. J. S. M. Moreno, L. Kollár and M. M. Pereira, *Chem. Lett.*, 2009, **38**, 844.
26. R. M. B. Carrilho, A. C. B. Neves, M. A. O. Lourenço, A. R. Abreu, M. T. S. Rosado, P. E. Abreu, M. E. S. Eusébio, L. Kollár, J. C. Bayón and M. M. Pereira, *J. Organomet. Chem.*, 2012, **698**, 28.

27. G. N. Costa, R. M. B. Carrilho, L. D. Dias, J. C. Viana, G. L. B. Aquino, M. Pineiro and M. M. Pereira, *J. Mol. Catal. A: Chem.*, 2016, **416**, 73.
28. Z. Freixa, M. M. Pereira, J. C. Bayón, A. M. S. Silva, J. A. R. Salvador, A. M. Beja, J. A. Paixão and M. Ramos, *Tetrahedron: Asymmetry*, 2001, **12**, 1083.
29. R. Lazzaroni, P. Pertici, S. Bertozzi and G. Fabrizi, *J. Mol. Catal.*, 1990, **58**, 75.
30. L. A. van der Veen, M. D. K. Boele, F. R. Bregman, P. C. J. Kamer, P. W. N. M. van Leeuwen, K. Goubitz, J. Fraanje, H. Schenk and C. Bo, *J. Am. Chem. Soc.*, 1998, **120**, 11616.
31. J. M. Brown and A. G. Kent, *J. Chem. Soc., Perkin Trans.*, 1987, **2**, 1597.
32. C. Claver, M. Diéguez, O. Pàmies and S. Castillón, *Top. Organomet. Chem.*, 2006, **18**, 35.
33. C. P. Casey, G. T. Whiteker, M. G. Melville, L. M. Petrovich, J. A. Gavney Jr. and D. R. Powel, *J. Am. Chem. Soc.*, 1992, **114**, 5535.
34. M. Kranenburg, Y. E. M. van der Burgt, P. C. J. Kamer, P. W. N. M. van Leeuwen, K. Goubitz and J. Fraanje, *Organometallics*, 1995, **14**, 3081.
35. B. Kreidler, D. Fridag, B. Schemmer, B. Wechsler, A. Christiansen and D. Neumann, Evonik Degussa Gmbh, 2011, US 9127030.
36. B. F. Perandones, C. Godard and C. Claver, *Top. Curr. Chem.*, 2013, **342**, 79.
37. M. Diéguez, M. M. Pereira, A. M. Masdeu-Bultó, C. Claver and J. C. Bayón, *J. Mol. Catal. A: Chem.*, 1999, **143**, 111.
38. J. E. Babin and G. T. Whiteker, Union Carbide Corp., 1993, WO1993003839.
39. M. R. Axet, J. Benet-Buchholz, C. Claver and S. Castillón, *Adv. Synth. Catal.*, 2007, **349**, 1983.
40. T. Horiuchi, T. Ohta, K. Nozaki and H. Takaya, *Chem. Commun.*, 1996, 155.
41. K. Nozaki, T. Nanno and H. Takaya, *J. Organomet. Chem.*, 1997, **527**, 103.
42. Y. Yan and X. Zhang, *J. Am. Chem. Soc.*, 2006, **128**, 7198.
43. X. Zhang, B. Cao, Y. Yan, S. Yu, B. Ji and X. Zhang, *Chem. Eur. J.*, 2010, **16**, 871.
44. T. P. Clark, C. R. Landis, S. L. Freed, J. Klosin and K. A. Abboud, *J. Am. Chem. Soc.*, 2005, **127**, 5040.
45. A. T. Axtell, J. Klosin, G. T. Whiteker, C. J. Cobley, M. E. Fox, M. Jackson and K. A. Abboud, *Organometallics*, 2009, **28**, 2993.
46. A. T. Axtell, J. Klosin and K. A. Abboud, *Organometallics*, 2006, **25**, 5003.
47. J. Pospech, I. Fleischer, R. Franke, S. Buchholz and M. Beller, *Angew. Chem. Int. Ed.*, 2013, **52**, 2852.
48. C. Rodrigues, F. G. Delolo, J. Norinder, A. Börner, A. L. Bogado and A. A. Batista, *J. Mol. Catal. A: Chem.*, 2017, **426**, 586.
49. K. Kartashova, S. Mallet-Ladeira and M. R. Axet, *J. Organomet. Chem.*, 2015, **799–800**, 226.

50. G. M. Torres, R. Frauenlob, R. Franke and A. Börner, *Catal. Sci. Technol.*, 2015, **5**, 34.
51. M. A. Moreno, M. Haukka and T. A. Pakkanen, *J. Catal.*, 2003, **215**, 326.
52. I. Piras, R. Jennerjahn, R. Jackstell, A. Spannenberg, R. Franke and M. Beller, *Angew. Chem. Int. Ed.*, 2011, **50**, 280.
53. L. Wu, Q. Liu, R. Jackstell and M. Beller, *Angew. Chem. Int. Ed.*, 2014, **53**, 6310.
54. K. Takahashi, M. Yamashita, Y. Tanaka and K. Nozaki, *Angew. Chem. Int. Ed.*, 2012, **51**, 4383.
55. I. Fleischer, K. M. Dyballa, R. Jennerjahn, R. Jackstell, R. Franke, A. Spannenberg and M. Beller, *Angew. Chem. Int. Ed.*, 2013, **52**, 2949.
56. A. Kämper, S. J. Warrelmann, K. Reiswich, R. Kuhlmann, R. Franke and A. Behr, *Chem. Eng. Sci.*, 2016, **144**, 364.
57. A. Kämper, P. Kucmierczyk, T. Seidensticker, A. J. Vorholt, R. Franke and A. Behr, *Catal. Sci. Technol.*, 2016, **6**, 8072.
58. A. C. B. Neves, M. J. F. Calvete, T. M. V. D. P. Melo and M. M. Pereira, *Eur. J. Org. Chem.*, 2012, 6309.
59. P. W. N. M. van Leeuwen, A. J. Sandee, J. N. H. Reek and P. C. J. Kamer, *J. Mol. Catal. A: Chem.*, 2002, **182–183**, 107.
60. M. Wei, X. Zhang, D. G. Evans and X. Duan, *AIChE J.*, 2007, **53**, 2916.
61. N. E. Leadbeater and M. Marco, *Chem. Rev.*, 2002, **102**, 3217.
62. P. Serp, M. Corrias and P. Kalck, *Appl. Catal. A: Gen.*, 2003, **253**, 337.
63. *Alternative Solvents for Green Chemistry: Edition 2*, ed. F. M. Kerton and R. Marriott, RSC Green Chemistry Book Series, RSC Publishing, Cambridge, 2013, vol. **20**.
64. A. Behr, A. Kämper, R. Kuhlmann, A. J. Vorholt and R. Franke, *Catal. Sci. Technol.*, 2016, **6**, 208.
65. C. A. Henriques, F. M. S. Rodrigues, R. M. B. Carrilho, M. Silva, A. Fernandes, M. F. Ribeiro and M. M. Pereira, *Curr. Org. Chem.*, 2016, **20**, 1445.
66. M. Haumann and A. Riisager, *Chem. Rev.*, 2008, **108**, 1474.
67. E. G. Hope, R. Simayi and A. M. Stuart, *Top. Organomet. Chem.*, 2015, **52**, 217.
68. *Aqueous-Phase Organometallic Catalysis: Concepts and Applications, 2nd, Completely Revised and Enlarged Edition*, ed. B. Cornils and W. A. Herrmann, Wiley-VCH Verlag GmbH, Weinheim, 2006.
69. M. G. Pedrós, A. M. Masdeu-Bultó, J. Bayardon and D. Sinou, *Catal. Lett.*, 2006, **107**, 205.
70. C. W. Kohlpaintner, R. W. Fischer and B. Cornils, *Appl. Catal. A: Gen.*, 2001, **221**, 219.
71. H. Bahrmann, K. Bergrath, H.-J. Kleiner, P. Lappe, C. Naumann, D. Peters and D. Regnat, *J. Organomet. Chem.*, 1996, **520**, 97.
72. Q. Lin, W. Jiang, H. Fu, H. Chen and X. Li, *Appl. Catal. A: Gen.*, 2007, **328**, 83.
73. C. P. Mehnert, R. A. Cook, N. C. Dispenziere and M. Afeworki, *J. Am. Chem. Soc.*, 2002, **124**, 12932.

74. I. T. Horváth and J. Rábai, *Science*, 1994, **266**, 72.
75. D. Bonafoux, Z. Hua, B. Wang and I. Ojima, *J. Fluorine Chem.*, 2001, **112**, 101.
76. G. T. Whiteker and C. J. Cobley, *Top Organomet Chem.*, 2012, **42**, 35.
77. *Hydroformylation: Fundamentals, Processes, and Applications in Organic Synthesis*, ed. A. Börner and R. Franke, Wiley-VCH Verlag GmbH, Weinheim, 2016.
78. G. L. Parker, L. K. Smith and I. R. Baxendale, *Tetrahedron*, 2016, **72**, 1645.
79. W. Sarnecki and H. Pommer, BASF Ag, 1957, DE 1060386.
80. H. Siegel and W. Himmele, *Angew. Chem, Int. Ed., Engl.*, 1980, **19**, 178.
81. P. P. Fintton and F. H. Moffet, Hoffmann-La Roche, 1978, US 4124619.
82. R. Chansarkar, A. A. Kelkar and R. V. Chaudhari, *Ind. Eng. Chem. Res.*, 2007, **46**, 8629.
83. C. J. Cobley, C. H. Hanson, M. C. Lloyd, S. Simmond and W. J. Peng, *Org. Proc. Res. Dev.*, 2011, **15**, 284.
84. G. Masuyer, S. L. U. Schwager, E. D. Sturrock, R. E. Isaac and K. R. Acharya, *Sci. Rep.*, 2012, **2**, nr 717.
85. J. Gu, T. Storz, F. Vyverberg, C. Wu, R. J. Varsolona and K. Sutherland, *Org. Proc. Res. Dev.*, 2011, **15**, 942.
86. C. Holzapfel, E. Dasilva, L. D. Drijver and T. Bredenkamp, *ChemCatChem.*, 2016, **8**, 2912.
87. M. Marchetti, S. Paganelli, D. Carboni, F. Ulgheri and G. del Ponte, *J. Mol. Catal., A: Chem.*, 2008, **288**, 103.
88. W. L. Chen, Y.-R. Gao, S. Mao, Y.-L. Zhang, Y.-F. Wang and Y.-Q. Wang, *Org. Lett.*, 2012, **14**, 5920.
89. C. Cobley and C. L. Rand, Dow Global Technologies, 2008, WO2008134327.
90. W.-H. Chiou, C.-L. Kao, J.-C. Tsai and Y.-M. Chang, *Chem. Commun.*, 2013, **49**, 8232.
91. M. S. Morales-Ríos, N. F. Santos-Sánchez and P. Joseph-Nathan, *J. Nat. Prod.*, 2002, **65**, 136.
92. V. Aureggi, V. Ehmke, J. Wieland, W. B. Schweizer, B. Bernet, D. Bur, S. Meyer, M. Rottmann, C. Freymond, R. Brun, B. Breit and F. Diederich, *Chem Eur. J.*, 2013, **19**, 155.
93. P. Sun, C. Sun and S. M. Weinreb, *J. Org. Chem.*, 2002, **67**, 4337.
94. J. F. Biard, S. Guyot, C. Roussakis, J. F. Verbist, J. Vercauteren, J. F. Weber and K. Boukef, *Tetrahedron Lett.*, 1994, **35**, 2691.
95. M. Jugé, N. Grimaud, J.-F. Biard, M.-P. Sauviat, M. Nabil, J.-F. Verbist and J.-Y. Petit, *Toxicon*, 2001, **39**, 1231.
96. L. O. Piccolo, D. E. Valoti and M. G. Visenti, Blaschim S.p.A., 1988, US 4736061.
97. P. J. Harrington and E. Lodewijk, *Org. Process Res. Dev.*, 1997, **1**, 72.
98. J. M. Brown and S. J. Cook, *Tetrahedron*, 1986, **42**, 5105.
99. G. Parrinello and J. K. Stille, *J. Am. Chem. Soc.*, 1987, **109**, 7122.
100. K. B. Rajurkar, S. S. Tonde, M. R. Didgikar, S. S. Joshi and R. V. Chaudhari, *Ind. Eng. Chem. Res.*, 2007, **46**, 8480.

101. R. V. Chaudhari, *Top. Catal.*, 2012, **55**, 439.
102. M. L. Abrams, J. Buser, J. R. Calvin, M. D. Johnson, B. R. Jones, G. Lambertus, C. R. Landis, J. R. Martinelli, S. A. May, A. McFarland and J. Stout, *Org. Process Res. Dev.*, 2016, **20**, 901.
103. R. M. Risi and S. D. Burke, *Org. Lett.*, 2012, **14**, 1180.
104. E. K. Hoegenauer and E. J. Thomas, *Org. Biomol. Chem.*, 2012, **10**, 6995.
105. D. Rodphaya, J. Sekiguchi and Y. Yamada, *J. Antibiot.*, 1986, **39**, 629.
106. R. M. Risi, A. M. Maza and S. D. Burke, *J. Org. Chem.*, 2015, **80**, 204.
107. P. Liu and E. N. Jacobsen, *J. Am. Chem. Soc.*, 2001, **123**, 10772.
108. S. M. Ringel, R. C. Greenough, S. Roemer, D. Connor, A. L. Gutt, B. Blair, G. Kanter and M. von Strandtmann, *J. Antibiot.*, 1977, **30**, 371.
109. K. Gerth, P. Washausen, G. Höfle, H. Irschik and H. Reichenbach, *J. Antibiot.*, 1996, **49**, 71.
110. T. Spangenberg, B. Breit and A. Mann, *Org. Lett.*, 2009, **11**, 261.
111. E. Airiau, T. Spangenberg, N. Girard, B. Breit and A. Mann, *Org. Lett.*, 2010, **12**, 528.
112. Z. A. Kasun, X. Gao, R. M. Lipinski and M. J. Krische, *J. Am. Chem. Soc.*, 2015, **137**, 8900.
113. H. A. Brooks, D. Gardner, J. P. Poyser and T. J. King, *J. Antibiot.*, 1984, **37**, 1501.
114. D. A. Kevin, II, D. A. F. Meujo and M. T. Hamann, *Exp. Op. Drug Disc.*, 2009, **4**, 109.
115. C. Rein, P. Demel, R. A. Outten, T. Netscher and B. Breit, *Angew. Chem.*, 2007, **119**, 8824.
116. K. M. Laemmerhold and B. Breit, *Angew. Chem. Int. Ed.*, 2010, **49**, 2367.
117. X. Ma and D. R. Gang, *Nat. Prod. Rep.*, 2004, **21**, 752.

CHAPTER 6

Transfer Hydrogenation with Non-toxic Metals for Drug Synthesis

B. ROYO

ITQB NOVA, Instituto de Tecnologia Química e Biológica António Xavier, Universidade Nova de Lisboa, Av. da República, 2780-157 Oeiras, Portugal
Email: broyo@itqb.unl.pt

6.1 Introduction

The aim of this chapter is to provide an overview of the application of transfer hydrogenation reactions in the preparation of drugs. It covers two processes: transfer hydrogenation and hydrogen borrowing reactions, which are another version of transfer hydrogenation processes.

The synthesis of enantiopure alcohols and amines is of vital importance in the pharmaceutical industry. They are commonly accessed through selective catalytic hydrogenation of the corresponding prochiral ketones and imines through direct reaction with molecular hydrogen (Chapter 5) or by metal-catalysed asymmetric transfer hydrogenation (ATH). The last method uses an alcohol as a hydrogen donor instead of hazardous hydrogen gas, making this process safe and environmentally friendly. Usually, high-cost transition metals are used as catalysts for ATH. However, cheap, non-toxic, and environmentally benign catalysts have recently emerged as an alternative to precious metals. This chapter presents the most common catalytic systems developed for ATH giving special emphasis to the recent developments using cheap metal catalysts and their application in the pharmaceutical industry.

Green Chemistry Series No. 54
Sustainable Synthesis of Pharmaceuticals: Using Transition Metal Complexes as Catalysts
Edited by Mariette M. Pereira and Mário J. F. Calvete

Published by the Royal Society of Chemistry, www.rsc.org

Selected examples to illustrate the application of ATH processes in the preparation of bioactive molecules are discussed.

In addition, borrowing hydrogen processes for the formation of C–N and C–C bonds are presented. This process exploits the use of alcohols as alkylating agents by temporary removal of hydrogen. In this chemistry, the metal catalyst borrows hydrogen from an alcohol to give an intermediate ketone or aldehyde, which can be transformed into an imine or alkene. The return of hydrogen from the catalyst leads to the formation of new C–N and C–C bonds. This approach represents a green technology for the creation of C–N and C–C bonds, avoiding the use of toxic, expensive alkyl halides used in the conventional synthetic approach, which generate stoichiometric amounts of salts as waste. Furthermore, alcohols are readily available from renewable feedstock, making this methodology especially suitable for the valorisation of biomass. The atom-efficient formation of C–N and C–C bonds represents a key step for the synthesis of a plethora of compounds widely used in the pharmaceutical industry.

6.2 Transfer Hydrogenation

Transfer hydrogenation (TH) is the addition of hydrogen to an unsaturated molecule, typically C=O and/or C=N, using a hydride source other than gaseous H_2, in the presence of a catalyst (Scheme 6.1). TH constitutes an alternative method of hydrogenation, offering significant benefits with respect to the conventional hydrogenation reaction because it avoids the use of highly flammable molecular hydrogen and the need for special pressure equipment. The hydrogen sources commonly employed are isopropanol or formic acid, which are non-toxic, stable, easy to handle, and inexpensive.

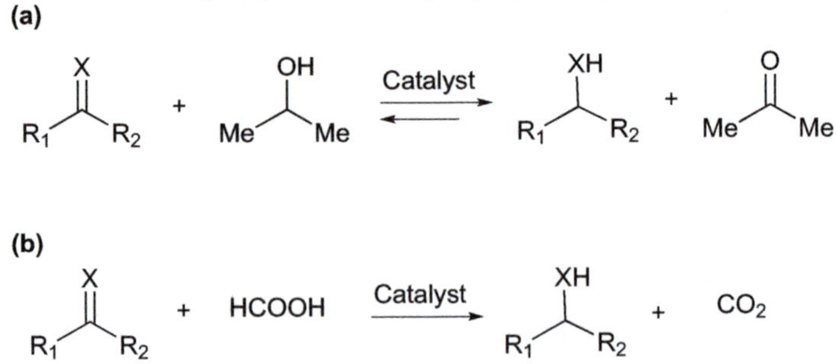

Scheme 6.1 TH of ketones and imines using isopropanol (a) and formic acid (b).

TH is an atom-efficient, environmentally benign method in which the alcohol acts as both the reaction solvent and the hydrogen source, and the catalysts are in general readily accessible and not sensitive. The operational simplicity of this process has made TH an elegant tool for hydrogenation of unsaturated compounds. The most versatile TH systems are:

(i) Isopropanol in the presence of catalytic amounts of base. In this case, acetone is formed as a side product (Scheme 6.1a). The major drawback is the reversibility of the reaction, often leading to incomplete conversion, and in ATH to racemisation. This limitation can be overcome by continuously distilling off the acetone formed.
(ii) Formic acid, usually as an azeotropic 5 : 2 mixture of HCO_2H and NEt_3 (Scheme 6.1b). This reaction is irreversible, affording high conversions without racemisation. The side product formed is CO_2; therefore reactions are performed in an open flask.

The use of transition metals as TH catalysts has been intensively studied during the last few decades.[1,2] Henbest described the first examples of metal-catalysed TH in the 1960s, using an iridium-based complex,[3] and later, Sasson and Blum[4] reported the first ruthenium complex in TH. These catalytic systems required high temperatures to get reasonable activities. The major advance in this area was the discovery that the addition of base dramatically increases the rate of hydrogen transfer reactions.[5] Since then, numerous catalytic systems operating under mild conditions have been developed using a base as a co-catalyst and employing iridium, ruthenium and rhodium complexes as catalysts. Most research efforts have focussed on the asymmetric version of the transfer hydrogenation reaction (ATH), which consists of the reduction of prochiral compounds ketones and imines in the presence of catalysts containing chiral ligands (Scheme 6.2).[6,7] Nowadays, ATH has been established as a powerful and elegant tool for enantioselective synthesis both in academia and industry.[7]

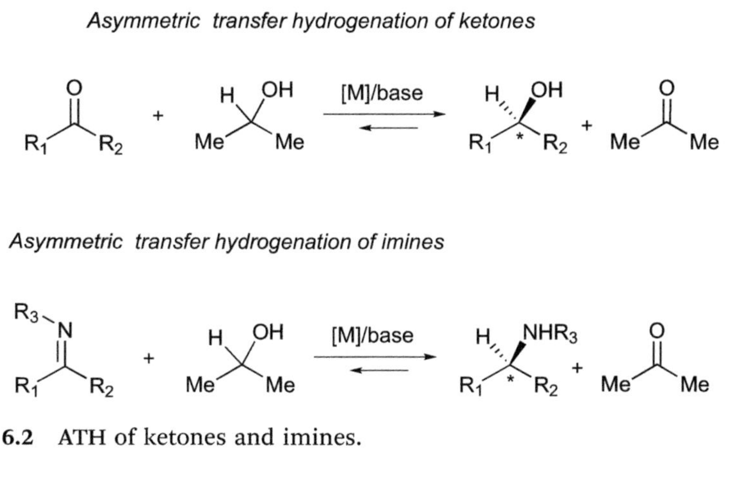

Scheme 6.2 ATH of ketones and imines.

The major breakthrough in ATH occurred when Noyori and co-workers disclosed the chiral catalyst Ru(II)-TsDPEN [**1**; TsDPEN = *N*-(*p*-toluenesulfonyl)-1,2-diphenylethylenediamine; Figure 6.1] bearing a chiral bidentate monotosylated diamine ligand, which showed excellent performance in the ATH of ketones and imines.[8–10] The robust Noyori catalyst was very effective in terms of stereoselectivity and activity for ATH using either *i*PrOH or formic acid as hydrogen donors. This significant discovery inspired intensive research in this area. The development of many structural variations of **1** has allowed the preparation of excellent catalysts capable to promote numerous applications from academic laboratory research to commercial-scale applications. Selected variants of the Noyori catalyst are shown in Figure 6.1. These include, the introduction of the NO ligands, in particular, amino alcohols (**2**),[11] chiral PP ligands (**3**), and tethered-type catalysts (**4**).[12] The presence of the Ru–NH linkage in the ruthenium catalysts was revealed to be essential to give excellent levels of catalytic efficiency in the TH of ketones. This effect is rationalised in terms of an outer-sphere mechanism, involving the N–H group of the supporting ligand (as discussed in Section 6.3). Research on transfer hydrogenation catalysis has extended to other metals, mainly to Rh(III) and Ir(III).[6] Half-sandwich Rh and Ir complexes bearing chiral bidentate NN or NO ligands (Figure 6.1, complexes **5–7**), along with Noyori's ruthenium(II) complexes are the best catalysts described in the literature so far. In all cases, the organometallic fragment stays intact in the catalytic cycle and *ee* values of 95–99% are obtained. The major challenge in ATH is increasing activity and productivity because TON values are usually

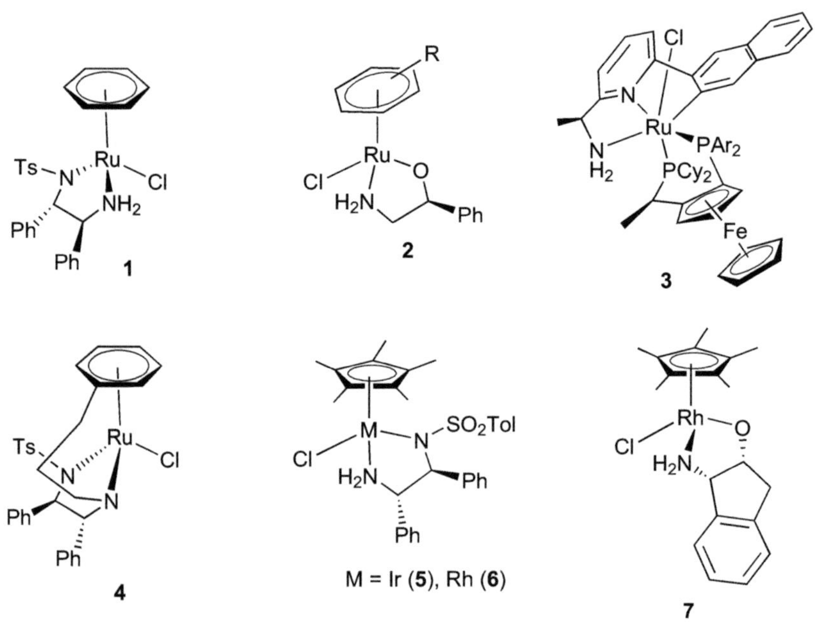

Figure 6.1 Selected catalysts for ATH.

below 1000 and TOFs are also moderate (<2000). Regarding the substrate scope, reduction of carbonyls and imines is well developed, while the transfer of C=C bonds is not competitive with hydrogenation methods.

6.2.1 Mechanistic Overview of Transfer Hydrogenation of Ketones

Concerning the general mechanism accepted for metal-catalysed transfer hydrogenation of ketones using a secondary alcohol as a hydrogen source, two possible variants are possible: inner-sphere mechanism, which implies substrate coordination to the catalyst, and outer-sphere mechanism, with no substrate coordination. In the inner-sphere mechanism, also denominated as metal hydride mechanism, the reaction takes place *via* a metal hydride intermediate, which is formed by the interaction of the catalyst with the hydrogen source. The unsaturated organic substrate binds directly to the catalyst, and the hydride is transferred from the metal hydride to the substrate. Therefore, there is a direct interaction between the donor and the acceptor molecules with the metal at different stages of the reaction. The involvement of mono- or di-hydride metal species depends on the surrounding ligands coordinated to the metal. In the mono-hydridic route, only one of the hydrogens from the alcohol, the α-C–H hydrogen, is transferred to the metal (Scheme 6.3a). The reaction starts with the generation of a metal alkoxide complex **A**, then β-hydride elimination from the alkoxide generates an intermediate metal monohydride species (M–H) **B** (Scheme 6.3a). Subsequent coordination of the substrate to the metal (ligand exchange), followed by the H-transfer from the metal to the coordinated ketone forms an intermediate alkoxide **C**. Reductive elimination of the product completes the catalytic cycle. In the dihydridic route, both the OH proton and the α-C–H hydrogen atom of the hydrogen donor are transferred to the metal, the former *via* oxidative addition and the latter *via* β-hydride elimination (Scheme 6.3b). Ligand exchange, H-transfer from the metal to the coordinate ketone, and reductive elimination yield the final product.[13]

In the outer-sphere mechanism, the reaction proceeds without coordination of the substrate to the metal so a "not-alkoxide" intermediate species is formed. This type of mechanism was described by Noyori for the chiral Ru(II)-TsDPEN catalyst (**1**), who proposed that the reaction occurred *via* six-membered transition states (TSs) in the outer coordination sphere of ruthenium.[14] The mechanism is outlined in Scheme 6.4. The first step is the transformation of the Ru–Cl species **1** to a hydride Ru–H (**C**). This transformation occurs in two consecutive steps: one is the generation of the coordinatively unsaturated ruthenium species **A** by abstraction of HCl, promoted by the base, and then, concerted hydride and proton transfer from isopropanol to **A** *via* a cyclic six-membered TS **B**, generating the Ru–H active species **C**. The second step is the conversion of the ketone to the chiral alcohol *via* a six-membered TS **D**, through simultaneous transfer of the

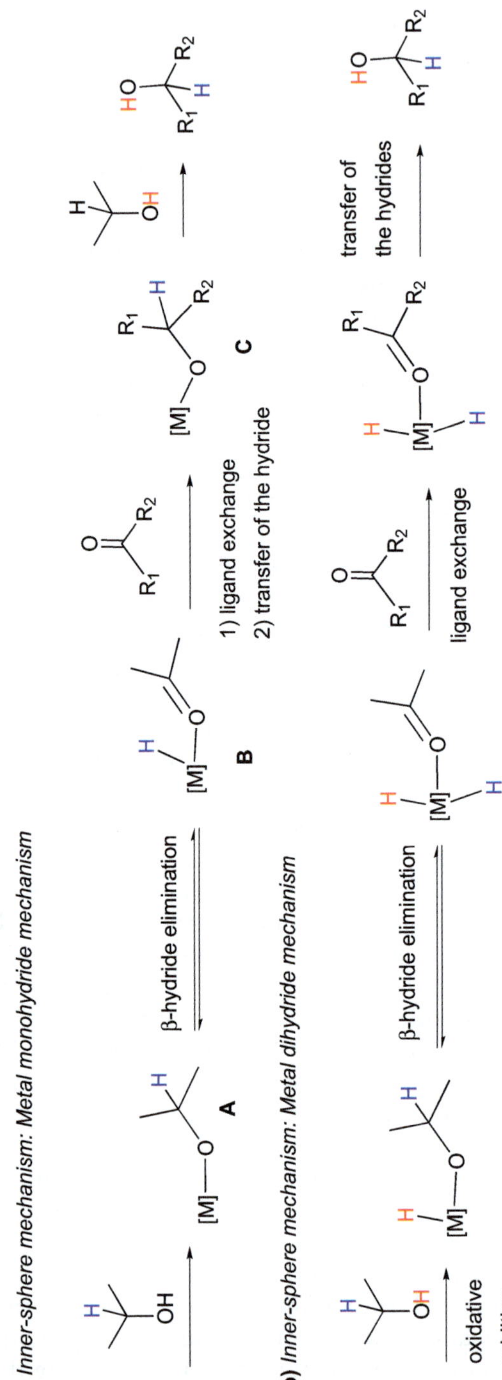

Scheme 6.3 TH mechanism *via* inner-sphere monohydride (a) and dihydride (b).

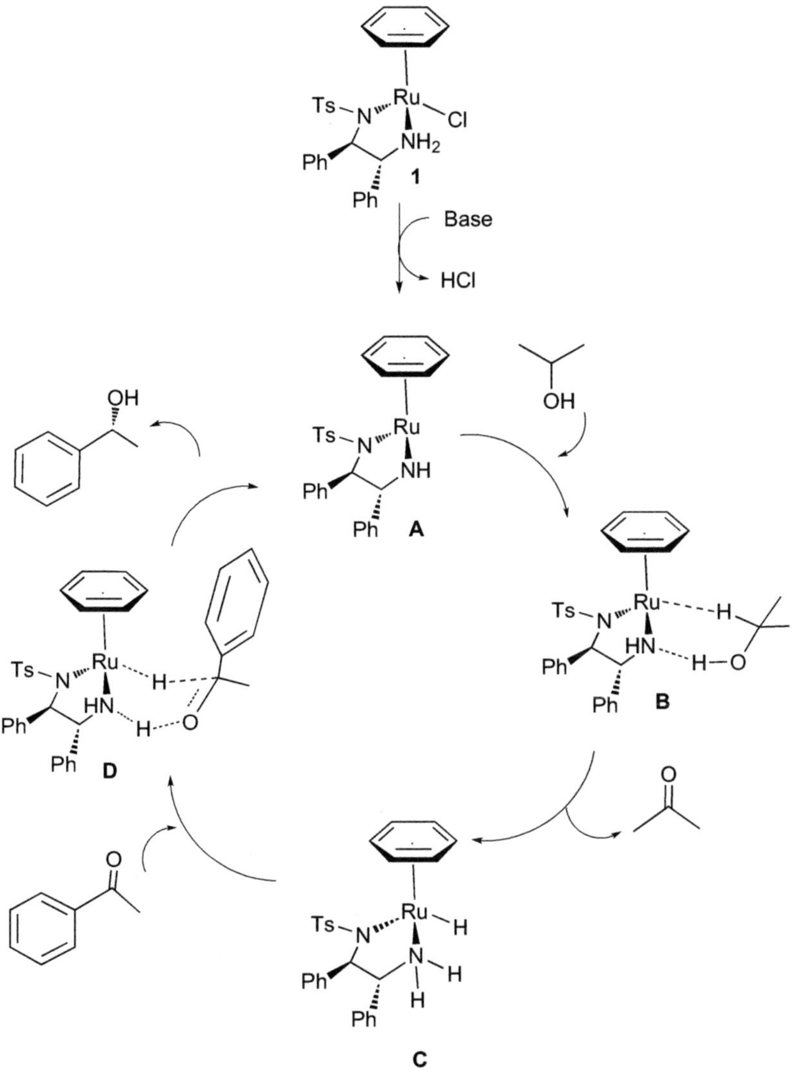

Scheme 6.4 Concerted outer-sphere mechanism of TH *via* six-membered cyclic transition state A.

hydride from ruthenium, and the proton from the amine ligand. This mechanism is often called metal–ligand bifunctional catalysis, indicating the active participation of the ligand in the catalytic reaction.

An ionic mechanism instead of the concerted mechanism has been proposed for different types of transition metals and substrates, *e.g.* imines. In the outer-sphere ionic mechanism, protonation of the imines precedes hydride transfer to the metal (Scheme 6.5). The hydride transfer from the formed iminium ion is suggested to occur without coordination of the imine to the metal centre.[15,16]

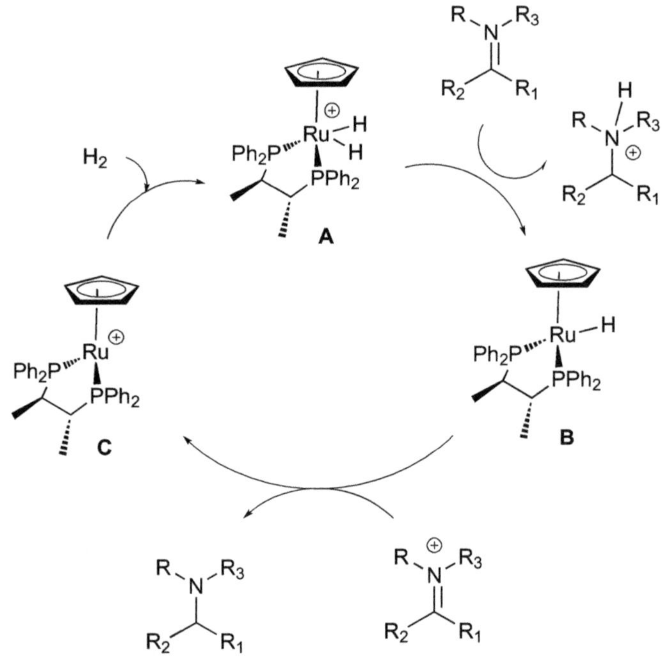

Scheme 6.5 Mechanism of the ionic hydrogenation of imines.

6.2.2 Transfer Hydrogenation with Cheap Metals

Today, ATH of carbonyl compounds and imines relies on the use of precious metals, such as Ru, Rh, and Ir. The problem with these metals is that they are expensive and raise toxicity concerns; they can be harmful to humans and to the environment. Therefore, their use in the preparation of drugs is subjected to strict regulatory control by the pharmaceutical industry in order to minimize the presence of residues of these metals in the final products. Earth-abundant first-row transition metals, such as iron, manganese, cobalt and nickel, are very attractive metals to replace precious metals in catalysis and sustainable pharmaceutical manufacturing. Their low cost, ready availability, low toxicity, and greater sustainability are their major advantages. Interestingly, living organisms use these first-row transition metals to catalyse their life-sustaining processes. For all these reasons, research on the use of Earth-abundant metals for catalysis, and in particular for transfer hydrogenation processes, has recently attracted considerable attention in the scientific community. Surely, breakthroughs in this area will provide the next generation of sustainable chemical methods for drug discovery.

Among first-row transition metals, iron, being one of the most abundant metals in the Earth's crust, is a fascinating alternative to precious metals for catalysis.[17–19] In the last decade, significant progress has been achieved in the development of sustainable catalytic processes with iron.[17–25] In the field of ATH, hydrogenation of polar bonds has been pioneered by Gao and Morris. The first example of iron-catalysed ATH of ketones was reported in 2004 by Gao using *in situ* generated catalysts from $[Et_3NH][HFe_3(CO)_{11}]$ and chiral PNNP ligands.[26] Later in 2008, Morris described the first example of a well-defined iron complex capable of catalysing ATH reactions (Figure 6.2).[27] Iron complexes of the general type [Fe(CO)Br(PNNP)] (**8**) bearing the tetradentate PNNP ligands already used by Gao for iron systems generated *in situ* displayed excellent catalytic activities in the ATH of acetophenone with *i*PrOH and KOBut under mild conditions.[28,29] A common structural feature in these catalysts is the presence of a carbonyl ligand completing the coordination sphere of the iron(II) centre, which was proven to be essential for the catalytic activity of these complexes.[30] Structural modification of the PNNP ligand by removal of the *ortho*-phenylene linkers between the phosphorous and imine moieties afforded the so-called second-generation catalysts with improved activities (TOF of 28 000 h^{-1}) and selectivities (up to 82% *ee*) for acetophenone. Further structural modifications of the PNNP ligand, including variations of the size of the substituents in the backbone[31] and change of the substituents of their phosphorous atoms,[32] allowed fine-tuning of the electronic and steric properties of the iron complexes. These studies revealed that the catalytic activity of the iron complexes **8–12** increased as the size of the substituents in the backbone of ligands increased (**10** performed better than **9**), and when the phosphorous atoms were substituted with phenyl groups over alkyl groups. Mechanistic investigations using the iron(II)–PNNP ligands demonstrated the active participation of the PNNP ligand through selective deprotonation of the imine group promoted by the base. It was proved that the partial ligand reduction is the key for the high catalytic activity of this iron complex, indicating that an outer-sphere mechanism is probably operating in these reactions.[33,34] Based on these results, a third generation of amine–imine P–N–NH–P ligands was developed by the group of Morris with the expectation that the presence of the amine group would provide more active iron catalysts. In fact, it was observed that complex **12** depicted in Figure 6.2 displayed excellent TOF in the hydrogenation of ketones (TOF 242 s^{-1}), surpassing the activity of ruthenium-based ATH catalysts.[35,36]

Beller's group recently disclosed the first iron-catalysed transfer hydrogenation of ketimines by facile *in situ* generation of the active catalyst from the iron precursor $[Et_3NH][HFe(CO)_{11}]$ and ligand **13** (Scheme 6.6) using 2-isopropanol as the hydrogen source. Up to 99% yield and 98% *ee* of the corresponding enantiomeric amines were obtained. A wide variety of chiral amines was accessed through this novel protocol, including aromatic, heteroaromatic and cyclic amines from the corresponding *N*-(diphenyl-phosphinyl)ketimines (Scheme 6.6).[37]

Apart from the tetradentate PNNP diimine ligands, chiral macrocyclic P_2N_4-type ligands disclosed by Gao have been successfully applied in ATH (Scheme 6.7a).[38] One of the most effective catalytic systems was obtained by combination of ligand **14** with the trinuclear iron(0) cluster

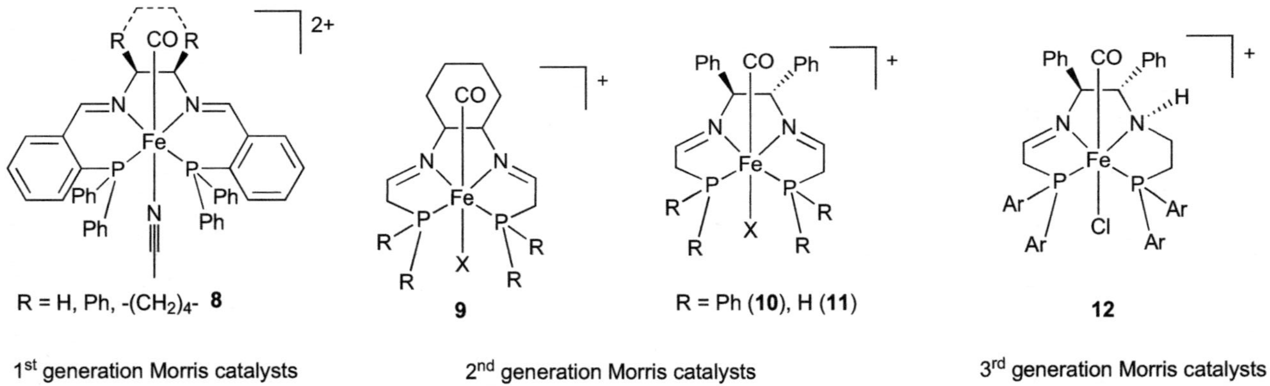

Figure 6.2 Morris's catalysts for ATH.

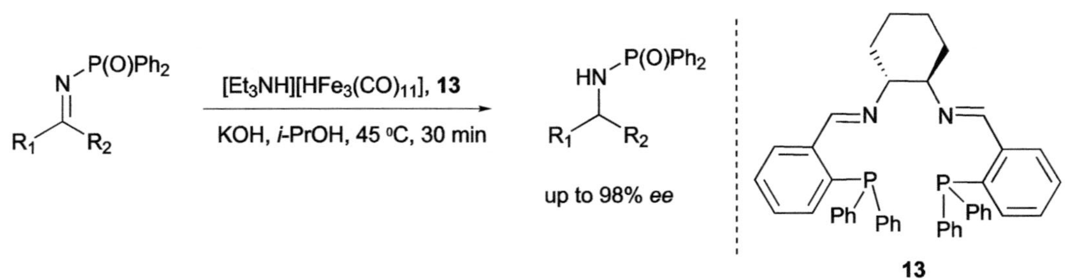

Scheme 6.6 ATH of imines catalysed by iron complexes.

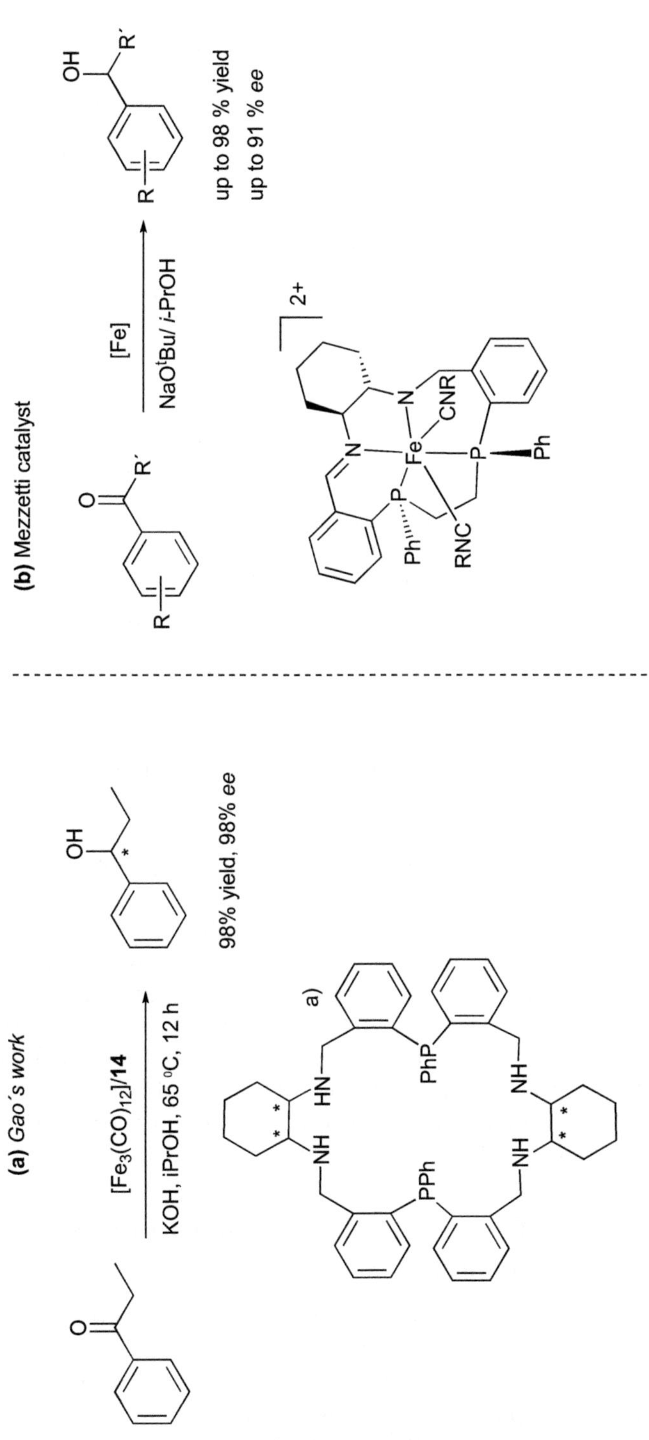

14

Scheme 6.7 Iron-catalysed ATH of ketones.

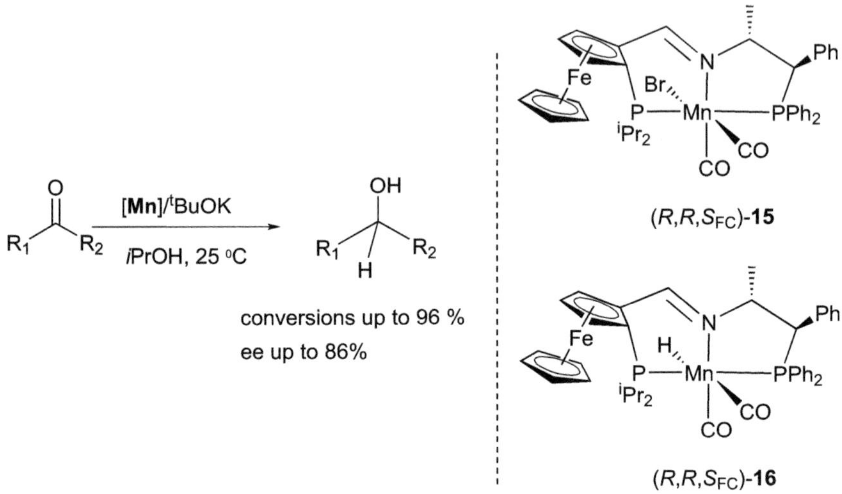

Scheme 6.8 First example of iron-catalysed ATH of imines.

$Fe_3(CO)_{12}$, which exhibited extraordinary enantioselectivities (up to 98% *ee*) in the reduction of a broad range of ketones, including aromatic, hetero-aromatic, and β-ketoesters. However, modest activities (TOF up to 40 h^{-1}) compared with Morrison's catalysts were obtained. More recently, Mezzetti described a successful ATH of a broad scope of substrates, including ketones, enones, and imines, using well-defined bis(isonitrile) iron(II) complexes bearing a C2-symmetric diamino $(NH)_2P_2$ macrocyclic system (Scheme 6.7b). Excellent enantioselectivities up to 99% *ee* and high activities were achieved for a broad scope of ketones.[39–41]

Other first-row transition metals, such as manganese, cobalt, and nickel, are elusive in ATH. The use of a tridentate NPO-type ligand in combination with $Ni(PPh_3)_2Cl_2$ allowed the formation of catalytically active Ni species capable to perform the reduction of a variety of ketones with a 90% yield and *ee* of 80%.[42] In addition, manganese complexes are emerging as an interesting class of catalysts for TH. Recently, Kirchner described the first examples of manganese complexes, $[Mn(PNP')(Br)(CO)_2]$ (**15**) and $[Mn(PNP')(H)(CO)_2]$ (**16**) containing a tridentate ligand with a planar chiral ferrocene, active for the ATH of ketones; conversions of up to 96% and *ee* values of up to 86% were obtained (Scheme 6.8).[43]

6.2.3 Asymmetric Transfer Hydrogenation in the Synthesis of Bioactive Molecules

Asymmetric transfer hydrogenation technology has found interesting applications in the total synthesis of bioactive molecules.[44] Many chiral β-amino aryl ethanols have been found to be potential synthetic precursors of pharmaceutically important molecules. For example, the synthesis of (*R*)-(−)-tembamide (**17**) and (*R*)-(−)-aegeline (**18**), α- and β-adrenergic blockers and agonists in the treatment of cardiovascular diseases, have been efficiently obtained by ATH of the corresponding ketones using the Rh complex [RhCp*/**19**] (**20**, Scheme 6.9). Using sodium formate as the hydride donor, **17** and **18** were obtained with yields of 95% and *ee* values of 99%.[45]

Another useful application of ATH includes the synthesis of ladostigil (**21**), which is a drug for the treatment of Alzheimer's disease. The only stereocentre present in **21** was generated by a ruthenium-catalysed ATH using the catalyst Ru–TsDPEN in combination with aqueous formate (Scheme 6.10).[46]

A ruthenium catalyst was also successfully used in the synthesis of eslicarbazepine acetate (**22**), which is an antiepileptic drug used as an add-on treatment with other antiepileptic medicines. ATH of the corresponding ketone precursor with formic acid as a hydrogen source and using a weak anionic ion exchange resin instead of trimethylamine as the base afforded a high yield and good enantiomeric excess of **22** (Scheme 6.11).[47]

Another interesting example to illustrate the applicability of ATH is the synthesis of (*R*)-salmeterol (**24**), a potent and long-acting β_2 adrenoceptor agonist used as a bronchodilator for the prevention of bronchospasm in patients with asthma and chronic obstructive pulmonary disease (Scheme 6.12).[48] The (*R*)-enantiomer is more active as a β-agonist and has fewer side effects than the racemate. Therefore, the enantioselective synthesis of **24** is of great interest for the pharmaceutical industry. (*R*)-Salmeterol was synthesised in eight steps using salicylaldehyde as the starting material. The key chiral intermediate α-bromoalcohol was prepared *via* Rh-catalysed ATH of the prochiral α-bromoketone **23** using sodium formate as the hydrogen donor under mild conditions. Under these conditions, excellent enantioselectivity (98% *ee*) of the α-bromoalcohol was obtained.

Many other practical applications of ATH for the preparation of bioactive molecules with pharmacologic interest can be found in the literature including, for example, the synthesis of (*S*)-panaxjapyne A (**25**),[49] (+)-yashabushitriol (**26**),[50] (−)-aglaroxin (**27**),[51] (*S*)-duloxetine (**28**),[52] and (*R*)-fluoxetine (**29**)[52] (Figure 6.3).

An elegant approach to the synthesis of relevant biologically active compounds is the combination of the ATH strategy with a dynamic kinetic resolution (DKR) process. This method has been successfully applied to the synthesis of important pharmaceutical intermediates. The synthesis of tesaglitazar (AZ-242) (**30**), a drug candidate used for the treatment of type II diabetes, is an interesting example to illustrate the efficient combination of ATH with DKR processes.[53] Ratovelomanana-Vidal and co-workers in collaboration with Hoffmann-La Roche developed a highly efficient procedure for ATH of prochiral α-alkoxy-β-ketoesters coupled with a DKR process under mild conditions with HCO_2H/NEt_3 as the hydrogen source and using the well-defined, commercially available chiral catalyst Ru^{II}-TsDPEN.[54] The ATH/DKR of a racemic mixture of the α-ethoxy-β-ketoester **31** was performed on a gram scale by using the Noyori Ru(II) catalyst **1** in CH_2Cl_2 and an azeotropic mixture (5 : 2) of HCOOH–NEt_3 at 20 °C, yielding the key intermediate ethyl ester (*S*)-**32** with > 95% *ee* (Scheme 6.13).[56,57]

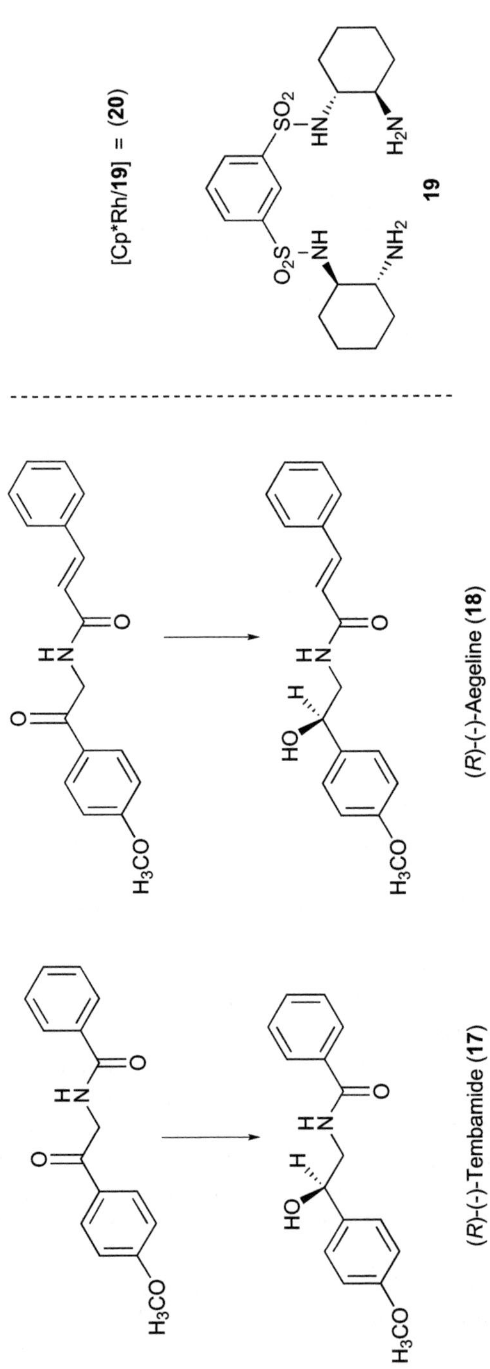

Scheme 6.9 Synthesis of bioactive molecules through ATH processes.

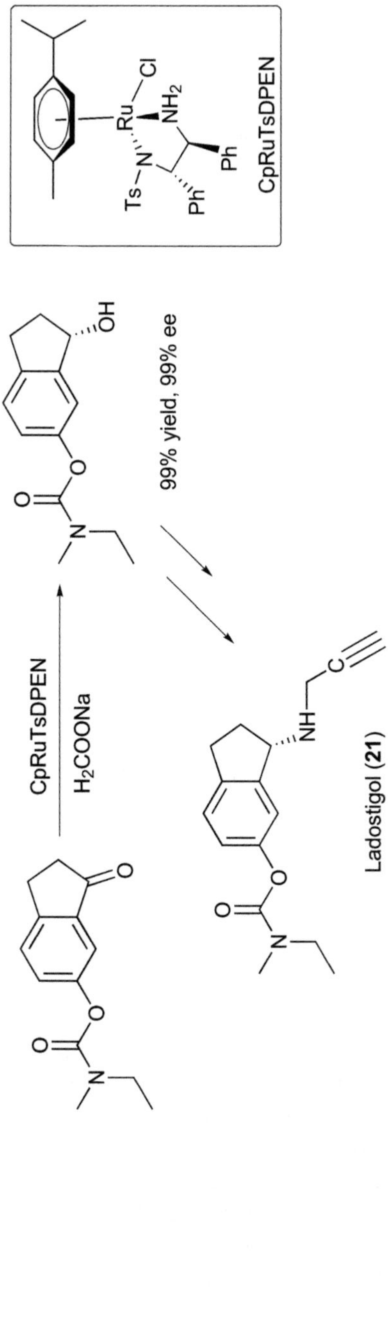

Scheme 6.10 Synthesis of ladostigil.

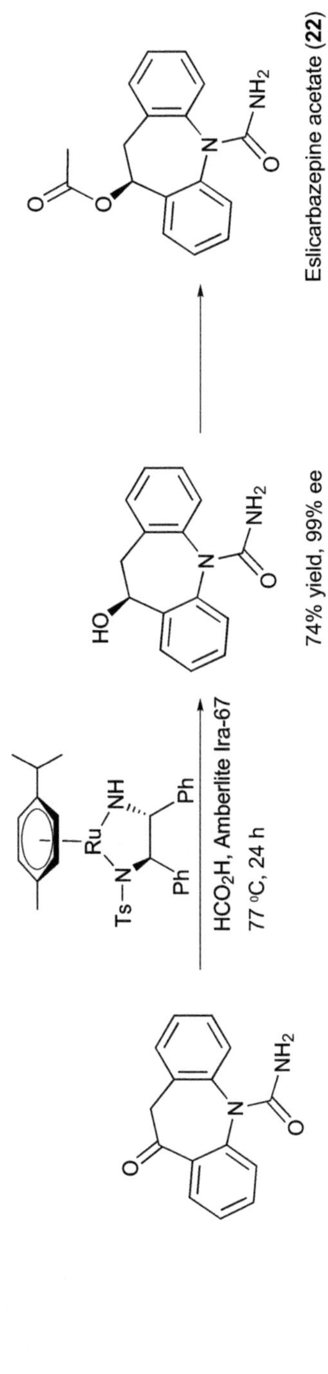

Scheme 6.11 Synthesis of eslicarbazepine acetate.

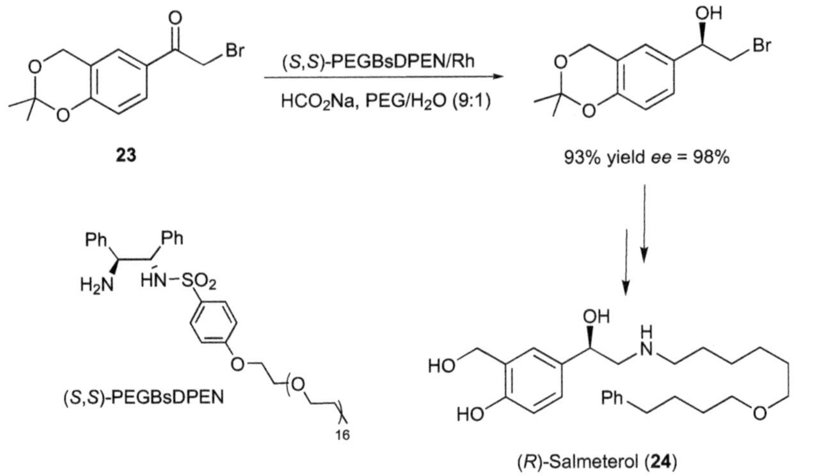

Scheme 6.12 Synthesis of salmeterol.

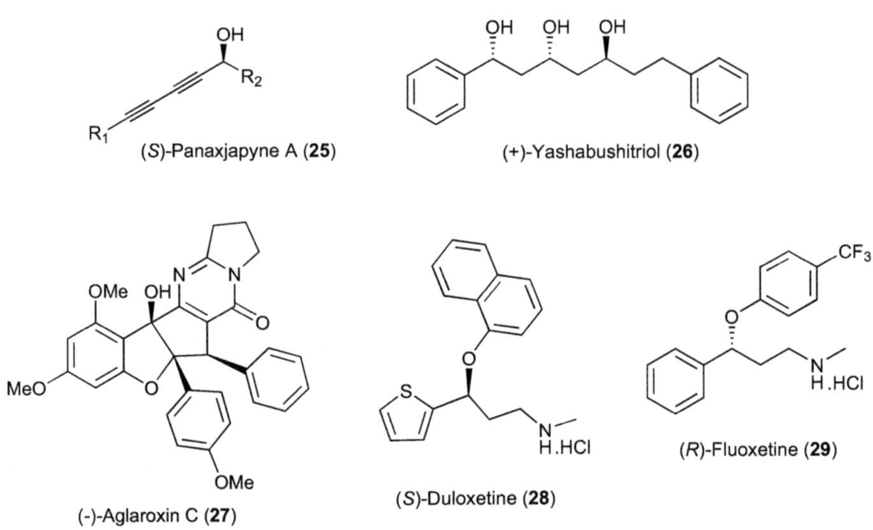

Figure 6.3 Pharmaceutical relevant molecules obtained *via* ATH.

Another example to illustrate the excellent performance and potential application of Noyori catalyst **1** in the preparation of bioactive molecules is the large-scale synthesis of (1′*S*,3*R*,4*R*)-4-acetoxy-3-(2′-fluoro-1′-trimethylsilyloxyethyl)-2-azedinone (**34**) as a new fluorine-containing intermediate that constitutes the core structure of β-lactam antibiotics, such as **35** in certain cholesterol drugs.[55] The synthetic key step relies again on a combination of a DKR/ATH processes of ethyl 2-benzamidomethyl-4-fluoro-3-oxo-butanoate (**33**), performed using HCO_2H–TEA (5 : 2) (TEA = triethylamine) as the

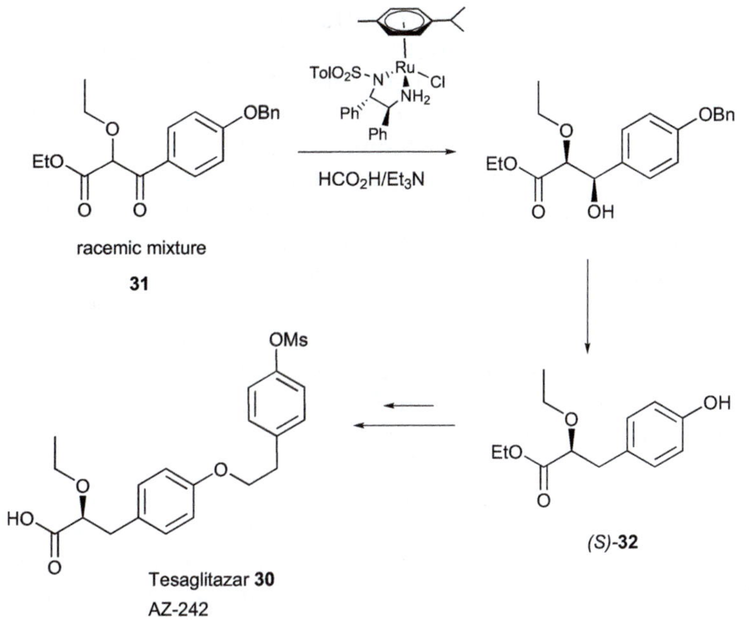

Scheme 6.13 Synthesis of Tesaglitazar.

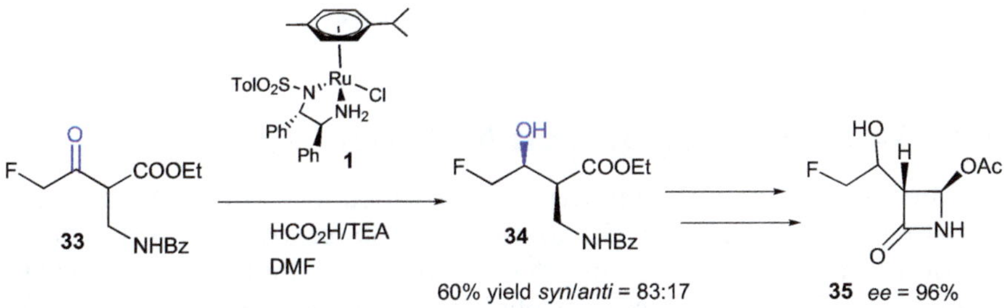

Scheme 6.14 Synthesis of β-lactams through combination of DKR/ATH processes.

hydrogen source at room temperature in the presence of a ruthenium catalyst. Under these conditions, the corresponding alcohol was obtained in quantitative yield with 83 : 17 diastereomeric ratio (*syn*/*anti*) (Scheme 6.14). This alcohol was further transformed into the expected fluorinated 2-azetidinone **35** in three steps (overall yield 21%).[58]

Asymmetric transfer hydrogenation method for the reduction of imines has also been applied for the synthesis of drugs. An example is the convenient preparation of an interesting family of 1-aryl tetrahydroisoquinoline (THIQ) derivatives of pharmaceutical interest. Compounds **36** and **37**, which are known to modulate glutamate neurotransmission in the central nervous system, were obtained in good yields and enantioselectivities using the ruthenium complex [RuCl(η^6-benzene)TsDPEN] (**1**) as the catalyst and the azeotropic mixture 5 : 2 $HCO_2H–NEt_3$ as the hydrogen source (Scheme 6.15). The corresponding valuable amines were isolated with high atom economy in high yields and excellent selectivities. The same approach led to the preparation of **38**, a potent noncompetitive AMPA (α-amino-3-hydroxy-5-methyl-4-isoxazolepropionic acid) receptor antagonist being investigated in phase III trials as an antiepileptic agent. Interestingly, it was demonstrated that the presence of the two electron-rich methoxy groups on the substrate was essential to stabilize the favoured TS species. This fact was explained through the existence of a strong C–H/π attractive interaction between a C(sp^2)–H substituent on the η^6-benzene ligand of the Ru complex and the aromatic ring of the isoquinoline skeleton when two methoxy groups are present in the substrate (TS **A**, in Scheme 6.15). In fact, significant lower yields and enantioselectivities were obtained in the ATH of substrates containing just one methoxy group.[59]

Another example of the efficient use of a transfer hydrogenation process for the reduction of imines is the large-scale synthesis of the 5HT2B receptor antagonist LY414197 (**40**; Scheme 6.16).[61] The synthesis of **40** has been developed using a ruthenium-based ATH. Without using any optimisation, a 95% yield of **40** with 96% *ee* was obtained in the reduction of the imine **39**. The method was particularly simple and convenient because of the easy isolation of the chiral amine **40** by filtration. This procedure replaced the well-known $NaBH_4$/CBZ-(*S*)-proline method previously used for the preparation of similar drugs, in which it was necessary to use an excess of an expensive chiral ligand.

6.3 Borrowing Hydrogen Methodology

The borrowing hydrogen approach, also called hydrogen autotransfer or dehydrogenative activation, is another version of transfer hydrogenation reaction and consists of a recently developed method for the formation of C–N and C–C bonds by activation of alcohols in the presence of a metal catalyst.[60]

The borrowing hydrogen methodology exploits the interconversion of alcohols to carbonyl compounds, and the transfer of hydrogen from the alcohol to the final product, mediated by the metal catalyst. The formation of a new C–N bond proceeds through a series of discrete steps involving sequential dehydrogenation and hydrogenation reactions (Scheme 6.17). Initially, the catalyst temporarily removes hydrogen (dehydrogenation process) from the alcohol (**41**) to form the more reactive intermediate aldehyde (**42**). This aldehyde reacts with the amine (**43**) to form the corresponding intermediate imine (**44**). The hydrogen from the catalyst is then incorporated into the imine (hydrogenation process), forming the final amine product (**45**). It is important to note that the overall process is a redox-neutral process, consisting of the coupling of an alcohol and an amine with the formation of the corresponding alkylated amine and water. The metal catalyst acts as a hydride shuttle between the starting material and the final product.

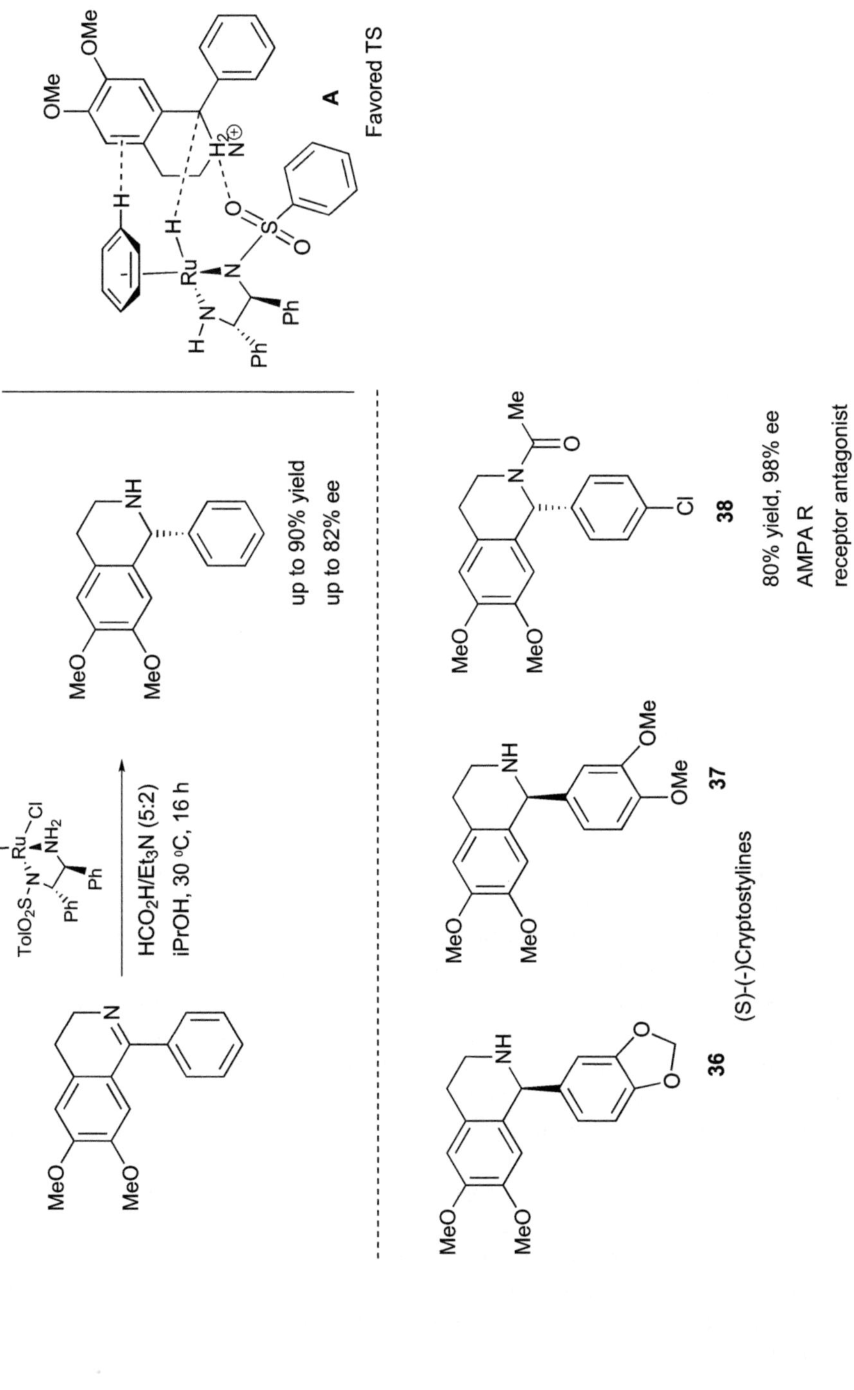

Scheme 6.15 Synthesis of bioactive molecules through ATH of imines.

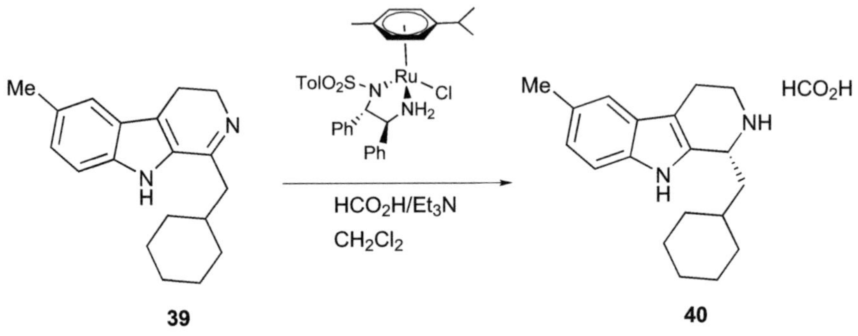

Scheme 6.16 Large-scale synthesis of 5HT2B receptor antagonist LY414197, by ATH.

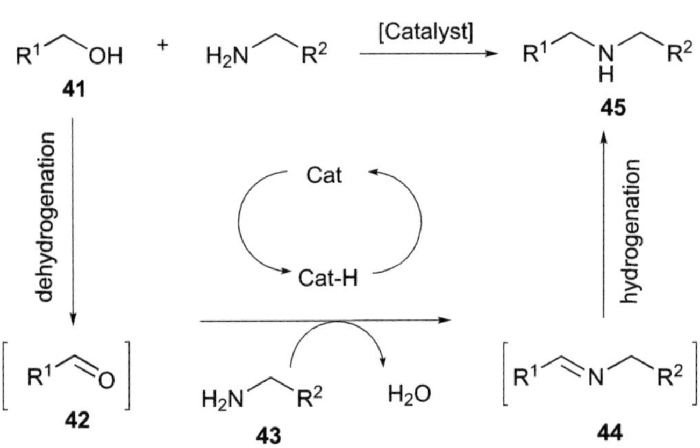

Scheme 6.17 C–N bond formation by borrowing hydrogen.

The borrowing hydrogen methodology represents a straightforward and sustainable method for *N*-alkylation since direct reaction of an alcohol with an amine is an atom-economic reaction, in which water is the only byproduct. Moreover, alcohols are readily available from renewable feedstock, as well as being inexpensive and non-toxic alkylating agents. The direct reaction of an alcohol with an amine was first disclosed independently by Watanabe[60] and Grigg,[61] and since then extraordinary progress has been made in this field, mainly using precious metal catalysts containing ruthenium and iridium. Comprehensive reviews on recent advances in this area are available in the literature.[62–65]

This new approach for the formation of C–N bonds offers interesting advantages over traditional methods, such as direct alkylation or reductive elimination, owing to their operational simplicity and the significant improvement in atom economy. Interestingly, it can be applied to a large number of substrates and appears to have tolerance for a wide variety of functional groups.

Recently, more sustainable catalytic systems for C–N bond formation based on eco-friendly and widely abundant metals have been described in the literature. Pioneering reports by Feringa and Barta described the first alkylation of amines with alcohols performed by an iron complex through a hydrogen borrowing approach.[66,67] The crucial feature of the catalytic system is the bifunctional nature of the iron catalyst, which is capable of temporarily storing the hydrogen, allowing the dehydrogenation of the alcohol in the first step, and transferring the hydrogen to the intermediate imine, forming the final product (Scheme 6.18).

This work inspired further developments in this area and a variety of new catalytic systems based on the use of manganese,[68,69] iron,[70–72] and cobalt[73,74] are now available (Figure 6.4). The scope of this new approach for the synthesis of amines includes the alkylation of a variety of aliphatic and aromatic amines, both secondary and tertiary, with a wide range of alcohols, and the use of diols in the formation of five-, six-, and seven-membered nitrogen heterocycles, which are privileged structures in the synthesis of pharmaceutically relevant molecules (Figure 6.5). Despite these recent achievements using non-precious metals, this field of research is still in its infancy. The development of more efficient catalytic systems in terms of activity, scope, and reaction conditions is a challenging task. Further development of new hydrogen borrowing processes for green and efficient chemical synthesis will surely displace conventional and industrial synthetic methods. Such processes will allow the development of

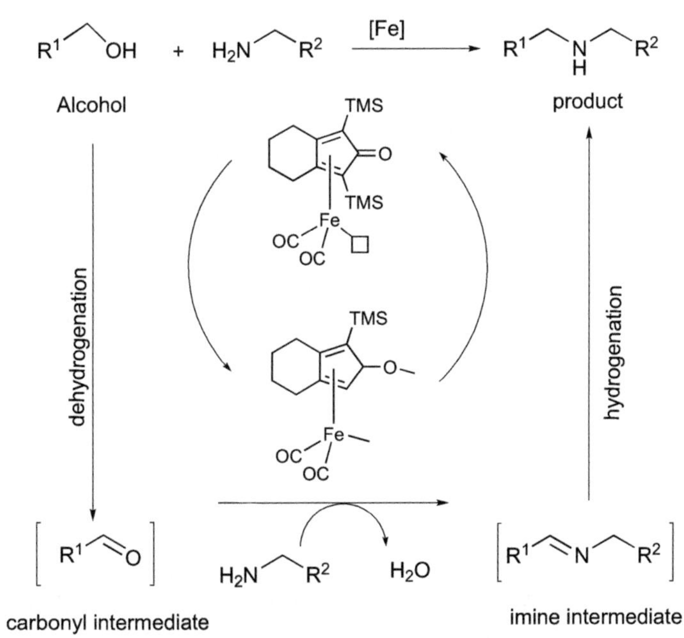

Scheme 6.18 Iron-catalysed N-alkylation of amines with alcohols.

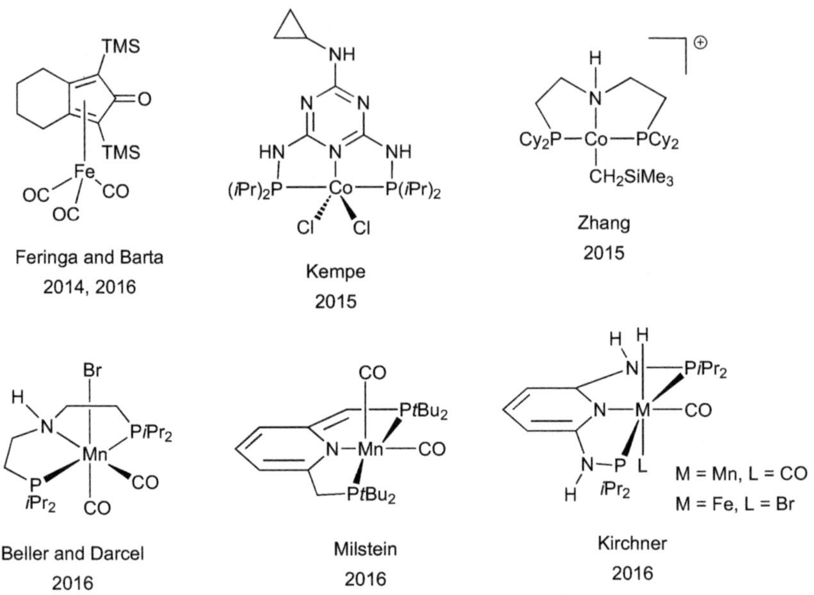

Figure 6.4 Selected iron, cobalt and manganese for hydrogen borrowing processes.

atom-economic synthetic strategies avoiding toxic reagents and with no waste generation.

To showcase the utility of this methodology in the pharmaceutical industry, the convenient synthesis of PF 03463275 (**46**), a GlyT1 inhibitor developed for the treatment of schizophrenia, is depicted in Scheme 6.19. Berliner recently described the first kilogram-scale application of direct alkylation of amines using the iridium catalyst $(Cp^{*}IrCl_2)_2$ in low loading (0.033 mol%).[75] The presence of water and a tertiary amine is crucial to achieving efficient and robust catalyst performance on both small and large scale. The significant operational and environmental advantages of this process over traditional oxidative–reductive amination sequences demonstrate the great potential of this new methodology.

Preparation of natural products such as noranabasamine (**47**) can also be performed using the hydrogen borrowing methodology by selective diamination of isosorbide, which is obtained from D-glucose.[76] Both the *R* and *S* enantiomers of the amphibian alkaloid **47** were prepared in an overall yield >30% with 89% and 86% *ee*, respectively. An enantioselective iridium *N*-heterocyclisation reaction with either (*R*)- or (*S*)-1-phenylethylamine and 1-(5-methoxypyridin-3-yl)-1,5-pentanediol was employed to generate the 2-(pyridine-3-yl)-piperidine ring system in 69–72% yield (Scheme 6.20).

There are many pharmaceuticals that contain the dimethylamino group, for example, the anti-inflammatory agents **48–50** (Scheme 6.21). Antergan (**48**) was among the first antihistamine drug to be sold and its structure provided the framework for the development of alternative antihistaminic agents, such as chlorphenamine (**49**) and tripelennamine (**50**), which are

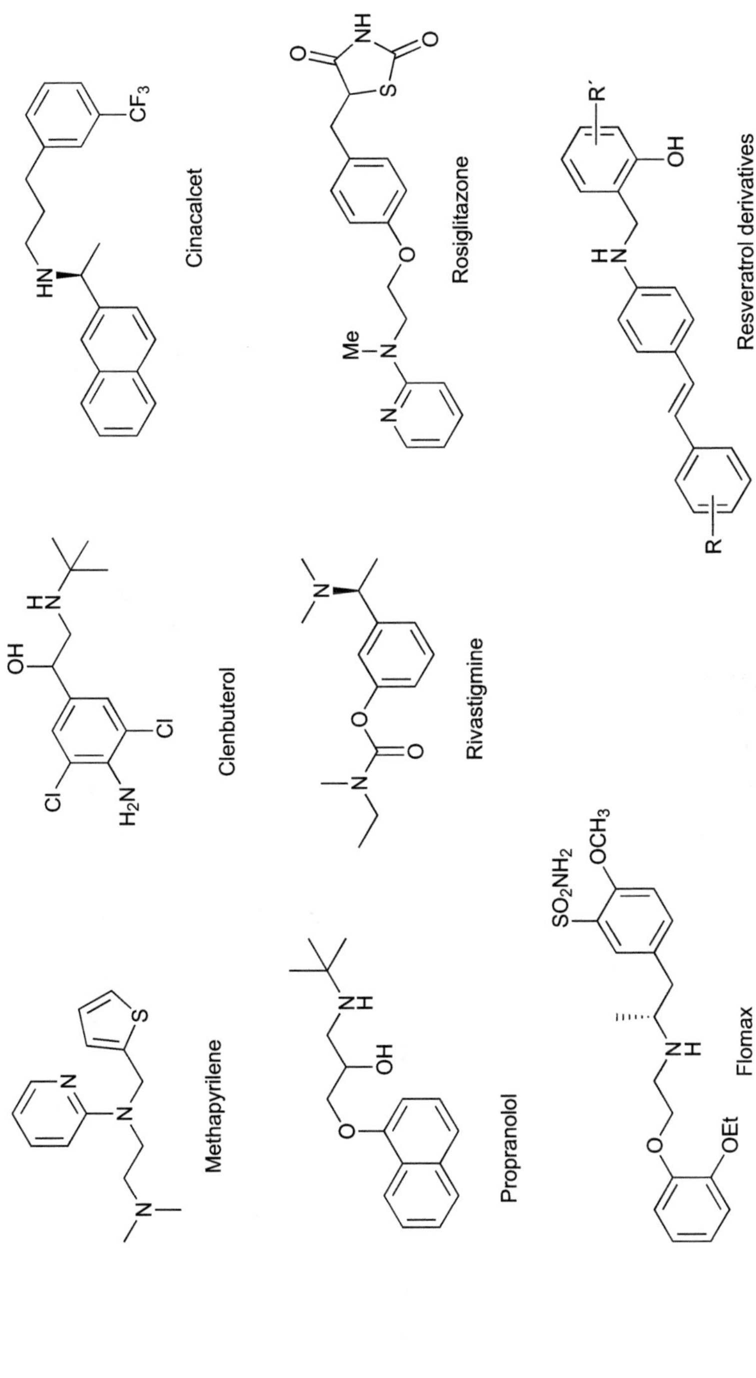

Figure 6.5 Pharmaceutical relevant molecules obtained *via* HBP.

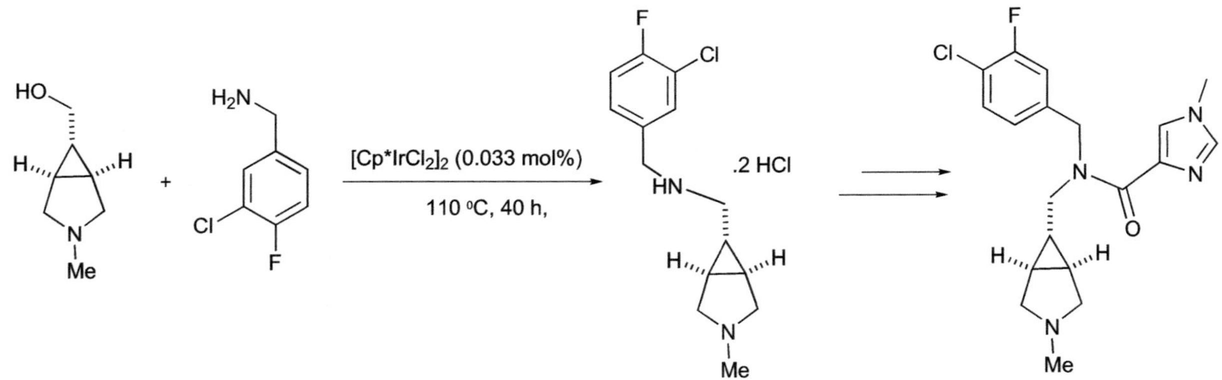

PF 03463275 (**46**)

Scheme 6.19 Synthesis of PF 03463275.

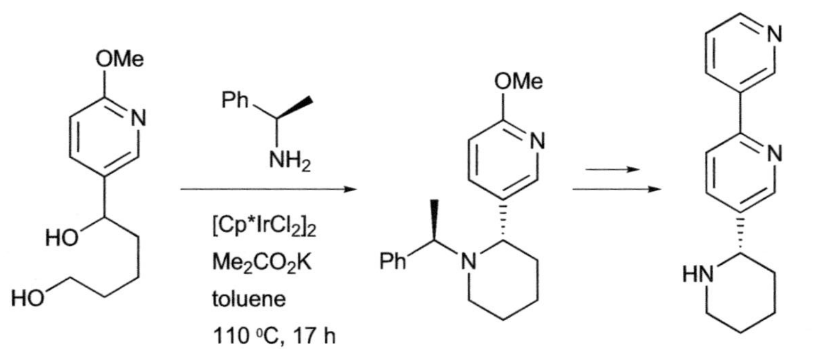

Noranabasamine (**47**)

Scheme 6.20 Synthesis of Noranabasamine.

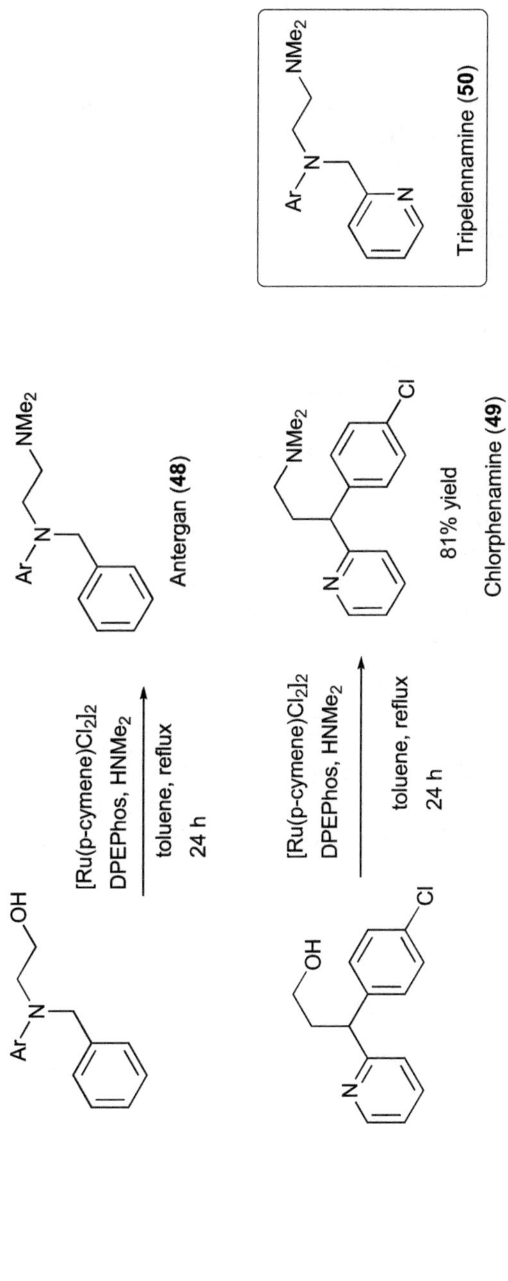

Scheme 6.21 Synthesis of pharmaceutical drugs through hydrogen borrowing processes.

used in the prevention of the symptoms of allergic conditions. Williams described the synthesis of these pharmaceutical drugs through ruthenium-catalysed hydrogen borrowing processes using a bidentate phosphine as the co-ligand.[77]

So far, the majority of the applications of hydrogen borrowing methodology in the pharmaceutical industry are developed using noble metals. An interesting survey of the borrowing hydrogen approach to the synthesis of some pharmaceutically relevant intermediates has been recently reported, in which the scope and limitations of this methodology are discussed.[78] Promising results using cheap and non-toxic metals for the synthesis of drugs have been recently disclosed. In this section, selected examples using iron catalysts are examined.

Iron-catalysed direct amine alkylation with an alcohol has been successfully applied in a key step for the preparation of the drug piribedil (**51**), a dopamine antagonist used in the treatment of Parkinson's disease (Scheme 6.22). The reaction of commercially available 1-(2-pyrimidyl)-piperazine (**52**) and piperonyl alcohol (**53**) afforded a 54% isolated yield of **51** in the presence of catalytic amounts of the iron catalyst **54**. Notably, no auxiliary reagents were required and the green solvent cyclopentyl methyl ether (CPME) was used in the synthesis (Scheme 6.22).[66]

Another interesting example of the application of **54** is the synthesis of a key intermediate to muscarinic agonist *N*-[5-[(1′-substituted-acetoxy)methyl]-2-furfuryl]dialkylamines (**55**). It is known that furanic compounds are an important class of pharmaceutically active compounds with potential antimuscarinic activity. Feringa and Barta described an elegant and fully sustainable synthesis of **55** intermediates from cellulose derived platform chemicals, through direct reaction of a phenylamine derivative and furan-2,5-diyldimethanol, using **54** as the catalyst (Scheme 6.23).[67]

Piperidine derivatives are privileged structural motifs present in many drugs used in medical practice (Figure 6.6). They include drugs used as local anaesthetics, analgesics, antiarrhythmics, and antihypertensives.[79] Interestingly, hydrogen borrowing provides an interesting approach for their

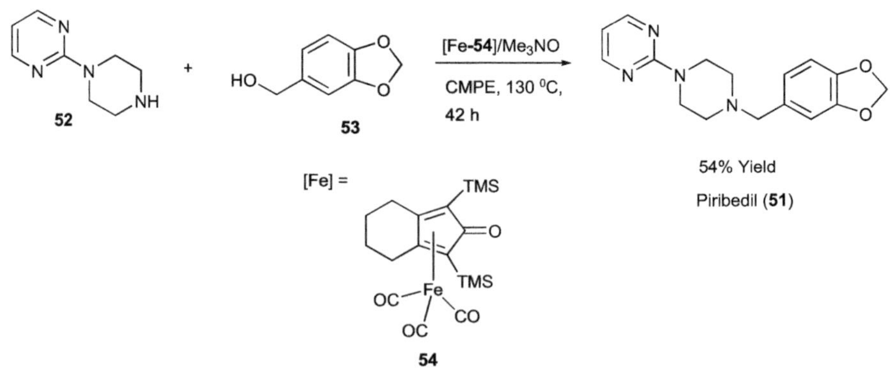

Scheme 6.22 Synthesis of piribedil.

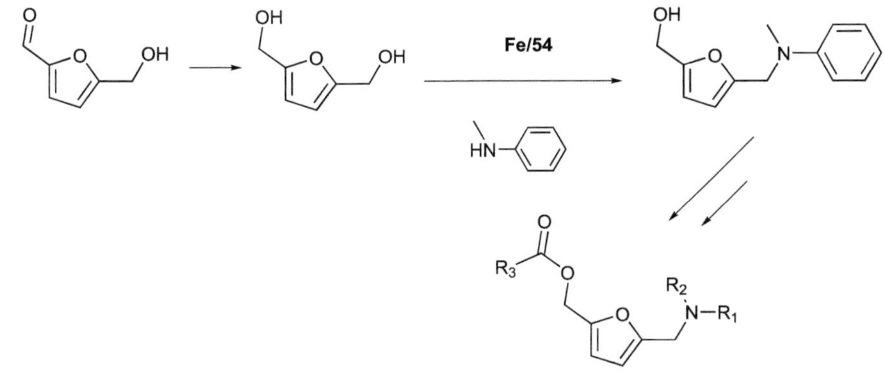

Scheme 6.23 Synthesis of pharmaceutical active building blocks from lignocellulose.

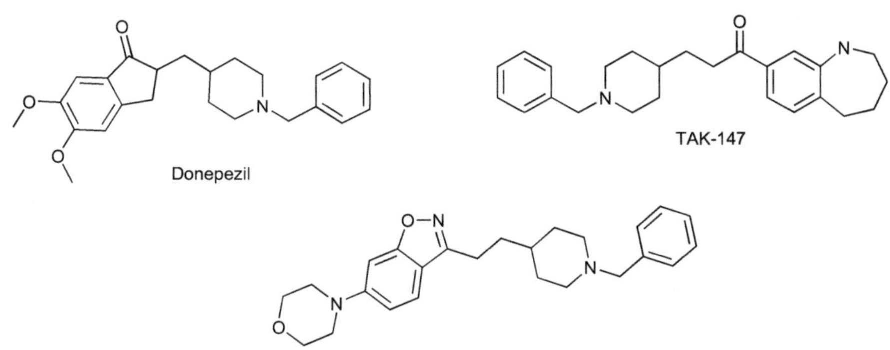

Figure 6.6 Piperidine bioactive molecules.

synthesis. As depicted in Scheme 6.24, a wide variety of *N*-benzylpiperidines can be prepared by reacting benzyl-protected five-, six- and seven-membered-heterocycles from benzylamines with diols (Scheme 6.24a) or by reacting benzyl alcohols with piperidines in the presence of the iron catalyst **54** (Scheme 6.24b).[67]

Notable, the hydrogen borrowing methodology can also be applied to C–C bond-forming reactions. An interesting example of these reactions is the α-alkylation of ketones with alcohols. A general pathway is depicted in Scheme 6.25. Initially, the catalyst, indicated as [M], temporarily removes hydrogen from the alcohol (**56**) to form the more reactive aldehyde (**57**). This aldehyde is then converted into an alkene (**58**), by *in situ* aldol condensation, and the "borrowed" hydrogen is transferred to the alkene to form the final product (**59**), with the overall formation of a new C–C bond.[80] Krische's group disclosed a beautiful synthetic methodology for C–C bond formation under transfer hydrogenation conditions using iridium-based catalysts.[81]

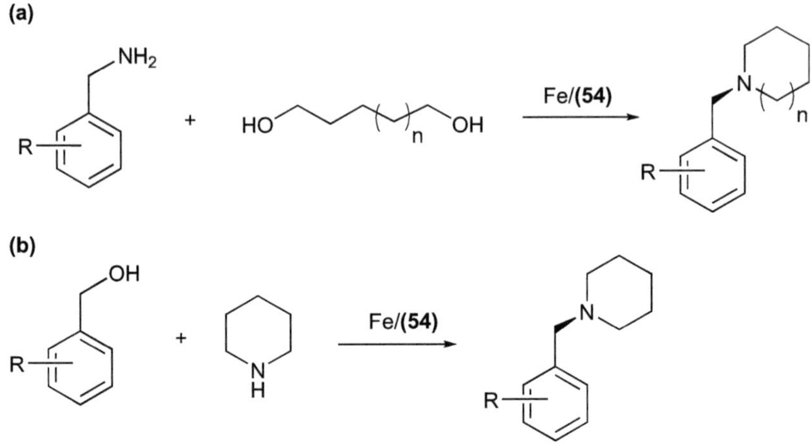

Scheme 6.24 Synthesis of piperidine derivatives by N-alkylation of amines with alcohols.

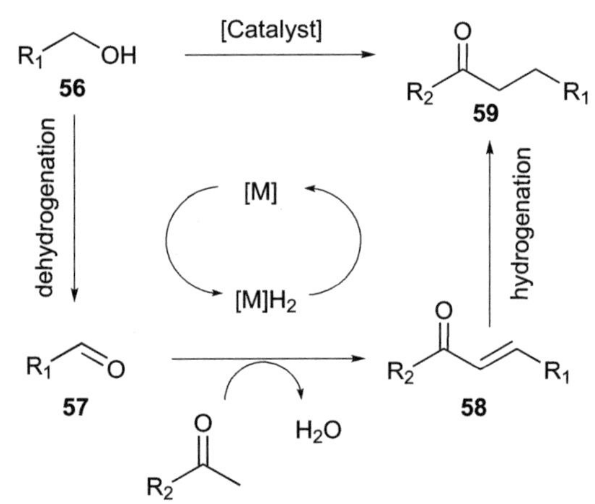

Scheme 6.25 C–C bond formation by hydrogen borrowing: α-alkylation of ketones with alcohols.

Other substrates than ketones can be alkylated with alcohols using the borrowing hydrogen strategy. Some representative examples of C–C bond-forming reactions include the synthesis of arylacetonitriles (**60**) and indole derivatives (**61**, Scheme 6.26).[82]

In addition, the hydrogen borrowing methodology can be applied to the activation of two alcohols in a one-pot reaction. Both alcohols are converted into the corresponding carbonyl compounds by the temporary removal of hydrogen. Subsequent aldol condensation leads to the formation of an α,β-unsaturated ketone, which undergoes hydrogenation giving the saturated alcohol as the final product (Scheme 6.27). Self-coupling of alcohols to construct new C–C bonds is a greatly attractive approach that avoids the use

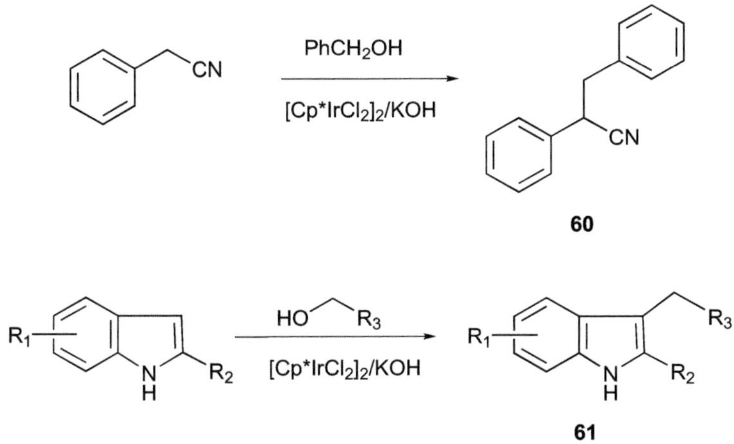

Scheme 6.26 Alkylation of arylacetonitriles and indole through hydrogen borrowing methodology.

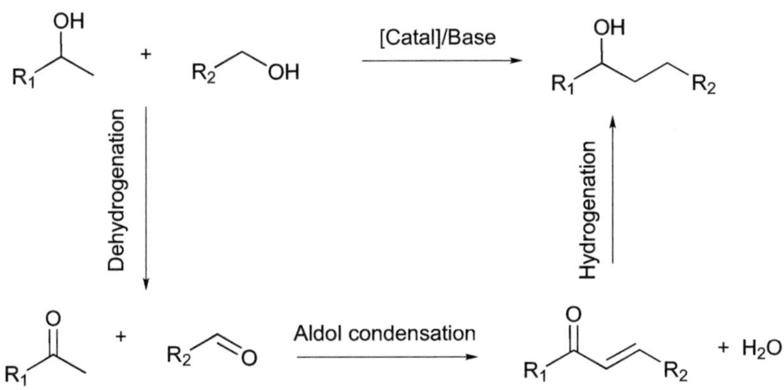

Scheme 6.27 One-pot self-coupling of alcohols to construct new C–C bonds through hydrogen borrowing.

of expensive alkyl halides and strong bases used in the traditional synthesis and in addition does not generate stoichiometric amounts of salts and waste. In the last decade, important efforts have been invested in this area of research and a wide range of catalytic systems based on the use of Ir,[83–85] Pd,[86] Rh,[87] Ru,[88–90] Ni,[91] Fe,[92] and Cu[93] are available. Nevertheless, despite noteworthy progress, the effectiveness of the catalysts is not sufficiently high in terms of selectivity and yield for their applicability in an industrial context. The design of improved catalysts is of great interest. Exploiting metal–ligand cooperativity has been demonstrated to be a useful approach to enhance the effectiveness of the catalytic systems. As an example, a bifunctional 2-(2-pyridyl-2-ol)-1,10-phenanthroline (phenyl-OH)-based Ru(II) complex (**62**) was found to be a highly efficient catalyst for the one-pot β-alkylation of secondary alcohols with primary alcohols (Scheme 6.28).

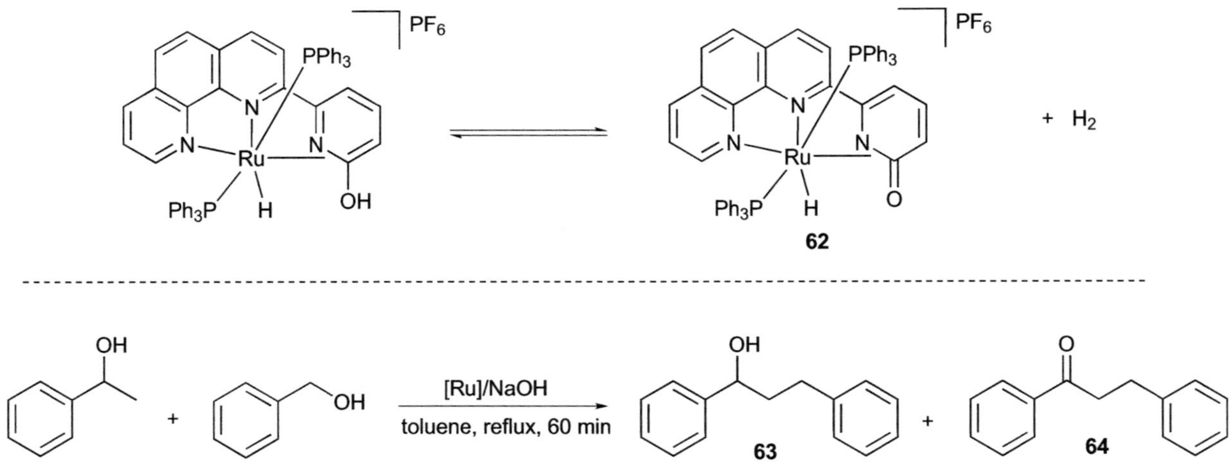

Scheme 6.28 Bifunctional ruthenium(II) complex as catalyst for N-alkylation of amines with alcohols.

Several aromatic, aliphatic, and heteroatom-substituted alcohols were selectively cross-coupled in high yields, surpassing the activity of reported ruthenium and iridium catalysts. It was demonstrated that the metal–ligand cooperativity in complex **62** was responsible for the remarkable high catalytic activity in the cross-coupling reaction of alcohols.[94]

6.4 Conclusion

In summary, ATH has evolved into a key technology for small- and large-scale manufacture of pharmaceuticals. Development of catalysts based on Earth-abundant metals to replace precious metals for ATH has rapidly attracted the interest of the scientific community. This field of research is evolving fast and surely there will be important breakthroughs in the near future. The borrowing hydrogen methodology that exploits the use of alcohols as alkylating agents provides a green and atom-economic pathway for the synthesis of bioactive molecules. Such methodology would improve the safety, cost-effectiveness, and scalability of the transformations employed in medicinal chemistry laboratories. Further development of improved catalytic systems using readily available reagents, with broad substrate scope, high selectivity and efficiency under milder conditions will have a direct impact in the pharmaceutical industry.

Acknowledgements

B.R. thanks FCT for consolidation contract IF/00346/2013 and research unit Green-it "Bioresources for Sustainability" (UDI/Multi/04551/2013).

References

1. G. Brieger and T. J. Nestrick, *Chem. Rev.*, 1974, **74**, 567.
2. F. Foubelo, C. Nájera and M. Yus, *Tetrahedron: Asymmetry*, 2015, **26**, 769.
3. J. Trocha-Grimshaw and H. B. Henbest, *Chem. Commun.*, 1967, 544.
4. Y. Sasson and J. Blum, *Tetrahedron Lett.*, 1971, **12**, 2167.
5. R. L. Chowdhury and J. E. Bäckvall, *J. Chem. Soc., Chem. Commun.*, 1991, 1063.
6. D. Wang and D. Astruc, *Chem. Rev.*, 2015, **115**, 6621.
7. C.-A. Bussaca, D. R. Fandrick, J. J. Song and C. H. Senanayake, *Adv. Synth. Catal.*, 2011, **353**, 1825.
8. S. Hashiguchi, A. Fuji, J. Takehara, T. Ikariya and R. Noyori, *J. Am. Chem. Soc.*, 1995, **117**, 7562.
9. A. Fujii, S. Hashiguchi, N. Uematsu, T. Ikariya and R. Noyori, *J. Am. Chem. Soc.*, 1996, **118**, 2521.
10. I. Yamada and R. Noyori, *Org. Lett.*, 2000, 3425.
11. J. Takehara, S. Hashiguchi, A. Fujii, I. Shin-ichi, T. Ikariya and R. Noyori, *Chem. Commun.*, 1996, 233.
12. J. B. Åberg, J. S. M. Samec and J. E. Bäckvall, *Chem. Commun.*, 2006, 2771.

13. J. S. M. Samec, J.-E. Bäckvall, P. G. Andersson and P. Brandt, *Chem. Soc. Rev.*, 2006, **35**, 237.
14. R. Noyori, M. Yamakawa and S. Hashiguchi, *J. Org. Chem.*, 2001, **66**, 7931.
15. M. Schlaf, P. Ghosh, P. J. Fagan, E. Hauptmn and R. M. Bullock, *Angew. Chem., Int. Ed.*, 2001, **40**, 3887.
16. H. Guan, M. Iimura, M. P. Magee, J. R. Norton and G. Zhu, *J. Am. Chem. Soc.*, 2005, **127**, 7805.
17. I. Bauer and H.-J. Knölker, *Chem. Rev.*, 2015, **115**, 3170.
18. W. M. Czaplik, M. Mayer, J. Cvengros and A. J. von Wangelin, *ChemSusChem*, 2009, **2**, 396.
19. K. Junge, K. Schröder and M. Beller, *Chem. Commun.*, 2011, **47**, 4849.
20. M. Albrecht, R. Bedford and B. Plietker, *Organometallics*, 2014, **33**, 5619.
21. P. J. Chirik and T. B. Gunnoe, *ACS Catal.*, 2015, **5**, 5584.
22. B. Su, Z. C. Cao and Z. J. Shi, *Acc. Chem. Res.*, 2015, **48**, 886.
23. S. Enthaler, K. Junge and M. Beller, *Angew. Chem., Int. Ed.*, 2008, **47**, 3317.
24. *Catalysis without Precious Metals*, ed. R. M. Bullock, Wiley-VCH, Weinheim, Germany, 2010.
25. C. Bolm, *Nat. Chem.*, 2009, **1**, 420.
26. J. S. Chen, L. L. Chen, Y. Xing, G. Chen, W. Y. Shen, Z. R. Dong, Y. Y. Li and J. X. Gao, *Acta Chim. Sin.*, 2004, **62**, 1745.
27. C. Sui-Seng, F. Freutel, A. J. Lough and R. H. Morris, *Angew. Chem., Int. Ed.*, 2008, **47**, 940.
28. A. Mikhailine, A. J. Lough and R. H. Morris, *J. Am. Chem. Soc.*, 2009, **131**, 1394.
29. R. H. Morris, *Chem. Soc. Rev.*, 2009, **38**, 2282.
30. P. E. Sues, A. J. Lough and R. H. Morris, *Organometallics*, 2011, **30**, 4418.
31. A. A. Mikhailine and R. H. Morris, *Inorg. Chem.*, 2010, **49**, 11039.
32. P. E. Sues, K. Z. Demmans and R. H. Morris, *Dalton Trans.*, 2014, **43**, 7650.
33. D. E. Prokopchuk, J. F. Sonnenberg, N. Meuer, M. Z. D. Iuliis, A. J. Lough and R. H. Morris, *Organometallics*, 2012, **31**, 3056.
34. P. O. Lagaditis, P. E. Sues, J. F. Sonnenberg, K. Y. Wan, A. J. Lough and R. H. Morris, *J. Am. Chem. Soc.*, 2014, **136**, 1367.
35. W. W. Zuo, A. J. Lough, Y. F. Li and R. H. Morris, *Science*, 2013, **342**, 1080.
36. W. Zuo and R. H. Morris, *Nat. Protoc.*, 2015, **10**, 241.
37. S. Zhou, S. Fleischer, K. Junge, S. Das, D. Addis and M. Beller, *Angew. Chem., Int. Ed.*, 2010, **49**, 8121.
38. Y. Y. Li, S. L. Yu, W. Y. Shen and J. X. Gao, *Acc. Chem. Res.*, 2015, **48**, 2587.
39. R. Bigler and A. Mezzetti, *Org. Lett.*, 2014, **16**, 6460.
40. R. Bigler, E. Otth and A. Mezzetti, *Organometallics*, 2014, **33**, 4086.
41. R. Bigler, R. Huber and A. Mezzetti, *Angew. Chem., Int. Ed.*, 2015, **54**, 5171.
42. N. A. Eberhardt and H. Guan, *Chem. Rev.*, 2016, **116**, 8373.

43. A. Zirakzadeh, S. R. M. M. de Aguiar, B. Stöger, M. Widhalm and K. Kirchner, *ChemCatChem*, 2017, **9**, 1.
44. D. J. Ager, A. H. M. de Vries and J. G. de Vries, *Chem. Soc. Rev.*, 2012, **41**, 3340.
45. N. A. Cortez, G. Aguirre, M. Parra-Hake and R. Somanathan, *Tetrahedron: Asymmetry*, 2013, **24**, 1297.
46. Z. Luo, F. Qin, S. Yan and X. Li, *Tetrahedron: Asymmetry*, 2012, **23**, 333.
47. R. Wisdom, J. Jung and A. Meudt, EP2,383,261 A1, 2011.
48. J. Liu, D. Zhou, X. Jia, L. Huang, X. Li and A. S. C. Chan, *Tetrahedron: Asymmetry*, 2008, **19**, 1824.
49. Z. Fang and M. Wills, *Org. Lett.*, 2014, **16**, 374.
50. Z. Fang, G. J. Clarkson and M. Wills, *Tetrahedron Lett.*, 2013, **54**, 6834.
51. S. D. Stone, N. J. Lajkiewicz, L. Whitesell and A. Hilmy, *Chem. Soc.*, 2015, **137**, 525.
52. E. Buitrago, H. Lundberg, H. Andersson, P. Ryberg and H. Adolfsson, *ChemCatChem*, 2012, **4**, 2082.
53. L. Monnereau, D. Cartigny, M. Scalone, T. Ayad and V. Ratovelomanana-Vidal, *Chem. – Eur. J.*, 2015, **21**, 11799.
54. D. Cartigny, K. Püntener, T. Ayad, M. Scalone and V. Ratovelomanana-Vidal, *Org. Lett.*, 2010, **12**, 3788.
55. Y. Wang, H. Zhang, W. Huang, J. Kong, J. Zhou and B. Zhang, *Eur. J. Med. Chem.*, 2009, **44**, 1638.
56. I. Plantan, M. Stephan, U. Urleb and B. Mohar, *Tetrahedron Lett.*, 2009, **50**, 2676.
57. M. Perez, Z. Wu, M. Scalone, T. Ayad and V. Ratovelomanana-Vidal, *Eur. J. Org. Chem.*, 2015, 6503.
58. K. Yamada, M. Takeda and T. Iwakuma, *J. Chem. Soc. Perkin Trans. I*, 1983, 265.
59. Y. Watanabe, Y. Tsuji, Y. Ohsugi and J. M. J. Williams, *Tetrahedron Lett.*, 1981, **22**, 2667.
60. T. D. Nixon, M. K. Whittlesey and J. M. J. Williams, *Dalton Trans.*, 2009, 753.
61. R. Grigg, T. R. B. Mitchell, S. Sutthivaiyakit and N. Tongpenyai, *J. Chem. Soc., Chem. Commun.*, 1981, 611.
62. G. Guillen, D. J. Ramon and M. Yus, *Chem. Rev.*, 2010, **110**, 1611.
63. M. H. S. A. Hamid, P. A. Slatford and J. M. J. Williams, *Adv. Synth. Catal.*, 2007, **349**, 1555.
64. C. Gunanathan and D. Milstein, *Science*, 2013, **341**, 1229712.
65. G. E. Dobereiner and R. H. Crabtree, *Chem. Rev.*, 2010, **110**, 681.
66. T. Yan, B. L. Feringa and K. Barta, *Nat. Commun.*, 2014, **5**, 5602.
67. T. Yan, B. L. Feringa and K. Barta, *ACS Catal.*, 2016, **6**, 381.
68. A. Mukherjee, A. Nerush, G. Leitus, L. J. W. Shimon, Y. Ben-David, N. A. E. Jalapa and D. Milstein, *J. Am. Chem. Soc.*, 2016, **138**, 4298.
69. M. Mastalir, M. Glatz, N. Gorgas, B. Stöger, E. Pittenauer, G. Allmaier, L. F. Veiros and K. Kirchner, *Chem. – Eur. J.*, 2016, **22**, 1.
70. A. J. Rawlings, L. J. Diorazio and M. Wills, *Org. Lett.*, 2015, **17**, 1086.

71. H.-J. Pan, T. W. Ng and Y. Zhao, *Chem. Commun.*, 2015, **51**, 11907.
72. G. Zhang, Z. Yin and S. Zheng, *Org. Lett.*, 2016, **18**, 300.
73. S. Rösler, M. Ertl, T. Irrgang and R. Kempe, *Angew. Chem., Int. Ed.*, 2016, **54**, 15046.
74. M. Mastalir, G. Tomsu, E. Pittenauer, G. Allmaier and K. Kirchner, *Org. Lett.*, 2016, **18**, 3462.
75. M. A. Berliner, S. P. A. Dubant, T. Makowski, K. Ng, B. Sitter, C. Wager and Y. Zhang, *Org. Process Res. Dev.*, 2011, **15**, 1052.
76. L. Miao, S. C. DiMaggio, H. Shu and M. L. Trudell, *Org. Lett.*, 2009, **11**, 1579.
77. M. H. S. A. Hamid, C. L. Allen, G. W. Lamb, A. C. Maxwell, H. C. Maytum, A. J. A. Watson and J. M. J. Williams, *J. Am. Chem. Soc.*, 2009, **131**, 1766.
78. J. Leonard, A. J. Blacker, S. P. Marsden, M. F. Jones, K. R. Mulholland and R. Newton, *Org. Process Res. Dev.*, 2015, **19**, 1400.
79. A. K. Rathi, R. Syed, H.-S. Shin and R. V. Patel, *Expert Opin. Ther. Pat.*, 2016, **26**, 1744.
80. J. F. Bower and M. J. Krische, *Top. Organomet. Chem.*, 2011, **34**, 107.
81. J. F. Bower, I. S. Kim, R. L. Patman and M. J. Krische, *Angew. Chem., Int. Ed.*, 2009, **48**, 34.
82. S. Whitney, R. Grigg, A. Derrick and A. Keep, *Org. Lett.*, 2007, **9**, 3299.
83. D. Wang, K. Zhao, C. Xu, H. Miao and Y. Ding, *ACS Catal.*, 2014, **4**, 3910.
84. K. I. Fujita, C. Asai, T. Yamaguchi, F. Hanasaka and R. Yamaguchi, *Org. Lett.*, 2005, **7**, 4017.
85. D. Gnanamgari, E. L. O. Sauer, N. D. Schley, C. Butler, C. D. Incarvito and R. H. Crabtree, *Organometallics*, 2009, **28**, 321.
86. O. Kose and S. Saito, *Org. Biomol. Chem.*, 2010, **8**, 896.
87. P. Satyanarayana, G. M. Reddy, H. Maheswaran and M. L. Kantam, *Adv. Synth. Catal.*, 2013, **355**, 1859.
88. S. Musa, L. Ackermann and D. Gelman, *Adv. Synth. Catal.*, 2013, **355**, 3077.
89. A. Prades, M. Viciano, M. Sanaú and E. Peris, *Organometallics*, 2008, **27**, 4254.
90. C. Schlepphorst, B. Maji and F. Glorius, *ACS Catal.*, 2016, **6**, 4184.
91. G. Tang and C. H. Cheng, *Adv. Synth. Catal.*, 2011, **353**, 1918.
92. J. Yang, X. Liu, D. L. Meng, H. Y. Chen, Z. H. Zong, T. T. Feng and K. Sun, *Adv. Synth. Catal.*, 2012, **354**, 328.
93. G. E. Dobereiner and R. H. Crabtree, *Chem. Rev.*, 2010, **110**, 681.
94. K. Chakrabarti, B. Paul, M. Maji, B. C. Roy, S. Shee and S. Kundu, *Org. Biomol. Chem.*, 2016, **14**, 10988.

CHAPTER 7

Green Metal-catalysed Synthesis of Pharmaceutically Useful Asymmetric Epoxides and Sulfoxides

C. C. ROMÃO

ITQB-UNL, Av. da República, Estação Agronómica Nacional, 2780-157 Oeiras, Portugal
Email: ccr@itqb.unl.pt

7.1 Epoxidation and Sulfoxidation: Introduction

Epoxidation and sulfoxidation are two oxidative processes in organic chemistry that share several common features. Both correspond to the formal addition of an oxygen atom, [O], to an electron pair: (i) to the double C=C bond of an olefin (alkene) and (ii) to a lone pair of the sulfur atom, [S], in organic sulfides (SR_2). In both cases, the transfer of the [O] atom from an oxidant requires the use of a catalyst. The addition of the [O] atom to the C=C double bond may take place at either face of the olefin while the addition of the [O] atom to SR_2 may take place on any of its two lone electron pairs. If either the olefin or the sulfide is prochiral, such addition leads to racemic mixtures of either epoxide or sulfoxide since, in practice, oxidants are often achiral and unable to distinguish between enantiofaces (Scheme 7.1). Thus, in both cases, chiral catalysts capable of enantiotopic

Green Chemistry Series No. 54
Sustainable Synthesis of Pharmaceuticals: Using Transition Metal Complexes as Catalysts
Edited by Mariette M. Pereira and Mário J. F. Calvete

Published by the Royal Society of Chemistry, www.rsc.org

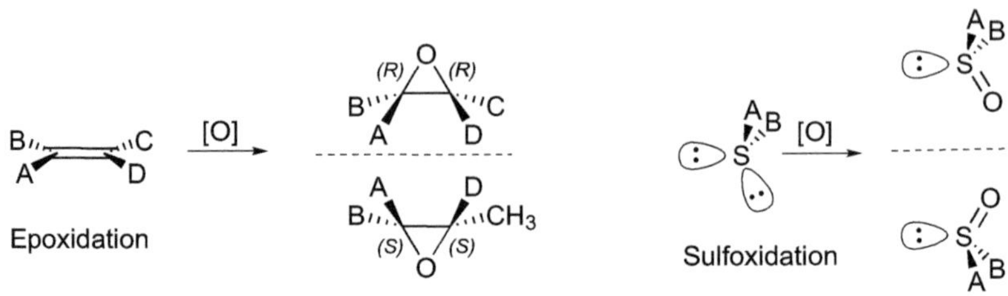

Scheme 7.1 Enantiomers from epoxidation and sulfoxidation.

recognition are required for the selective synthesis of enantiomerically pure epoxides or sulfoxides (or highly enantiomerically enriched mixtures).

Since both chiral epoxides and chiral sulfoxides are crucial functions in many pharmacologically active molecules, and are extremely versatile intermediates in organic synthesis, the quest for efficient enantioselective and more sustainable epoxidation and sulfoxidation catalysts has spurred an enormous amount of research effort over the years.[1] It then turns out that several types of catalysts that have been developed for asymmetric epoxidation (AE) can be used or adapted to perform asymmetric sulfoxidation (ASO), thus justifying the contents of this chapter.

Synthetic and industrial chemistry has been under sustained pressure to adopt safer and environmentally sustainable practices, which have been rationalised under the general umbrella of "green chemistry". This concept is an envelope where all sorts of inputs considering the decrease of the environmental footprint of each chemical process are embedded: saving energy and resources, eliminating toxicity, environmental poisoning and chemical waste are the goals. Catalysis, atom economy, high reaction selectivity and efficiency, cheaper and renewable chemicals are the concepts and tools that chemists should use to achieve such goals. AE and ASO will be discussed with such principles in mind because they involve potentially dangerous oxidants, toxic solvents and metals, and because they can maximise resources by producing 100% of the desired enantiomer instead of racemic mixtures with 50% product waste.

Although historically useful, hazardous oxidants like percarboxylic acids [RC(O)OOH] or dioxiranes (R_2CO_2) can only be accepted under very special situations. Shi's epoxidation is a remarkable exception, given its metal-free catalytic operation with *in situ* generation of chiral dioxiranes.[2–4] Other less aggressive oxidants like iodosyl arenes (ArIO), hypochlorite (NaOCl) or oxone ($KHSO_5$) are still acceptable under controlled conditions but are far from being environmentally harmless.

Molecular oxygen, O_2, might be considered the ultimate environmentally friendly oxidant, since it is abundant and 100% oxygen active, that is, its byproducts are H_2O or nothing. In practice, the reactivity of O_2 is difficult to control and concentrated O_2 in the presence of organic matter can lead to dangerous situations. Therefore, hydrogen peroxide, H_2O_2, is usually considered the most environmentally acceptable terminal oxidant (TO) with

H_2O as its sole by-product. However, *per se*, H_2O_2 is a weak epoxidation agent and needs to be activated by metal catalysts. Alkyl hydroperoxides, ROOH, are the next most environmentally friendly oxidants, but they also need to be activated like H_2O_2. Particularly useful are ^tBuOOH (*t*-butyl hydroperoxide; TBHP) and $PhC(Me)_2OOH$ (cumyl hydroperoxide; CHP), which form ^tBuOH and $PhC(Me)_2OH$, respectively, as end reaction products. Under proper conditions these alcohols can be reoxidised to the parent ROOH species without significant environmental problems. Progress in the use of both O_2 and H_2O_2 as environmentally sustainable oxidants has been comprehensively reviewed recently by Bryliakov.[5]

One final word is to simply acknowledge the crucial role of the late Professor Tsutomu Katsuki in the development of concepts and practice of enantioselective epoxidation and sulfoxidation, which has been recently put in historical perspective by some of his close collaborators.[6]

7.2 Asymmetric Transition Metal-catalysed Epoxidation of Olefins

From the point of view of practical synthetic methods, there are two main catalytic systems that successfully carry out the asymmetric olefin epoxidation reaction (AOE): the Katsuki–Sharpless or Sharpless Asymmetric Epoxidation (SAE) of allylic alcohols and the Jacobsen–Katsuki asymmetric epoxidation (JKAE) of unfunctionalised olefins. Both methods have achieved high levels of efficiency in terms of yields and enantiomeric excesses obtained, and are now widely employed in multistep synthetic pathways of important natural and pharmaceutical products where chirality is a key property. Both methods, though, have limitations from the green chemistry point of view, namely in their use of NaOCl and organic hydroperoxides as the TO. New catalysts using H_2O_2 as the TO have been recently discovered that can obtain amazing selectivity. For example, bis-hydroxamic acid complexes and Ti(salalen) complexes can take AOE to substrates that are not responsive to either SAE or JKAE. However, the fact that many types of olefins cannot undergo enantioselective epoxidation with any of these catalysts, but are amenable to racemic epoxidation with small, highly active achiral catalysts, coupled with the enhanced reactivity of epoxides, led the way to the use of the powerful kinetic resolution of racemic epoxide mixtures, namely in the case of the hydrolytic kinetic resolution technology.[7]

7.2.1 The Katsuki–Sharpless Asymmetric Epoxidation of Allylic Alcohols

After a few initial reports showing the possibility of inducing enantioselectivity in alkene epoxidation with simple chiral Mo and V complexes in conjunction with chiral auxiliaries, the breakthrough appeared in 1980 at the hands of Katsuki and Sharpless in a paper appropriately entitled

"The first practical method for asymmetric epoxidation".[8] This reaction is known as Sharpless Asymmetric Epoxidation (SAE, Scheme 7.2) and certainly contributed to his Nobel Prize in Chemistry, awarded in 2001.

The reaction uses very simple, commercially available reagents: $Ti(O^iPr)_4$, (+)-diethyl L-tartrate [L-(+)-DET] or (−)-diethyl D-tartrate [D-(−)-DET], and ^tBuOOH (TBHP) as the oxidant. A very large range of prochiral allylic alcohols ($R^3R^2C{=}CR^1CHR^4OH$) can be epoxidised with a readily predictable stereochemistry and excellent enantioselectivity, often >90% *ee*.

The predictability of the outcome of this reaction is sometimes considered "magic" or even "miraculous". Indeed, the stereochemistry of the major enantiomer obtained depends on the chirality of the DET used and can be very easily predicted by the mnemonic summarised in Scheme 7.3.

First, we visualize the allylic alcohol with its C=C bond lying on a plane with the CH_2OH group in the lower-right quadrant. If SAE is carried out with the L-(+)-DET ligand, the epoxide obtained results from addition of the O atom from below the plane. Obviously, when D-(−)-DET is used as chiral ligand, the attack takes place from above the plane producing the other enantiomer. This enantiotopic selectivity is independent of the substitution pattern at the double bond. Further examples of the power of this prediction are shown in Scheme 7.4 and in Figure 7.1.

When the C1 position of the alcohol is substituted (CHR^4), the substituent R^4 can be accommodated above or below the plane of the allylic alcohol in the Sharpless mnemonic. In that case, the product that will be formed faster

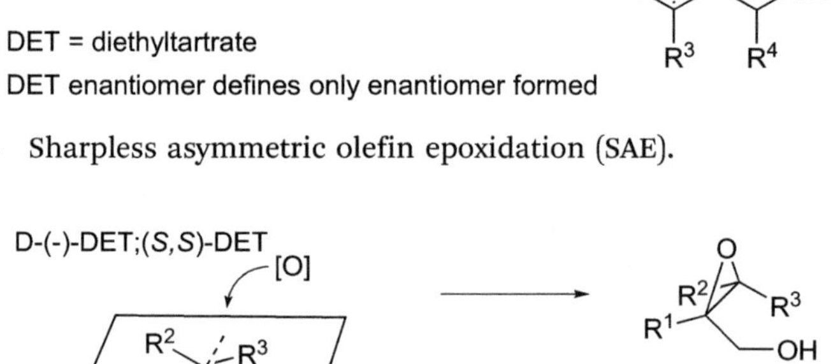

Scheme 7.2 Sharpless asymmetric olefin epoxidation (SAE).

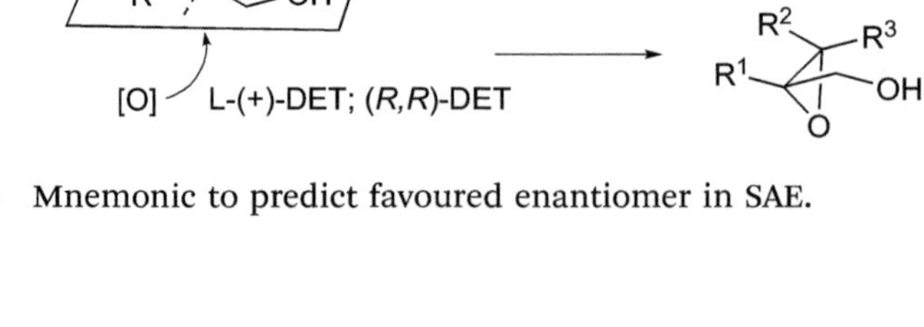

Scheme 7.3 Mnemonic to predict favoured enantiomer in SAE.

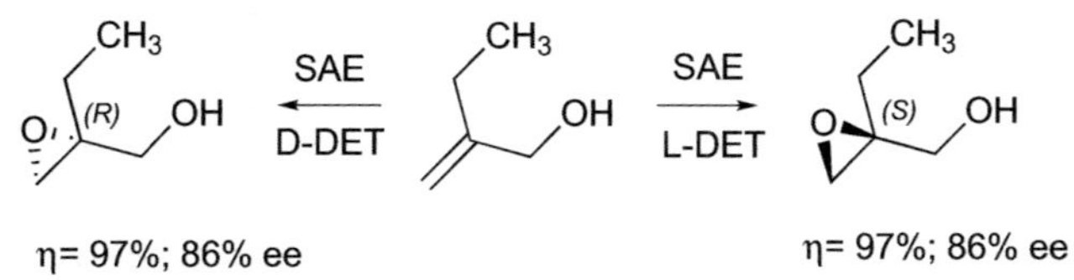

Scheme 7.4 Stereochemistry of asymmetric epoxidation of prochiral allylic alcohols.

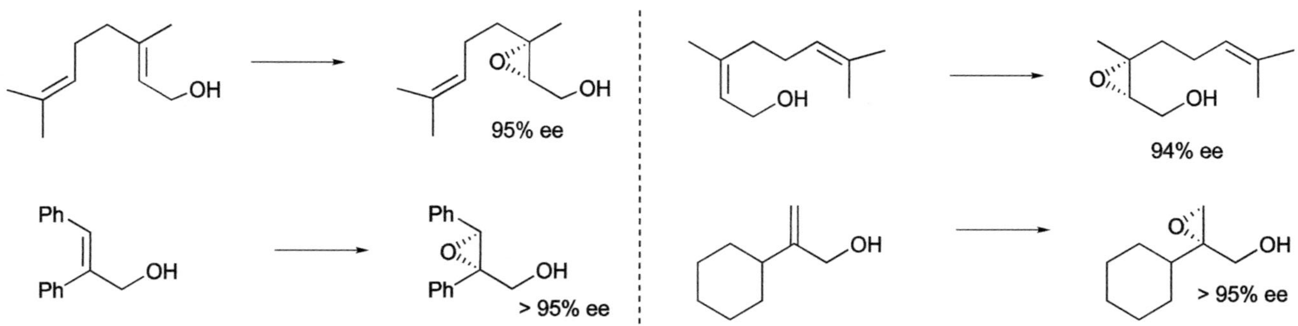

General reaction protocol:

$Ti(O^iPr)_4$/(+)-DET/^{t}BuOOH/ CH_2Cl_2/ -20 °C

DET = diethyltartrate

Figure 7.1 Early epoxidation reactions reported by Katsuki and Sharpless.[8]

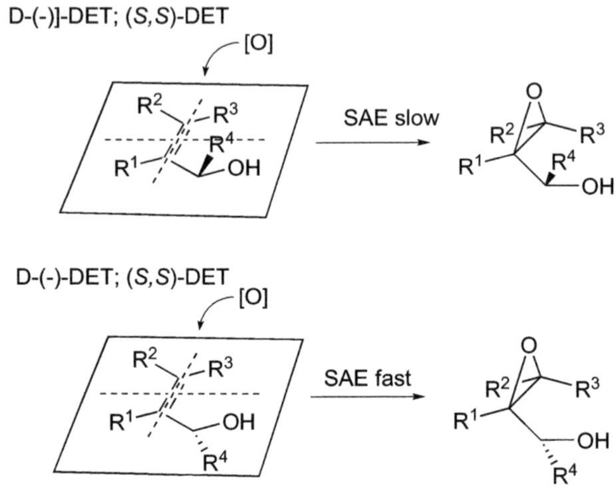

Scheme 7.5 Enantiomer prediction for SAE of allylic alcohols chiral at C1 (kinetic resolution).

is the one that results from minimising the contact between the tartrate scaffold and R^4. This kind of kinetic resolution is illustrated in Scheme 7.5. Allylic alcohols disubstituted at C1 (CR^4R^5) do not undergo selective SAE.

When $Ti(O^iPr)_4$ is reacted with L-(+)-DET the complex $[Ti(\kappa^2\text{-}O_2C_2H_2(CO_2Et)_2(O^iPr)_2]$ is formed (**A** in Scheme 7.6). If D-(−)-DET is used, the opposite optical isomer is formed. The existing experimental evidence strongly suggests that these monomers dimerise under the reaction conditions to give the species **B** in Scheme 7.6.

^{i}PrOH is then eliminated from **B** upon deprotonation of both TBHP and the allylic alcohol forming the proposed and accepted catalytically active species **C** shown in Scheme 7.6. In **C**, one of the Ti atoms retains the original **A** coordination sphere whereas the other (on the right-hand side) now accommodates both the oxidant and the substrate with the dangling C=C double bond that is going to be oxidised. The C(O)OEt group of the tartrate ligand is dissociated from that Ti atom, creating an empty coordination position, which is easily filled by the coordination of the second O atom of the ^{t}BuOO$^-$ ligand. This dihapto coordination mode of the ^{t}BuOO$^-$ ligand originates a marked positive charge on the O atom that is not bound to the ^{t}Bu group. So, this electrophilic O atom may react with the nucleophilic uncoordinated double bond of the allyl alcohol leading to the formation of the epoxide. The dangling allylic C=C bond can approach the electrophilic O atom from either face. The chemical result will be an epoxide in both cases, yet each face will produce a different enantiomeric form of that epoxide when the allylic alcohol is a prochiral molecule. Key mechanistic studies were performed by Sharpless and collaborators.[9,10]

The origin of this enantioselectivity can be traced in the transition states in Scheme 7.6. The top TS_S (**D**) shows the C=C bond of the coordinated

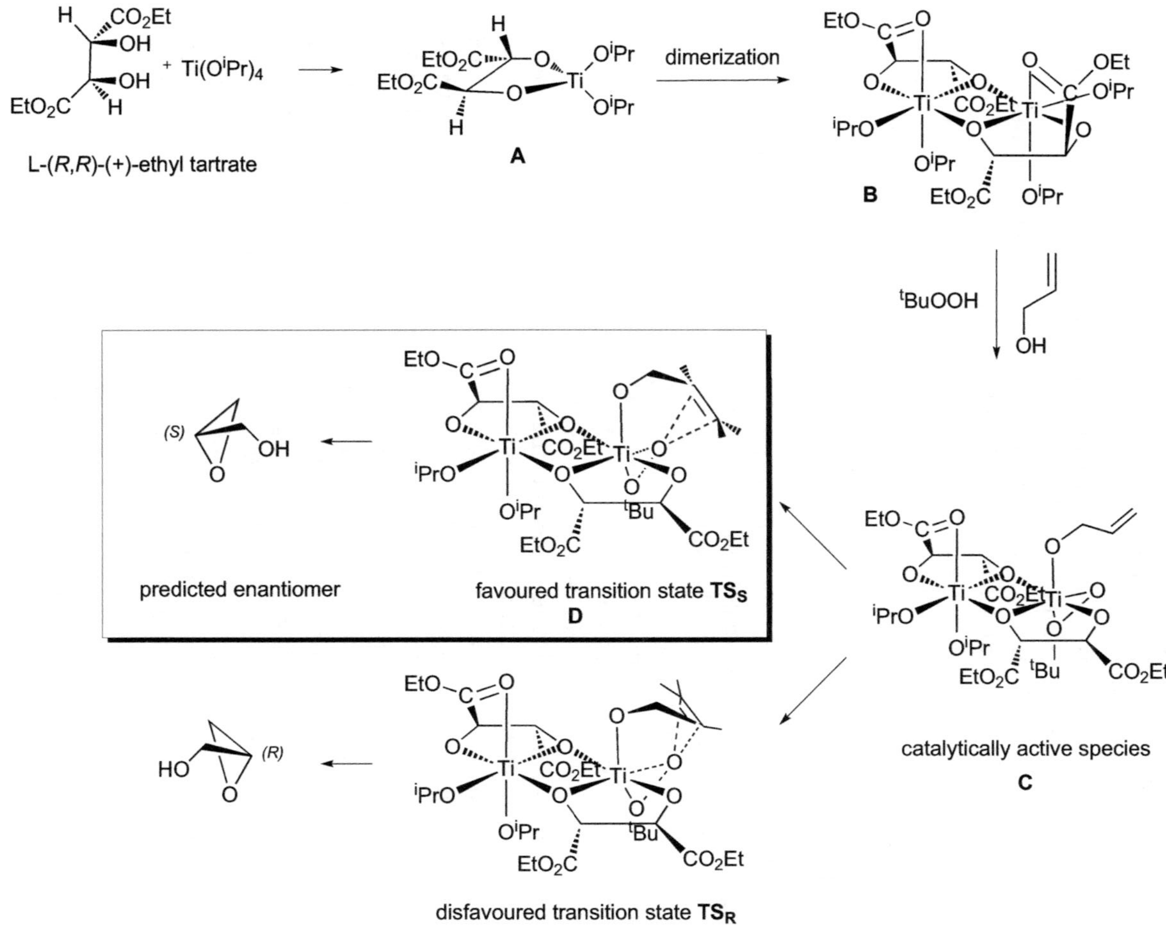

Scheme 7.6 Accepted model of the SAE catalytic precursor.[9,10]

allylic alcohol facing the transferrable O atom in a position that leads to the formation of the (*S*)-enantiomer of the epoxide, as predicted by the rule presented in Scheme 7.3, since the example uses (*R*,*R*)-DET as the ligand. The bottom TS_R presents the other enantioface to the electrophilic O atom leading to the opposite enantiomer, which is not favoured at all. A computational study by Rzepa concludes that the difference in activation energy between both transition states is only 3 kcal mol^{-1} in favour of the one predicted by the mnemonic and experimentally observed, yet such a small value already accounts for a 99% *ee* value.[11]

When the allylic alcohol bears a substituent at the C1 allylic position (R^4), this may be positioned above or below the plane of the mnemonic. The resulting prediction is now derived from Scheme 7.5. Approach of the allylic alcohol to a complex with the D-(−)-DET ligand is favoured when the R^4 group points away from the DET ligand (below the allyl plane), thus kinetically favouring the formation of the epoxide in the lower-right corner of Scheme 7.5. This is a kinetic resolution process that leads to the accumulation of one diastereomer over the other. Usually the relative rate constant of both processes $k_{rel} = k_{fast}/k_{slow}$ is greater than 25, which leads to a very high *ee* of the unreacted alcohol already at 60% conversion.

This simple steric rationale explains the fact that tertiary allylic alcohols do not undergo normal SAE: the presence of a putative R^5 at the C1 atom is quite destabilising of the interaction with the species **C**.

Normal SAE conditions mean that the reaction is usually carried out at ≈ -20 °C in CH_2Cl_2. Halogenated solvents are usually banned from the "green chemistry" catalogue. However, they are essentially irreplaceable in the SAE and actually seem to interact with the hydroperoxides weakening their O–O bonds and increasing their reactivity. The amount of $Ti(O^iPr)_4$ that is used in the normal SAE protocol is stoichiometric. This is far from being an economic solution and several attempts to improve this situation have been tested. One of the most useful is the addition of molecular sieves to the reaction mixture. These remove all water, which would destroy the catalyst and decompose the epoxides to diols, through acid catalysed ring-opening. In practice, molecular sieves improve the epoxide yield, lower the amounts of Ti catalyst and DET to *ca.* 10%.[12] Other possibilities to improve the sustainability of the catalytic system include the use of supported catalysts. Such processes facilitate the separation of the catalysts from the reaction mixtures and are beneficial for the purification process and waste minimisation. Examples of such supports are linear and branched poly(tartrate esters)[13] and mesoporous supports, like MCM-41.[14]

The negative temperatures at which the reactions take place increase the costs of industrial operations. Thus, some modified catalysts working at higher temperatures have been produced, yet their performances do not match that of the standard protocol.[15] Although the nature of the solvent and the oxidant, as well as the reaction conditions, are not the "greenest", the very high yields and exceptional enantioselectivity are highly favourable forms of atom economy and simplified purification processes, two very important factors in sustainable industrial epoxide synthesis. All types of mono-, di- and tri-substituted allylic alcohols are fine substrates for SAE. The reactivity and selectivity of SAE reactions are higher for *trans* (or E) disubstituted double bonds in allylic alcohols than for the isomeric *cis* (or Z) double bonds. Homoallylic alcohols also show a much smaller reactivity, as well as a lower and inverse enantioface selectivity when compared to that predicted for the normal SAE. Two examples are given in Scheme 7.7. In one case the epoxidation of the homoallylic alcohol **3** occurs very slowly to yield the "inverse" isomer **4**. In the other case, the tris-homoallylic alcohol **1** is epoxidised in an intermediate step of the venustratriol synthesis. The reaction using (−)-DET, is slow, requires trityl hydroperoxide (TrOOH) but yields the epoxide isomer **2** with high selectivity. However, a similar reaction using (+)-DET does not allow the clean synthesis of the epoxide with the opposite stereochemistry.[16]

Allylic diols also present less predictable reactivity and stereochemistry. For example, the diol **5** does not undergo SAE with (−)-DET, whereas using the opposite (+)-DET a quantitative yield is obtained as shown (**6**). On the other hand, when the terminal hydroxyl group of these diols was protected with $^{t}BuPh_2Si$ (TDS), the resulting siloxide **7** was epoxidised with SAE–(−)-DET to produce the predicted epoxide **8** in very high yield (Scheme 7.7).[17] This reaction was a crucial step in the process of the total synthesis of the highly unstable chromophore of (+)-neocarzinostatin the first example of an "enediyne" antitumoural agent to be characterised.[18]

As can be seen in many of its application examples below, the SAE reaction is highly chemoselective since only the allylic double bond terminated with the OH function is epoxidised even in the presence of other double bonds and several other functions in the same molecule. This chemoselectivity imparts a very high value to SAE in the synthesis of functionalised molecules, namely in biologically active and natural products.

7.2.1.1 Applications of SAE in organic synthesis and pharmaceutical industry

The big bonding strain of the epoxide ring resulting from the internal angles of the C–O–C triangle being much smaller than the sp_3 tetrahedral norm (109°) makes epoxides highly reactive species that are readily attacked by nucleophiles. Beyond their own chemical and biological properties, this varied reactivity makes epoxides unusually versatile intermediates in many organic transformations, producing alcohols, diols, amino alcohols, ketones, olefins and polymers (epoxy resins) as can be seen in Scheme 7.8.[19]

When this synthetic potential is coupled with the presence of stereogenic centres at the epoxide, which can actually be selectively installed and then transferred to the new scaffolds, the possibilities for the synthesis of a variety of high added value pharmaceuticals and natural products, either as intermediates or as end products, is immense. In fact, many natural products

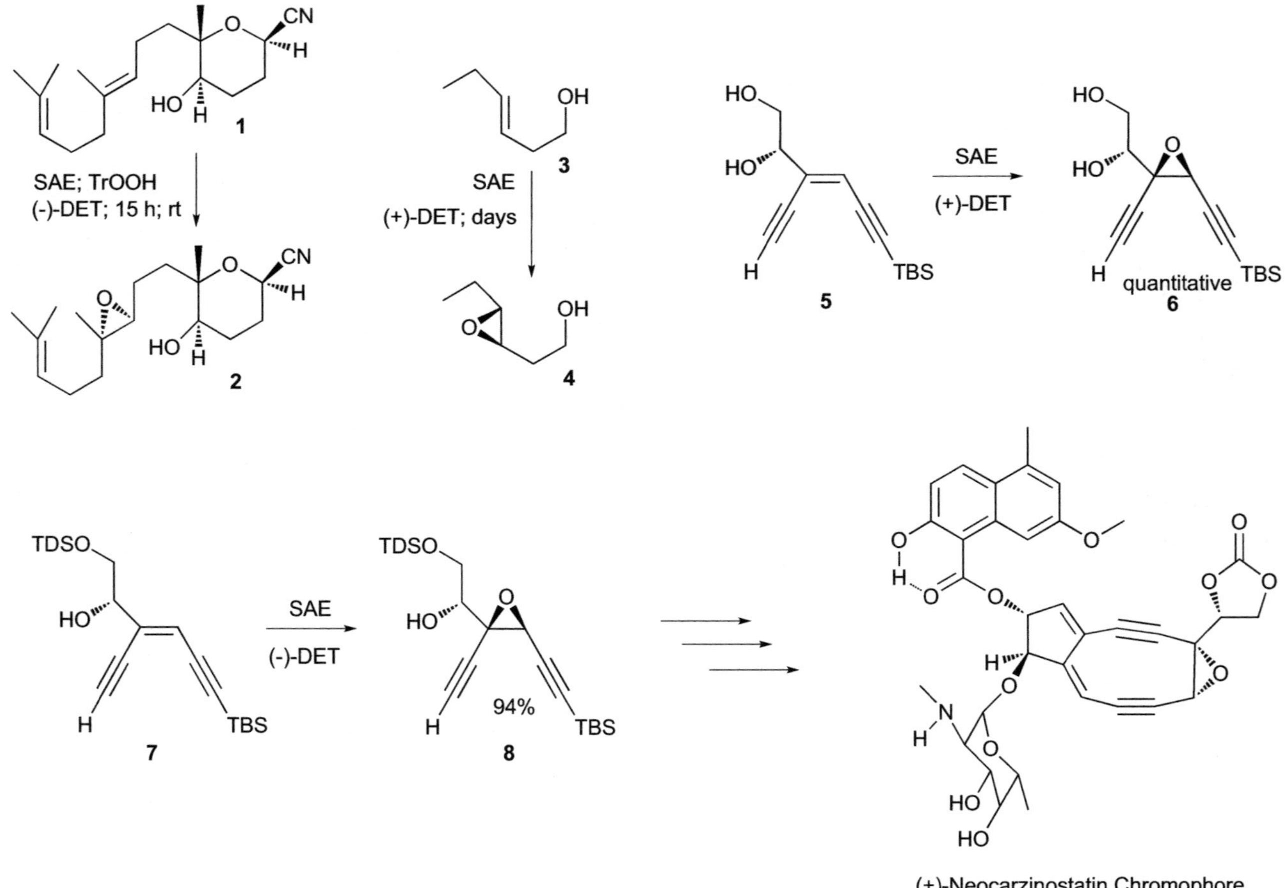

Scheme 7.7 Examples of SAE selectivity with diols and homoallylic alcohols.

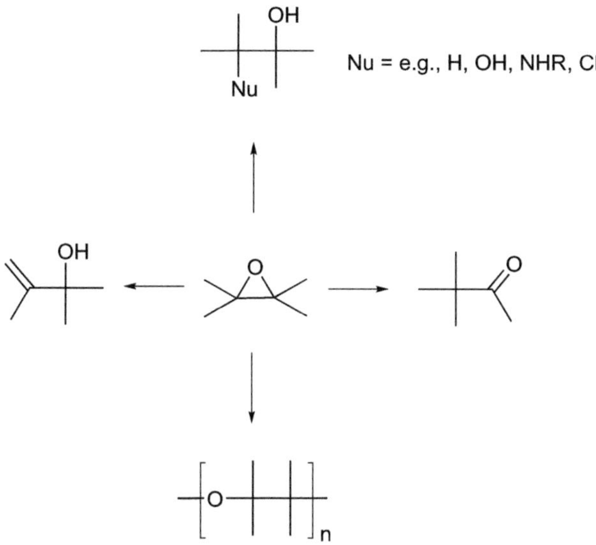

Scheme 7.8 Synthetic transformations of epoxides.

with very complex structures containing epoxide functions are known and can be prepared with the help of the SAE technology, as exemplified for neocarzinostatin (Scheme 7.7)[18] and venustatriol (Scheme 7.10).[16] The fact that an allylic epoxide can be opened to generate three OH functional groups at consecutive positions also opens large possibilities in the synthesis of polyalcohols. This possibility was masterfully demonstrated in the paper entitled "Total synthesis of (L)-hexoses",[20,21] which is briefly summarised in Scheme 7.9.

The SAE of compound **9** led to the 2,3-epoxy alcohol **10** in high yield and *ee*%. Further transformations of **10** enable the selective synthesis of both aldehydes **11** and **12** in high enantiomeric purity and excellent (>80%) yields. Olefination of those aldehydes implanted new allylic double bonds in **13** and **14**, which, upon new SAE reactions and adequate manipulation enabled the synthesis of all eight stereoisomers of L-hexoses: L-allose, L-altrose, L-mannose, L-glucose, L-gulose, L-idose, L-tolose and L-galactose. In the words of the authors, this academic exercise clearly demonstrates the power of the SAE reaction where the catalyst, not the substrate, imposes the selectivity.

Besides one allylic alcohol function, both farnesol and geraniol have other unfunctionalised double bonds, which are not substrates for SAE. Thus, following the SAE of (*E*,*E*)-farnesol and geraniol, Corey's group reported the very elegant and complex synthesis of venustatriol, a marine-derived product with antiviral and anti-inflammatory properties (Scheme 7.10).[16]

Besides the large number of examples of the use of SAE in organic synthesis[22] selected examples that take advantage of the combination of two very powerful transition metal-catalysed reactions, SAE and alkene (or alkyne) metathesis, are presented below.[23]

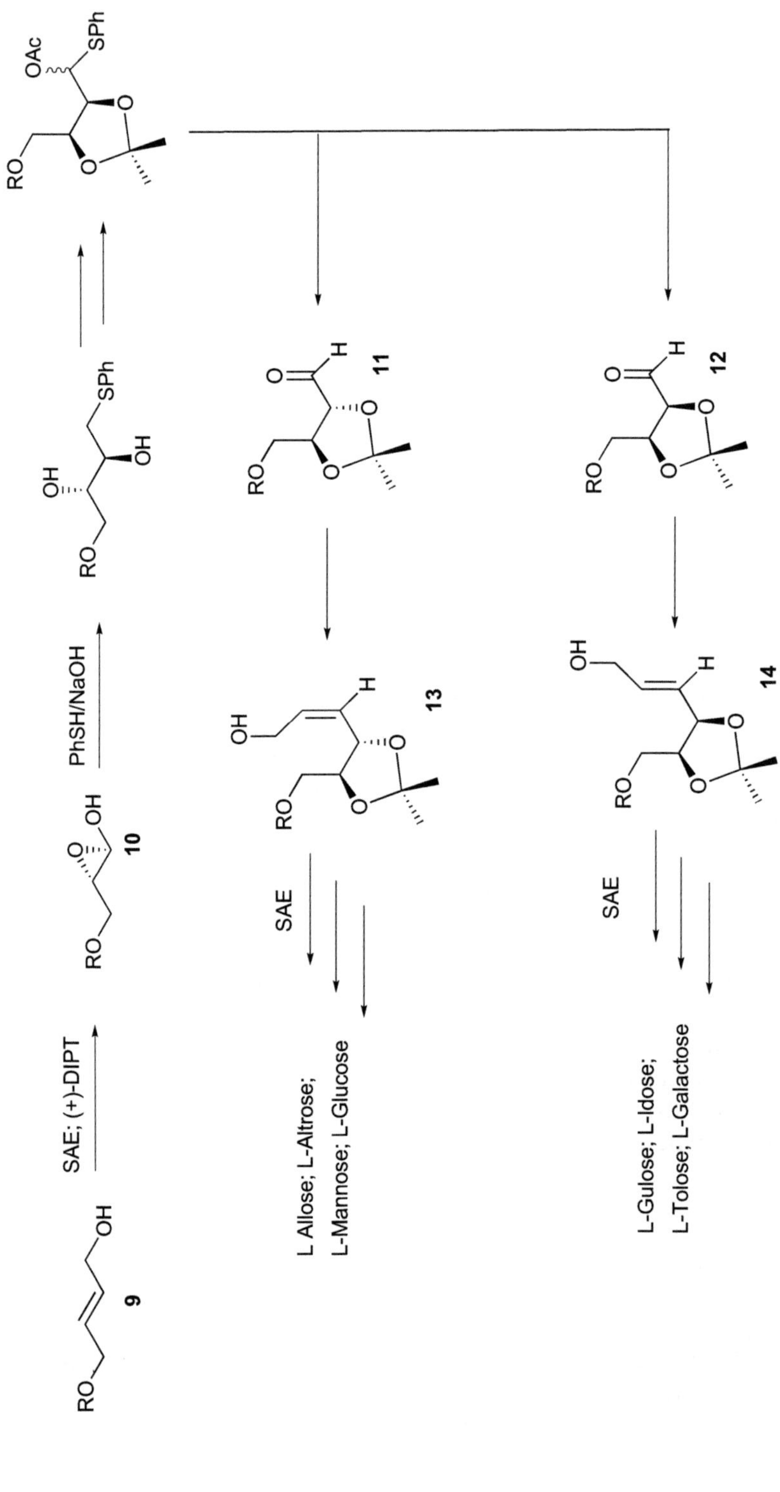

Scheme 7.9 Synthesis of L-hexoses.[20]

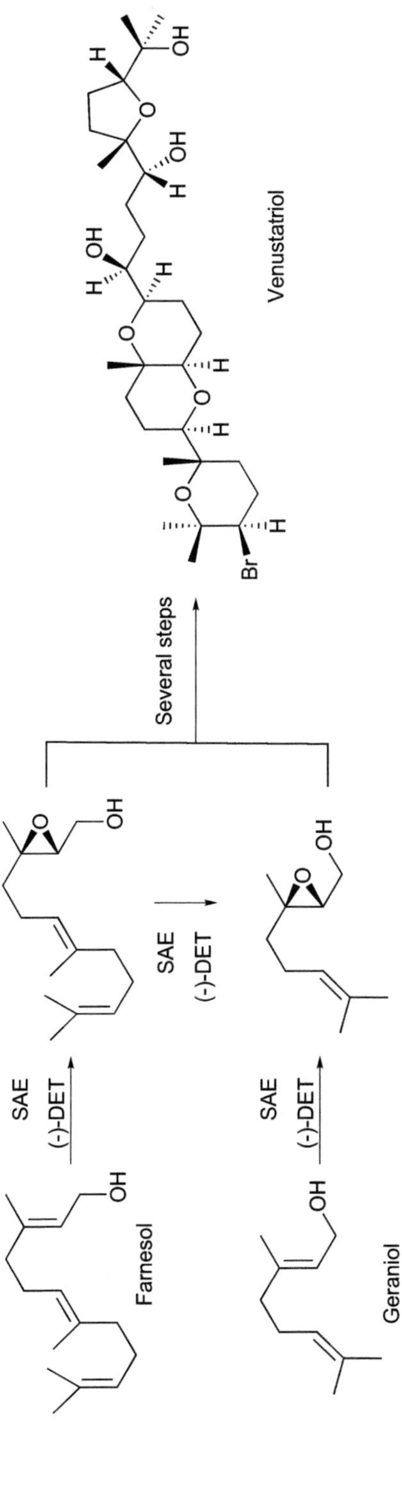

Scheme 7.10 SAE in the synthesis of venustatriol.[16]

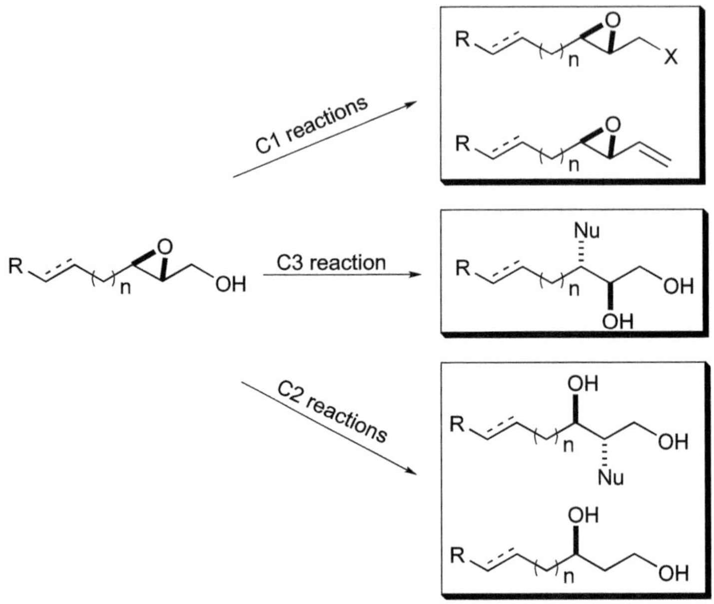

Scheme 7.11 Synthetic modifications of 2,3-epoxy alcohols containing C=C and C≡C bonds.[23]

The presence of terminal olefin or alkyne functions in the molecular frameworks shown in Scheme 7.11 suggests that a broad range of new molecules can be derived through the use of olefin or alkyne metathesis technologies, namely cross olefin metathesis and ring closure metathesis. In particular, the use of ring-closing metathesis (see Chapter 9) enables the preparation of cyclic molecules that are otherwise very difficult to assemble and are often rare natural products. The examples presented in Scheme 7.12 illustrate the potential of such transformations for the case of epoxy alcohols activated at C1 but other products can be also obtained by activation at C2 and C3.[23]

In terms of its industrial use, the SAE finds its place in the synthesis of drugs where a given enantiomer is responsible for the biological function. Some examples are gathered in Figure 7.2. The first success was the synthesis of the gipsy moth pheromone disparlure produced by the company J. T. Baker on a scale of a few kg year^{-1}.[24] At the Eisai Company Ltd, in Japan, the synthesis of a chiral version of a fragment of the calcium channel blocker emopamil left hand involves a crucial SAE reaction. The main advantage of this process is the simplification of the purification of the SAE reaction product by avoiding chromatography.[25] The epoxidation of cinnamyl alcohol is the first step in the long synthesis of both (*S*,*S*)-reboxetine succinate, a norepinephrine uptake inhibitor developed by Pfizer,[26] and the leukotriene antagonist S 0961 developed by Aventis.[27] In the former case, the original enantiomeric resolution process was replaced, leading to a 50% waste decrease, a big plus in terms of green chemistry. In the case of S 0961, the

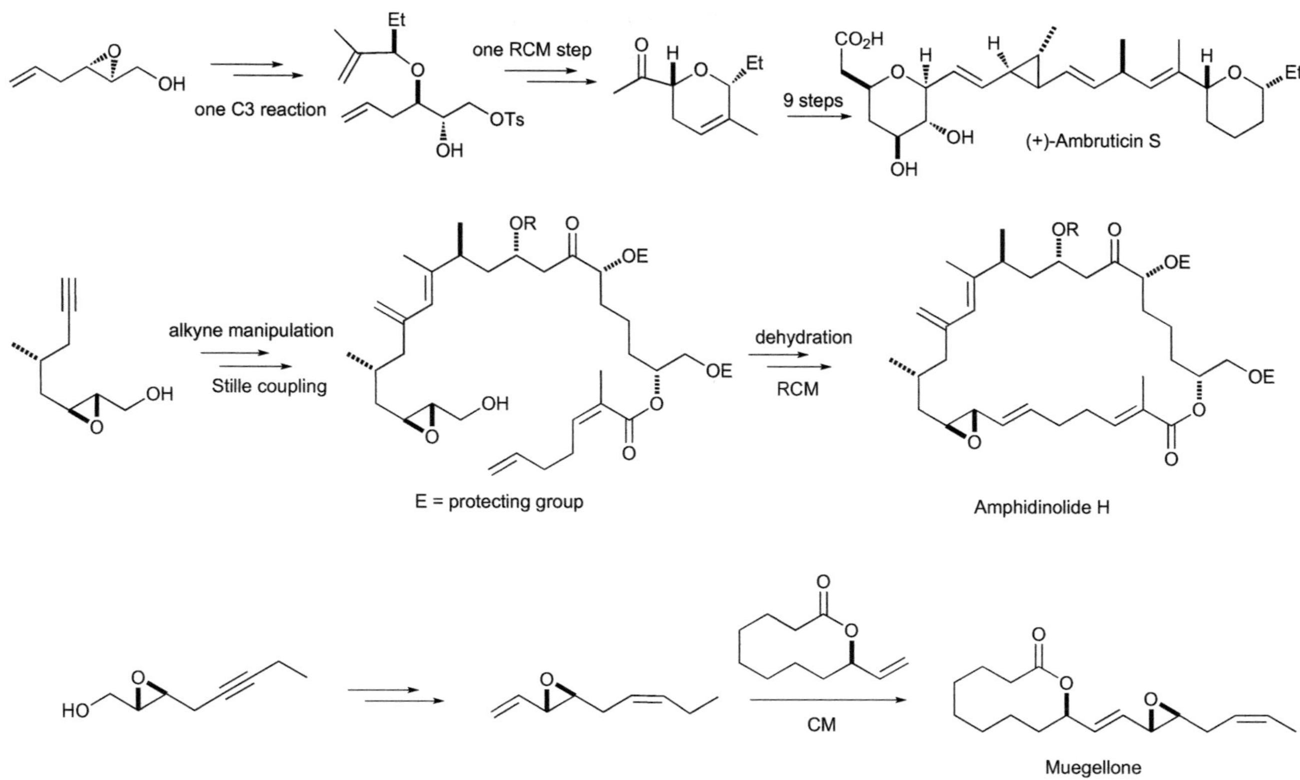

Scheme 7.12 Examples of natural products synthesised from 2,3-epoxy alcohols containing C=C and C≡C bonds.

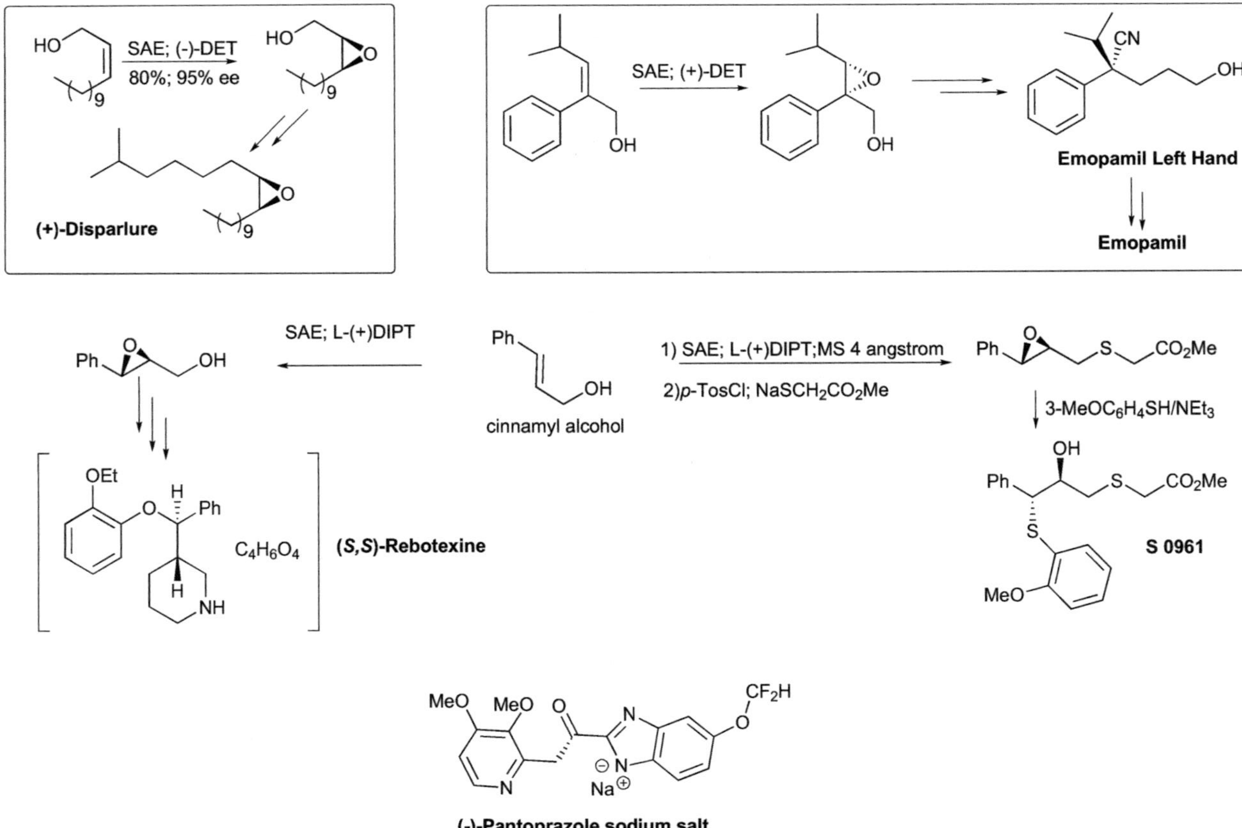

Figure 7.2 Some pharmaceutical, industrial synthesis involving SAE technology.

process is very simple, comprising three simple steps, as shown. To the best of our knowledge, the most recent example of the industrial use of SAE technology is the improved method for the synthesis of (−)-pantoprazole to produce the compound in 99.5% *ee* and 70% yield with a very simple recrystallisation procedure.[28]

Many other examples could be used to illustrate the power of the SAE technology in the pharmaceutical industry. In terms of its compliance with environmental sustainability practices, its main value rests in its high selectivity, which is key to savings in terms of resources and waste, as well as a remarkable simplification of lengthy, sometimes cumbersome, multistep synthetic processes.

7.2.2 The Jacobsen–Katsuki Epoxidation with M(salen) Complexes

In spite of its usefulness, the Sharpless epoxidation is limited to allylic alcohol substrates. Given the variety of non-allylic C=C double bonds that can be epoxidised on the way to products of interest, devising an enantioselective catalyst for the epoxidation of unfunctionalised and non-allylic olefins was a main goal for synthetic chemistry. The breakthrough appeared in 1990/91 with the so-called Mn(salen) catalytic system developed independently by Jacobsen[29] and Katsuki[30] (Figure 7.3). The commercially available Jacobsen catalyst is the hallmark of this technology enabling epoxidation of a range of *cis* olefins with >90% *ee*, some of which open wide avenues on the way to asymmetric synthesis of natural products and bioactive molecules. In addition, the second-generation Katsuki catalysts have also shown extraordinary results in many applications.

Their general structure is based on the tetradentate Schiff base derived from the condensation of two molecules of salicylaldehyde with ethylenediamine, abbreviated as salen, which is modified by design to produce chiral

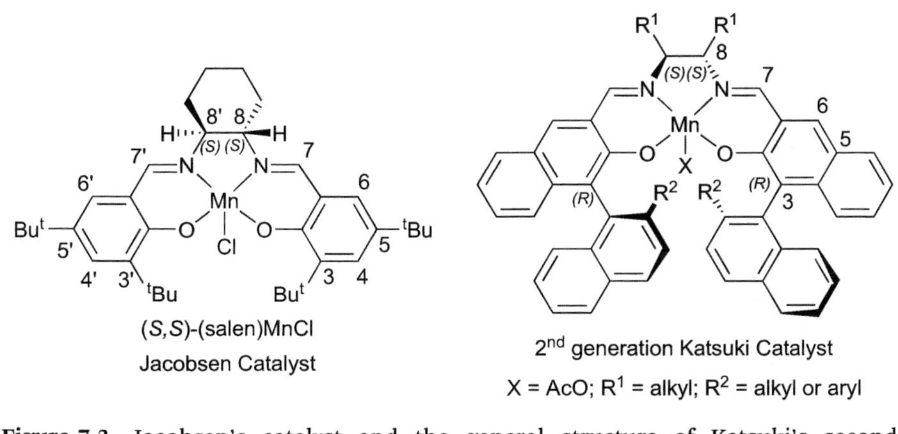

Figure 7.3 Jacobsen's catalyst and the general structure of Katsuki's second-generation catalysts.

salen ligands (salen*) as follows: the bridge connecting the two salen ligands is built from an enantiomerically pure diamine, *e.g.* $RHNCH_2CH_2NHR$ or 1,2-cyclohexane diamine; the 3,3′ positions are substituted with very bulky groups, *e.g.* ^{t}Bu in the Jacobsen catalysts and $PhEtHC^{-}$ in the first-generation Katsuki catalysts[30] and *R*-naphthyl in the case of the second-generation Katsuki catalysts,[31] and the 5,5′ positions can be occupied by several groups for stereochemical and electronic control. (Second-generation Katsuki catalysts are based on the naphthalene analogue of salicylaldehyde.) Finally, the central metal ion is always Mn(III) and the axial position below the N2O2 plane can be vacant or occupied either by anionic ligands X (*e.g.* AcO^{-}, Cl^{-}) or by neutral ligands, L, usually pyridine or amine *N*-oxides. The latter have a positive influence on the catalytic process, improving stability, reactivity and often selectivity.[32] Specific catalysts of this kind under the proper conditions can make the enantioselective epoxidation (>90% *ee*) of prochiral mono-, di-, tri- and tetra-substituted olefins to create an epoxide with two chiral centres and they can be stored indefinitely under air at room temperature.

The best substrates, that is, the ones that easily achieve high enantioselectivity, are *cis* olefins, namely those conjugated to a cyclic aromatic substituent. *cis*-ß-Methylstyrene (*cis*-PhHC=CHMe) is the most used model olefin.[33] The catalytic performance, in terms of the dependence of the enantio- and stereo-selectivity results, depends on the nature of different TOs, different co-ligands X or L, and different structural and stereochemical characteristics of the salen* ligand determined by their substituents at the C3,3′ and C5,5′ positions and the chiral diimine bridge. Importantly, the reaction is not stereoselective since *cis* alkenes produce mixtures of *cis* and *trans* epoxides. So, the only level of asymmetric induction is the selection of the more favourable enantioface of the prochiral olefin that interacts with the Mn=O fragment in the O-transfer step. The source of the JKAE technology goes back to 1983 when Kochi's group reported the reactions presented in eqn (7.1) and (7.2).[34]

$$[Cr^{III}(salen)(H_2O)_2]^{+} \underset{-PhI}{\overset{PhIO}{\longrightarrow}} [OCr^{V}(salen)]^{+} + 2H_2O \quad (7.1)$$

$$[OCr^{V}(salen)]^{+} \underset{H_2O}{\xrightarrow{R_2C{=}CR_2}} [Cr^{III}(salen)(H_2O)_2]^{+} + R_2C\overset{O}{\frown}CR_2 \quad (7.2)$$

The oxygen atom transfer from iodosylbenzene (PhIO) to Cr^{III} implies the formal oxidation of Cr^{III} to Cr^{V} in the final product. X-ray crystallography data on $[O{=}Cr^{V}(salen)]^{+}$ shows that the Cr^{V} ion rises slightly above the N2O2 plane and that the salen ligand adopts a stepped conformation. The strongly electrophilic oxo ligand in $[O{=}Cr^{V}(salen)]^{+}$ and substituted analogues is attacked by the electron-rich double bond of olefins which are epoxidised while $[Cr^{III}(salen)]^{+}$ is regenerated [eqn (7.2)]. Accordingly, electron-rich olefins are much more reactive than electron-poor ones.[35] Indeed *p*-methoxystyrene reacts *ca.* 1000 times faster than 1-octene a fact that limits the usefulness of the system $[Cr^{III}(salen)]^{+}$–PhIO as a catalyst since many alkenes are left almost untouched. On the contrary, the analogous $[Mn^{III}(salen)]^{+}$–PhIO system catalyses the epoxidation of a much broader range of substrates like styrenes, stilbenes, cyclic and acyclic olefins with 50–75% yields in minutes, at room temperature, in NCMe solvent, using PhIO as the TO. The first mechanistic investigation of this system was a landmark in the field.[36] The putative catalytically active species was considered to be $[O{=}Mn^{V}(salen)]^{+}$ in analogy with the $[O{=}Cr^{V}(salen)]^{+}$ system. Contrary to the Cr case, the highly reactive $[OMn^{V}(salen)]^{+}$ species has never been isolated and it is assumed that it rests in a dimer form, $[(salen)Mn^{IV}OMn^{IV}(salen)]^{2+}$, a hypothesis that was supported much later by mass spectrometry studies.[37] According to this information, the mechanism of catalytic epoxidation by the $[Mn^{III}(salen)L]^{+}$ system is given in Scheme 7.13.

Among the differences between the $[Cr^{III}(salen)]^{+}$ and the $[Mn^{III}(salen)]^{+}$ systems, two are rather important: one relates to the fact that the Mn system practically does not discriminate between electron-rich and electron-poor olefins; the other is that only the Mn system oxidises saturated hydrocarbons like cyclohexane, supporting the notion that its [O] transfer step is radical-based and not cationic.

Putting together all the information available in 6 years of frantic research, the controversial mechanism of the O-transfer step of the Mn(salen) system was described by Linker in 1997, as represented in Scheme 7.14.[38]

Pathway **A** is a concerted transfer leading exclusively to the *cis* epoxide, similar to that of the Cr(salen) system. Pathway **B** shows the formation of a radical intermediate, which, upon rotation of the C–C bond can lead to both the *cis* and/or the *trans* epoxide. The pathway represents the reversible formation of a metalloxetane, which can either lead to the *cis* epoxide or breakdown to the radical intermediate also forming the *trans* epoxide. In the Cr system, the radical intermediate is replaced by a cationic intermediate. This scheme agrees with the nature and stereochemistry of the reaction products and the side products, which may arise from radical

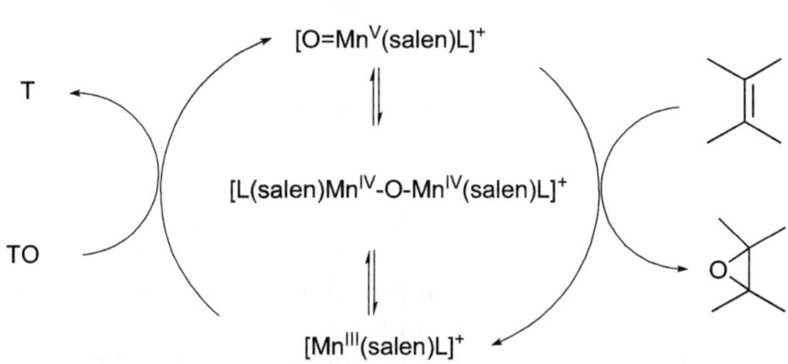

Scheme 7.13 Mechanism of catalytic epoxidation by the $[Mn^{III}(salen)L]^{+}$ system.

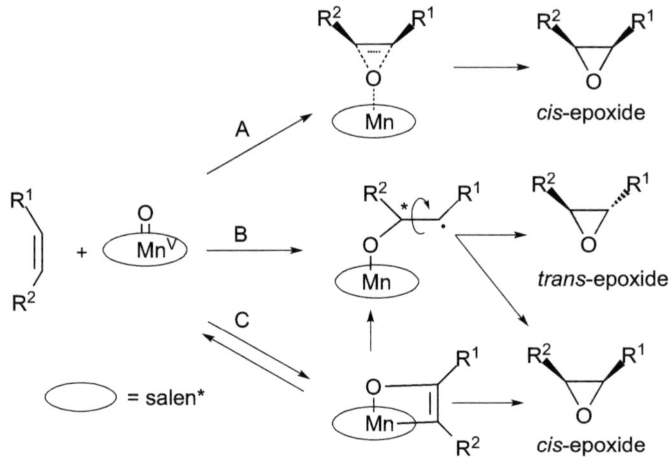

Scheme 7.14 Mechanistic pathways for Mn(salen)-catalysed asymmetric epoxidation JKAE.[38]

rearrangements.[36] This radical intermediate has actually been trapped by the TEMPO nitroxyl radical {(2,2,6,6-tetramethylpiperidin-1-yl)oxyl}, diverting the epoxidation towards the synthesis of alkoxyamines.[39] Still in 1997, Jacobsen's group strongly rejected pathway **C** in light of detailed experimental results.[32] In 2000, computational studies added support to such rejection.[40] In 2004, Abashkin proposed another intimate mechanism for the [O] transfer,[41] which was later reinterpreted by Corey bringing a fresh look at the JKAE mechanism (Scheme 7.20).[42]

In Kochi's salen ligand (chiral C_2 symmetry), the very low inversion barrier allows either enantioface of a C=C double bond to approach the metal centre and no enantioselectivity is attained. By imparting elements of chirality to the salen ligand, Jacobsen[29] and Katsuki[30] independently created complexes with fixed chirality that promoted the catalytic formation of epoxides with the highest enantioselectivities reported for non-enzymatic catalysis at that time. As shown in Figure 7.4, *cis* olefins were epoxidised with higher *ee* values than *trans* and terminal olefins. Using a bulkier -CPhEtH substituent at the 3,3′ positions, Katsuki obtained *ca.* 50% ee in the epoxidation of *trans*-PhCH=CHMe.[30]

Still, these *ee* values are unsatisfactory for practical use and improvement was sought through a better understanding of the origin of selectivity.[38,43,44] This search followed stereochemical arguments that are discussed in the light of Scheme 7.15. The side-on approach of the alkene, say *cis*-PhCH=CHMe, to the Mn=O bond was assumed with the olefin sliding over the salen plane, defined by the N2O2 atoms, towards the oxo ligand along path **p1**. This pathway favours the approach of the alkene with the larger Ph group (R_L) over the diamine bridge at position C8 to avoid steric clash with the bulky ^{t}Bu group at the C5 and C3 positions. The C3,3′ ^{t}Bu groups prevent the approach from the **p4** direction. *Trans* olefins have no favoured side-on

Olefin	Catalyst	ee (%)	Previous best ee %
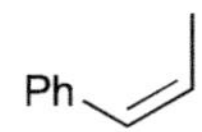	**16 (R,R)**	84	40
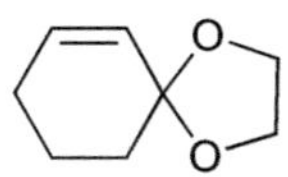	**16 (R,R)**	93	
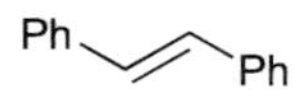	**15a (S,S)**	33	
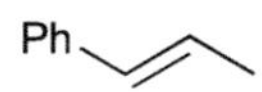	**15a (S,S)**	20	16
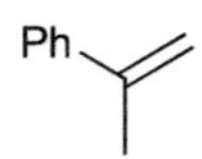	**16 (R,R)**	30	
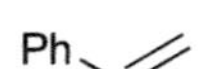	**16 (R,R)**	57	48

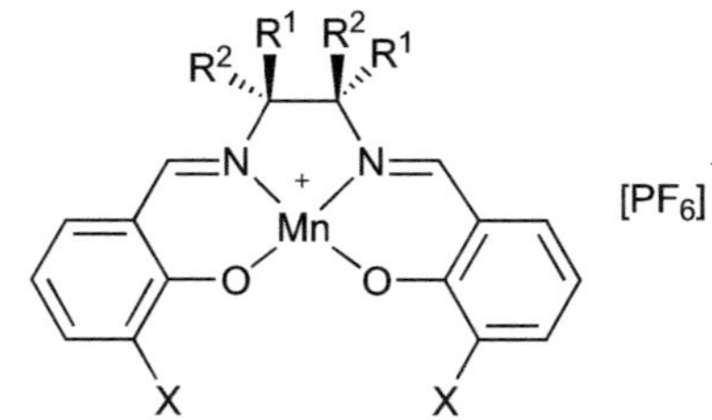

15a (S,S) R^1 = Ph; R^2 = H; X = H
15b (R,R) R^1 = H; R^2 = Ph; X = H
16 (R,R) R^1 = H; R^2 = Ph; X = ^{t}Bu

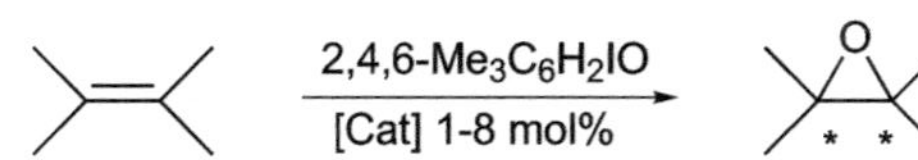

Figure 7.4 First enantioselective olefin epoxidation report with a chiral salen catalyst.[29]

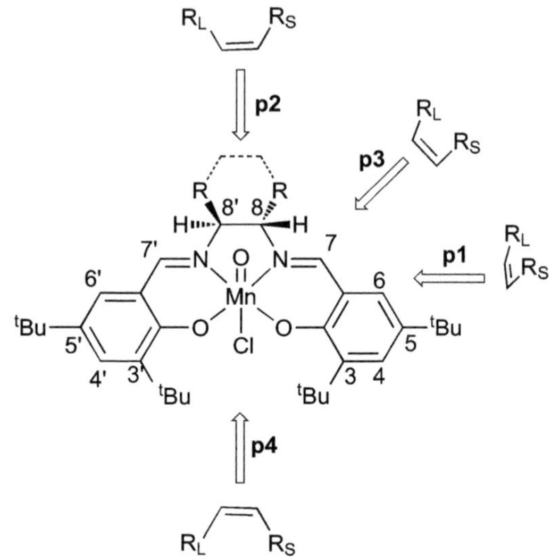

Scheme 7.15 Pathways proposed for the approximation of the olefin to the Mn=O function.

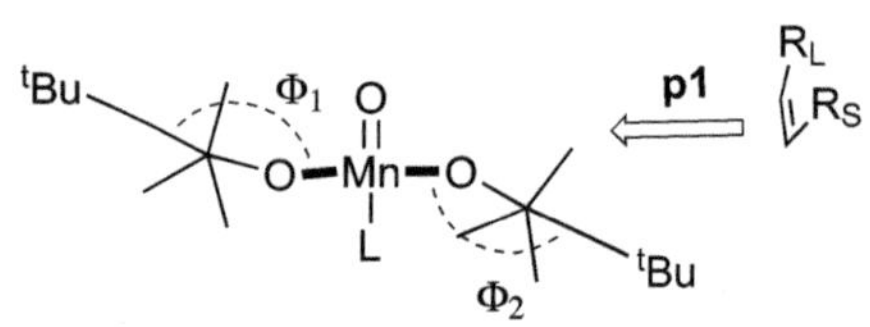

Scheme 7.16 Sketch of the salen ring when looking into the Mn=O bond, at the level of the N2O2 plane, along direction p4. Dotted lines define the out-of-planarity angles Φ_1 and Φ_2.[45]

approach route to the Mn=O bond, hence their low reactivity and enantioselectivity.

The amazing 92% *ee* obtained with the 1,2-diaminocyclohexane-based diimine bridge in Jacobsen's catalyst led this author to propose **p2** as the favoured path for alkene approach since it maximised the interaction between the substrate and the chiral part of the ligand (the bridge).[33] However, the optimisation of chiral substituents at the C3,3′ positions, in the so-called second-generation Katsuki catalysts (Figure 7.3), led this author to propose the **p3** olefin approach pathway as the most consistent with all experimental data.[43] These arguments lost their value in face of the recognition that the $[O{=}Mn(salen)L]^+$ complex is not planar as predicted by computational methods and sketched in Scheme 7.16.[45]

This stepped geometry immediately relieves the steric constraints along pathway **p1** because the ^{t}Bu groups at C3 and C5 lie well below the Mn=O bond. Of course, the chirality of the diimine bridge will reinforce and stabilise this non-planarity favouring pathway **p1** for the majority of the Mn-salen-catalysed epoxidations.[43,46]

We have mentioned above that ligands such as pyridine-*N*-oxide (PyO), 4-phenylpyridine-*N*-oxide (4-PPNO) or *N*-methylmorpholine-*N*-oxide are beneficial for the epoxidation. As an example, in the epoxidation of *cis*-PhCH=CHMe with PhIO using a catalyst like that in Scheme 7.15 with R = Ph and with Br instead of ^{t}Bu at the C5,5′ positions, addition of two equivalents of 4-PPNO increases the yield of epoxide from 2% to 76%, the conversion from 54% to 81%, the *cis* : *trans* ratio from 2 to 3 and the *ee* from 69% to 72%, in comparison to the same reaction in the absence of added 4-PPNO. Other striking examples were given by Jacobsen.[47] The increase in the reaction rate and catalytic turnover numbers were originally assigned to the stabilisation of the $[O{=}Mn^V(salen)]^+$ complex by formation of an octahedral species like $[O{=}Mn(OPy)(salen)]^+$ with both O ligands *trans* to each other. The strapped complex **18** in Scheme 7.17 does not change its activity upon addition of excess 4-PPNO. This strongly suggests that during the catalytic cycle the PyO co-ligand must be held in place for optimised results. This observation is very important in terms of the interpretation of the possible mechanisms of O-transfer as well as for the face selectivity question. In fact, computational methods have shown that the axial ligand increases the degree of distortion of the salen ring and the fold angles Φ_1 and Φ_2 lie between 155° and 165° depending on the nature of L.[48] This concept has recently been re-examined with remarkable success. By using chiral BINOL-phosphates as mononegative, O^- donor counter-ions, List's group obtained excellent epoxidation activity and selectivity over a wide range of olefin substrates using PhIO and an "achiral" Mn(salen) cation equipped with a simple ethylene diamine bridge (**17**, Scheme 7.17).[49] This alternative way of controlling both the chirality around the catalytic site and increase its reactivity was attributed to the formation of a "frustrated" Mn^+–(BINOL-phosphate)$^-$ ion-pair (**17**, Scheme 7.17).

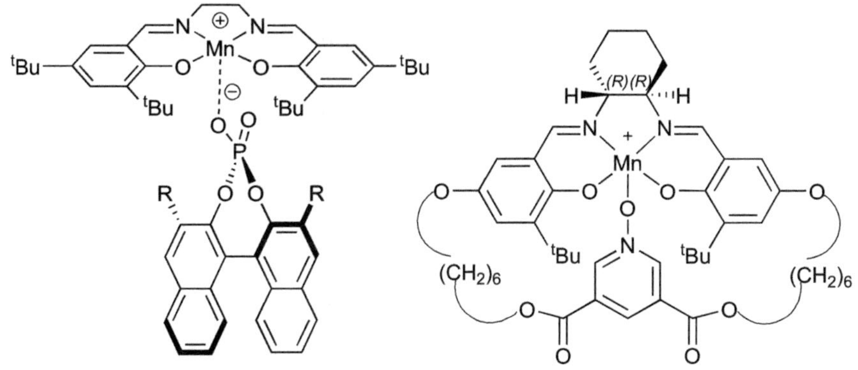

17 **18**

Scheme 7.17 Complexes designed to prove (**18**) or take advantage of (**17**) the role of O ligands *trans* to the Mn=O bond.

The nature and role of the TO influence the results of the asymmetric oxidation and have a central role when considering sustainability issues. With low solubility and high cost, PhIO and other iodosyl arenes are not ideal. Cheap aqueous NaOCl (sodium hypochlorite; common bleach) proved to be very efficient when the reagents and products are compatible with water. The reaction is carried out in a two-phase system with aqueous NaOCl buffered at *ca.* pH 11 and the olefin dissolved in CH_2Cl_2.[50] Many other oxidants can be used in non-aqueous systems, such as $KHSO_5$, magnesium monoperoxyphthalate (MMPP) and MCPBA (meta-chloroperoxybenzoic acid). The latter provides very good results when the reaction is performed at low temperature (−78 °C) and is particularly useful for the epoxidation of terminal olefins, *e.g.* styrene, and also to obtain *trans* epoxides from *cis* olefins enantioselectively.[51] TOs can be selected depending on many other factors, including safety, cost, substrate compatibility and environmental constraints. In the latter case, oxidants like O_2, TBHP and H_2O_2 are of great value.[52]

So far, we have considered $[O{=}Mn^V(salen]^+$ as the catalytic species responsible for the oxygen transfer from the oxidant to the alkene. However, this is a very labile species that was only observed by ESI (Electrospray Ionisation) mass spectrometry where it appeared together with the dimeric ion $[PhIO(salen)Mn–O–Mn(salen)OIPh]^{2+}$. MS experiments have shown that this dimer breaks down into the daughter ions $[PhIO(salen)Mn]^+$ and $[PhIO(salen)MnO]^+$. The latter, as well as several analogues with other TO ligands, were shown to form epoxides and sulfoxides by reaction with olefins and sulfides within the MS ion chamber.[53] It is likely that under normal catalytic conditions some TOs generate species other than $[O{=}Mn^V(salen]^+$, which are still capable of transferring oxygen atoms to the olefin. Several examples are provided in Figure 7.5 but $Mn^{IV}{=}O$ or Mn–O–Mn dimeric species may also be involved. The hydroperoxo species may arrive from the use of H_2O_2 or ROOH as oxidants. The TO adducts in Figure 7.5 represent the simple coordination of the terminal oxidant to the Mn^{III}salen complex without formal O-transfer to the Mn and the corresponding oxidation $Mn^{III} \rightarrow Mn^V$. This is a very likely situation, particularly in the case of the

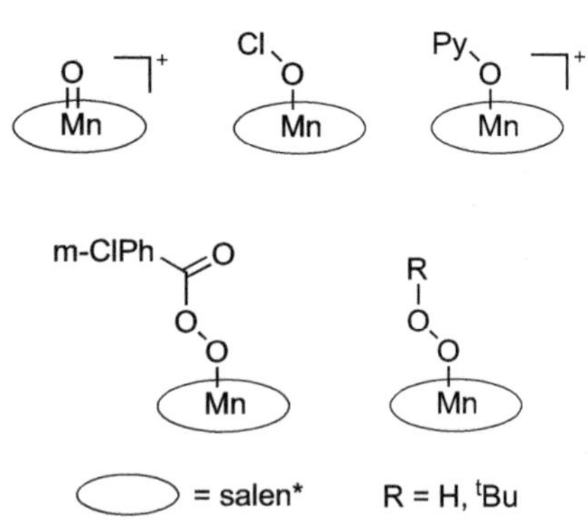

Figure 7.5 Coordination of several TOs to the Mn^{III}(salen) or Mn^V(salen) centres.

widely used oxidant ClO^-. MS studies have shown that both $[O{=}Mn^V salen)]^+$ and $[(ClO)Mn^{III}(salen)]$ species may coexist in the reaction mixture and their reactions with the olefinic substrates may follow independent pathways. Moreover, the basic (pH 11) conditions where these reactions are performed may favour the formation of $[HOMn^{III}(salen)]$ complexes, which can easily lead to μ-oxo dimers with different TO groups attached to each Mn ion.

Practical experience recommends that the oxidant should be the last reagent to be added to the epoxidation catalytic reaction. Spectroscopic studies show that adding it before the addition of other co-ligands or substrates leads to the formation of several mostly unproductive Mn species that contribute to a loss of selectivity and/or yield. Cases of multiple oxidation mechanisms operating simultaneously have been already observed experimentally, namely in the epoxidation of *cis* alkenes to give mixtures of both *cis* and *trans* epoxides in a manner that is dependent on the reagents used. The general scheme that has been proposed by several authors is presented in Scheme 7.18.

Collman's group showed that the diastereoselectivity and enantioselectivity in the epoxidation of (*Z*)-stilbene (*cis*-PhHC=CHPh) with a Jacobsen catalyst with four different counter-ions X^- carried out with three different iodosylbenzenes (TO = PhIO, C_6F_5IO, and MesIO) leads to results that depend strongly on the added co-ligand L (none, PyO, Ph_3PO, or N-Me-Imd) and the counter-ion (Cl^-, Br^-, BF_4^-, AcO^-). In this study the *cis* : *trans* ratio of the epoxide can be varied from 18 : 82 to 83 : 17 by varying the counter-ion and the iodosylarene. According to Scheme 7.18, the nature of the iodosylarene may control the importance of pathway **a** *vs.* pathway **b**, thus altering the final product ratio. The authors propose that in pathway **a** catalyst [(TO)Mn(salen)L] promotes *trans*-stilbene epoxide formation whereas pathway **b** promotes *cis*-stilbene epoxide through the "normal" $[O{=}Mn(salen)L]^+$ catalyst.[54] Although the type of mechanism raises a consensus, the selectivity assignment is not straightforward since other authors propose opposite selectivities from the same type of bifurcated mechanism in similar yet not identical experiments.[55]

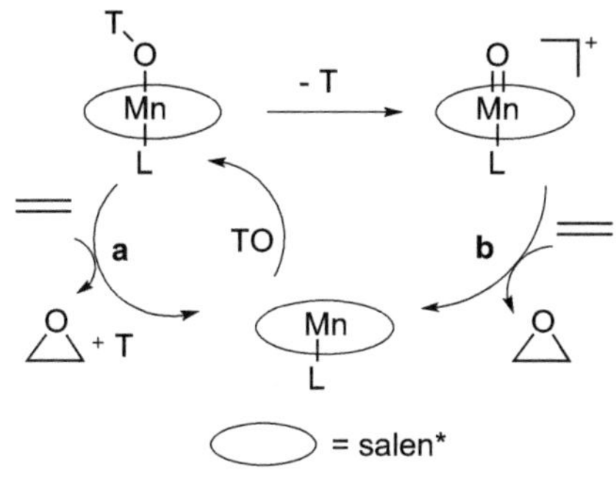

Scheme 7.18 Adaptation of Collman's proposal on the bifurcated O-transfer competing pathways in JKAE catalysis.[54]

If the concerted pathway dominates, stereoselection is preserved and *cis* olefins produce only *cis* epoxides. This is the case for non-conjugated, isolated olefins.[56] However, in the case of the radical pathway **B** in Scheme 7.14, *cis* olefins will lead to mixtures of *cis* and *trans* epoxides. A complete analysis of the stereochemical consequences of the rotation of the C–C bond in the radical intermediate of pathway **B** on the asymmetric induction of all the *cis* and *trans* epoxides obtained has been masterfully developed by Jacobsen's group.[51,57] The results are controlled by the enantioselectivity of the irreversible generation of the radical intermediate (two diastereomers with a chiral salen ring and a chiral C atom, Scheme 7.14) and the relative rates of collapse of each of these diastereomers to *cis* or *trans* epoxides. Such rates are essentially different because the radical intermediates are diastereomers. Generally, the process results in the enhancement of the *ee* of the major product in a so-called enantioselectivity refinement.[57] For example, the epoxidation of $C_6H_{13}C{\equiv}CCH{=}CHMe$ with a facial selectivity of 72% *ee* led to a value of 98% *ee* for the *trans* epoxide with a *cis* : *trans* ratio of 0.62.

So far we have concentrated mainly on the reactivity of *cis* olefins, of which the conjugated ones present the highest reactivity and enantioselectivity. This, as has been shown, is influenced not only by stereochemical reasons but also by co-ligands and TOs. However, electronic effects have to be taken into account in order to explain why *cis*-PhCH=CHMe is a much better substrate for epoxidation than CyCH=CHMe with roughly the same bulkiness, and also why the electronic nature of the substituents in positions C5,5′ influences the enantioselectivity of the reactions. In fact, analogues of the Jacobsen catalyst with donor ligands like OMe and OSi^iPr_3 replacing the ^{t}Bu groups in positions C5,5′ are very efficient catalysts for the synthesis of *trans* epoxides from *cis* olefins.[58,59]

A series of computational studies have looked into the influence of the spin state of $[O{=}Mn(salen)]^+$ on the reaction mechanism and product selectivity (*e.g.*[45]). In one of these studies, the participation of the O atoms of the salen ring in the reaction mechanism was proposed for the first time.[41] According to the authors, intermediate **C** in Scheme 7.14 was replaced by a five-membered ring (FMR), which originates exclusively *cis* epoxide as shown in Scheme 7.19. So, combining Schemes 7.14 and 7.19, competition between the three channels **A, B** and FMR would determine the outcome of the reaction in terms of product selectivity.

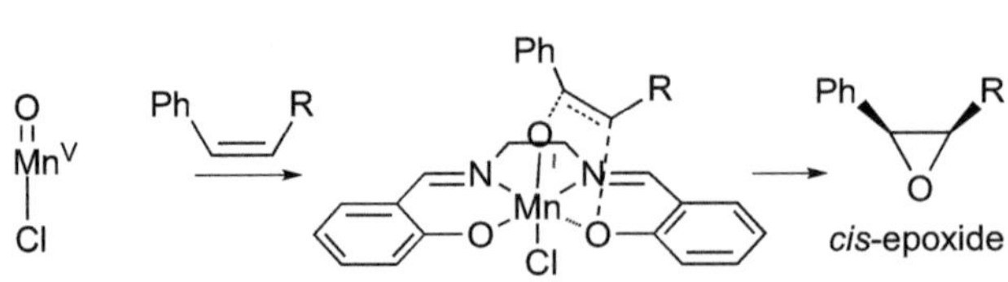

Scheme 7.19 Five-membered-ring type of interaction between Mn=O and incoming olefin.[41]

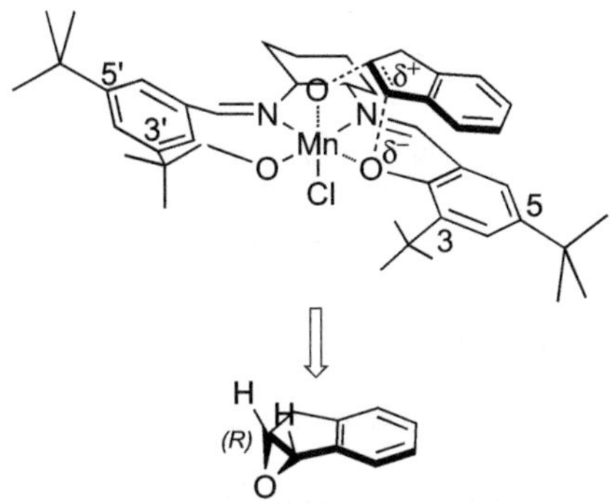

Scheme 7.20 Five-membered-ring transition state determines facial selectivity in the epoxidation of indene.[42]

In 2009, Corey's group produced a model that brings together the factors that influence the JKAE reaction into a simple and coherent picture.[42] This model takes into consideration that the conformation of the salen* ring in the Jacobsen catalyst is not only stepped as in Scheme 7.16 but also canted (both salen halves make a dihedral angle). As illustrated in Scheme 7.20, the ^{t}Bu group at C3′ points upwards from the N2O2 plane whereas the ^{t}Bu group at C3 points downwards from said plane. Such deformation allows an incoming *cis*-conjugated olefin, *e.g.* indene, to freely approach the Mn=O bond as in the **p1** pathway of Scheme 7.15. In principle, both indene faces could be presented looking down on the Mn=O fragment. However, *only one of these face orientations* places the benzylic (or equivalent) C atom of the *cis*-conjugated olefin above the O atom of the salen ligand. In this situation, the δ^{+} charge generated at this benzylic C atom by the interaction of the other end of the C=C bond with the electron-deficient oxo ligand will be stabilised by the δ^{-} charge on the salen O beneath. This interaction originates a FMR encompassing the olefin C=C and the catalyst O–Mn=O atoms. The collapse of this FMR leads to the epoxide with a very high face selectivity. Producing a similar FMR on the other enantioface requires the approach to the Mn=O function over the salen ring tilted upwards with the bulky C3′ ^{t}Bu group above its plane, which is obviously disfavoured on steric grounds. The same arguments support the low selectivity and reactivity of *trans* alkenes.

The authors compare the formation of this FMR to a [3 + 2] cycloaddition of the C=C bond to the O–Mn=O system. Besides presenting a rationale for enantioface selection that is not exclusively dependent on steric interactions, this model also explains other features that were never properly clarified. For example, donor substituents at the C5,5′ positions improve enantioselectivity owing to an increase of δ^{-} at the salen O atom and stabilisation of the FMR. Conversely, as also experimentally observed, electron withdrawing substituents at the C5,5′ positions decrease enantioselectivity. The presence of amine *N*-oxides and other co-ligands also increases δ^{-} at the salen O atom among all the other contributions to the overall reaction already mentioned.

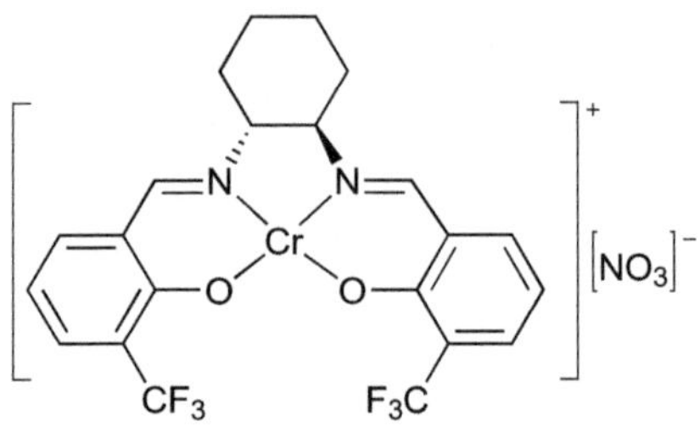

Scheme 7.21 Cr^{III}(salen) catalyst able to epoxidise *trans* olefins.

As pointed out by Corey, only a few kcal mol^{-1} are necessary to account for very high values of *ee*.

So far, the epoxidation of *trans* olefins has been left out of the proposed models. Such substrates were even considered by Jacobsen as essentially impossible to epoxidise selectively with salen type systems.[58] This group reported an indirect way to obtain *trans* epoxides in >90% *ee* through epoxidation of *cis* alkenes in the presence of both the Jacobsen catalyst with the $OSi^{i}Pr_3$ substituent at the salen ring C5,5′ positions and added cinchona alkaloid derivatives.[58] However, second-generation Katsuki catalysts, like the one in Figure 7.1, enable the epoxidation of *trans*-ß-alkyl-styrenes and similar molecules. For instance, *trans*-PhHC=CHMe is epoxidised with 77% yield, 91% *ee*, with PhIO as the TO at −30 °C.[60] This is possible because in these catalysts each of the biphenyl substituents at the positions C3,3′ stands perpendicular to the corresponding half of the salen ring in its stepped conformation thus allowing the *trans* olefins to access the Mo=O function. Another method uses $[Cr^{III}(salen)]^+$ catalysts: the Cr complex in Scheme 7.21 stoichiometrically epoxidises *trans*-PhHC=CHMe) with 92% *ee* and 45% yield. The yield is limited to 50% owing to the formation of the dimer $[(salen)CrOCr(salen)]^{2+}$.

7.2.2.1 Applications of JKAE Catalysis in the Pharmaceutical Industry

A long list of asymmetric epoxides produced with the JKAE technology can be found in a review by Gilheany.[44] Such epoxides can be part of a number of synthons used in many organic transformations. Beyond the examples presented on epoxidation of disubstituted *cis* and *trans* olefins, JKAE can be performed on terminal olefins, tri- and tetra-substituted olefins, as well as enol ethers. Dienes can also be epoxidised selectively at one double bond leaving the other untouched. The versatility of Jacobsen's catalyst and its low-cost synthesis[61] make it widely used in AE reactions at industrial scale and obviously in the synthesis of enantiopure drugs. The JKAE of two cinnamate esters, a chromene and indene, is exemplified in Scheme 7.22 as central steps in the syntheses of two anti-hypertensives (Diltiazem and Cromakalim), one anti-HIV agent (Crixivan) and one antitumoural agent (Taxol).

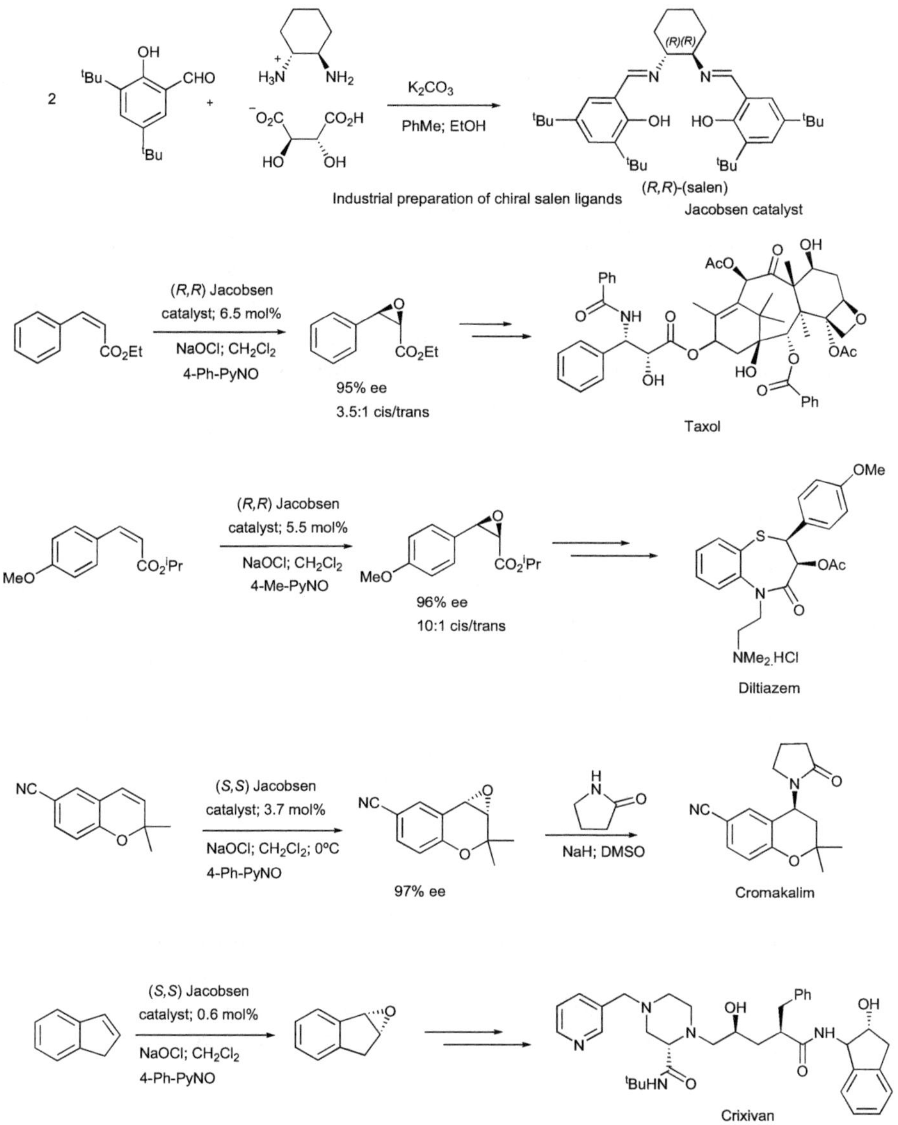

Scheme 7.22 Synthesis of Jacobsen's catalyst and examples of the use of JKAE in pharmaceutical chemistry.

7.2.2.2 *M(salen) Catalysis Using Environmentally Sustainable O_2 and H_2O_2 Oxidants*

The typical TOs used in the JKAE reactions described above do not fit within the ideal characteristics of a green catalytic system. In contrast to the SAE, which is incompatible with water, the JKAE can be performed with aqueous solvents and therefore, should be tuneable to the use of the two most environmentally benign oxidants, molecular oxygen (O_2) and hydrogen peroxide (H_2O_2). A great deal of effort has been put on this endeavour which has been comprehensively reviewed recently.[5] One of the important points to note is that the salen* ligands keep their role as "privileged catalysts". Metals other than Mn or Cr are needed for improved results.

The seminal report on epoxidation with O_2 came from Mukaiyama's group in 1992.[62] They used Jacobsen's catalyst and analogues to epoxidise dihydronaphthalene derivatives with *ee* values up to 77%. The enantioface selection was opposite to that of the normal Jacobsen–Katsuki-type catalysts with PhIO or NaOCl oxidants. The practical value is minimal given the relatively high catalyst load (12 mol%) and the need to use three equivalents of a co-catalyst (pivaldehyde; $Me_3CC(O)H$) that produces three equivalents of carboxylic acid Me_3CCO_2H by-product. By combining Ru^{II} ions with their second-generation salen ligands with axial chirality at the C3,3′ positions, Katsuki's group was able to epoxidise *cis* and *trans* styrene derivatives with O_2 (1 atm) under illumination with excellent selectivity but low efficacy (TON < 15).[63]

Further optimisation of the system led to complex **19** (R = H) in Figure 7.6, which catalyses the epoxidation of the typical range of JKAE substrates, including *trans*-PhHC=CHMe under air without the need for light. This complex reaches 94% *ee* and up to 99% yields for some epoxides.[64]

This very recent breakthrough suggests the possibility of extending the developments in the epoxidation with O_2 to other substrates. Jacobsen's catalyst was used in the first report of this reaction with modest results. On the contrary, the first-generation Katsuki's complex **20** in Figure 7.6 gives very high epoxide yields and *ee* values. The use of *N*-MeIm is also required for the catalysis, arguably because it breaks the Mn–O–OH bond (see Scheme 7.5) to form the $[O{=}Mn^{V}(salen)(N\text{-}MeIm)]^+$ catalyst. Unfortunately, the rather strong catalase type activity requires the use of an excess of 10 equivalents of H_2O_2. The use of the second-generation catalyst **21** in Figure 7.6 and an N-heterocyclic donor only requires 3 mol equivalents of H_2O_2 (from commercial aqueous 30% H_2O_2 solution) to achieve >97% *ee* of the epoxides. This system was further improved by replacing the N-heterocyclic ligands with ammonium acetate, which is likely oxidised to percarboxylic acid in the reaction. Although powerful, the system is limited to the oxidation of 2,2-dimethylchromene substrates.

Reduction of the salen imine function led to the salan* and salalen* ligands like **22**, **23**, and **24** in Figure 7.6. The μ-oxo Ti(salalen) dimer **22** and other salan analogues were proven to be very efficient using 30% H_2O_2 as the TO for the highly selective epoxidation (*ee* up to 99%) not only of the typical JKAE olefin substrates but also of terminal olefins. The catalyst charge is 1 mol% and only 1.01 equivalents of H_2O_2 are needed (Scheme 7.23).[65]

In one case, the reaction is carried out by generating the catalyst *in situ*, mixing 10 mol% of $Ti(O^iPr)_4$ and 10 mol% of chiral salan ligand **24** to epoxidise styrenes with >97% *ee* values.[66] The system has been extended to epoxidise other conjugated and non-conjugated olefins in very high yields

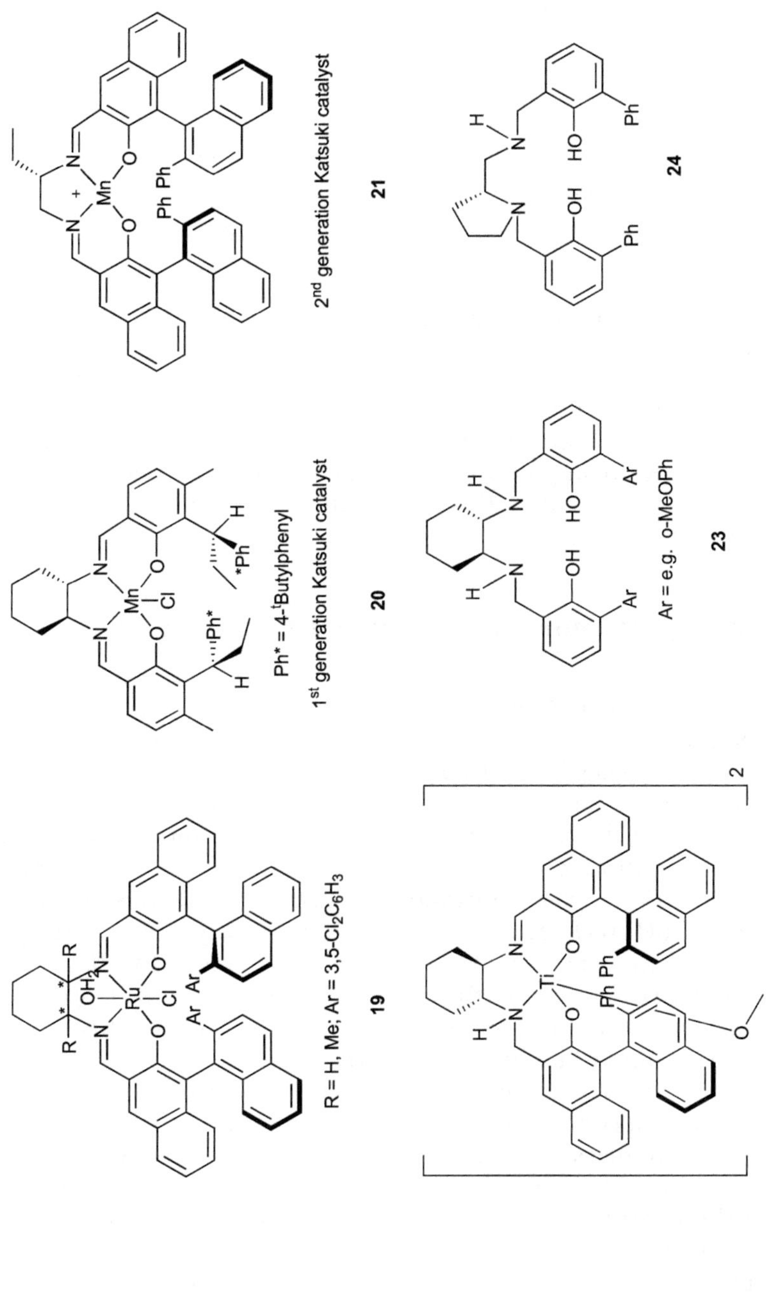

Figure 7.6 Examples of second-generation Katsuki-type catalysts and ligands for epoxidation with O_2 and H_2O_2 oxidants.

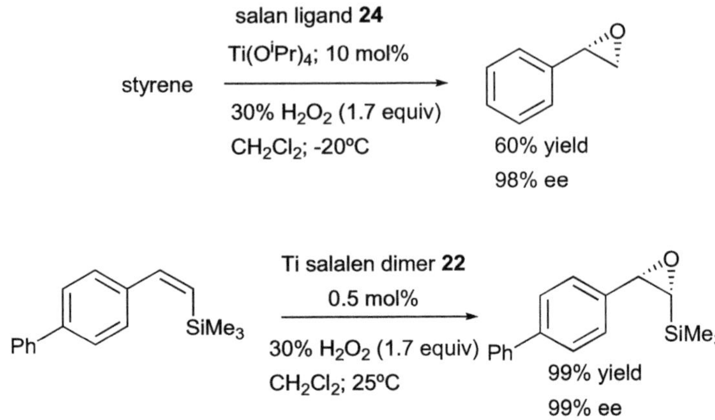

Scheme 7.23 Examples of "green" epoxidations using H_2O_2 as the TO.

and enantiomeric excesses. It is important to note that Ti(salen) catalysts prepared from Ti(O^{i}Pr)$_4$ and the salen ligands do not work. It is necessary to have at least one of the imine bonds reduced to form the salalen ligands in order to achieve epoxidising activity. The mechanism proposed by Katsuki involves the formation of a dimeric complex similar to that in the Sharpless epoxidation process.[65]

7.2.3 M(bis-hydroxamic acid)-catalysed Epoxidations

The value of hydroxamic acids (HA) as chiral ligands for AE was described by Sharpless's group even before their discovery of the Ti–DET system described in Section 7.2. However, the system was abandoned because its catalytically active species [(HA)VO(OR)$_2$] was not very stable and tended to decompose into either non-selective species like [VO(OR)$_3$] or inactive species like [(HA)$_2$VO(OR)]. More than 30 years later, Yamamoto re-examined this chemistry and introduced bis-hydroxamic acid (BHA) ligands as another kind of C_2-symmetric "privileged ligand".[67] These were able to stabilise the V active species [(BHA)VO(OR)$_2$] and achieve high activity and selectivity in the AE of almost all types of allylic alcohols (Scheme 7.24). The main advantages of these catalysts over the classic SAE process are the low catalyst concentration needed (*ca.* 1–2 mol%) and its use of aqueous TBHP as the oxidant, which are important green chemistry assets.

The kinetic resolution in the case of allylic alcohols with $R^4 \neq H$, as well as that of homoallylic alcohols, runs much more efficiently than in the SAE technology. The bulkiness of the ligands, some of which are exemplified in Scheme 7.24, controls their enantioselectivity. One important example is the synthesis of the intermediate **A** in the preparation of atorvastatin, a major hypolipidemic drug (Scheme 7.25).

The complexes formed between BHA acids and $MoO_2(acac)_2$ proved to be excellent catalysts for the enantioselective epoxidation of all sorts of olefins.

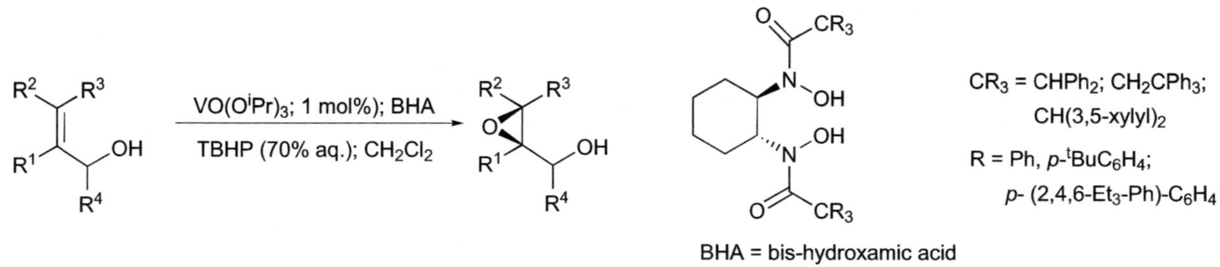

Scheme 7.24 Enantioselective epoxidation of allylic alcohols with bis-hydroxamic acid (BHA) ligands.[67]

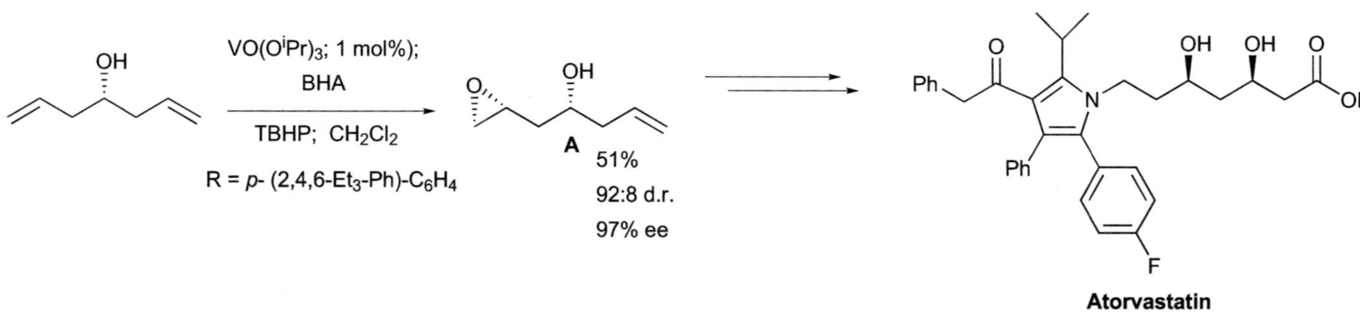

Scheme 7.25 Epoxidation of homoallylic alcohol in the synthesis of atorvastatin.[67]

In the case of polyolefins, the most electron-rich one was epoxidised faster. MoO_2X_2 derivatives have long been known to be excellent epoxidation catalysts. Yet, in spite of countless approaches using all sorts of ligands, the BHA system introduced by Yamamoto was the first to enable highly enantioselective epoxidation with $[Mo^{VI}O_2L_2]$ derivatives. This is certainly due to the large size of the CR_3 substituents used, which are necessary to form a chiral pocket close to the $[OMo(BHA)(OH)(\eta^2\text{-}O_2R)$ species involved in the catalytic process.[68] All previous attempts failed in providing such a stable and appropriate chiral pocket.[69]

Extending the use of these BHA ligands to other metals, like $Zr(O^tBu)_4$ or its Hf analogue, enabled the preparation of very powerful catalysts for the epoxidation of homoallylic and bis-homoallylic alcohols under anaerobic and water-free conditions very similar to those used in SAE protocols. This technology has a lot of potential for expansion through the use of other metal ions.[67] Indeed, use of $WO_2(acac)_2$ and BHA ligands has provided a very potent and environmentally benign epoxidation system for allylic and homoallylic alcohols using 30% H_2O_2 as the TO. The reaction requires NaCl as an additive to avoid epoxide ring-opening but does not require previous isolation of catalysts. All reagents can be added under air and normally at room temperature to achieve remarkable selectivity with a very broad range of allylic and homoallylic substrates. This is indeed a breakthrough in this area where H_2O_2 had never played a role as a TO.[70]

The very complete understanding of AOE with several systems led Yamamoto's group to combine SAE with W/BHA technologies to achieve the "Synthesis of virtually enantiopure aminodiols with three adjacent stereogenic centres by epoxidation and ring-opening".[71] Other lines of progress in this field were described by Mehrman.[72] This extraordinary achievement is a perfect example of the very advanced state of AOE technologies, which open the doors to the synthesis of a broad palette of high-value synthons in pharmaceutical and natural product chemistry.

7.2.4 M(aminopyridine)-catalysed Epoxidations

Inspired by the C_2-symmetric tetradentate coordination of the salen ligand, several research groups investigated the epoxidation activity of Mn^{II} complexes with C_2-symmetric tetradentate aminopyridine N4 ligands and H_2O_2 as the TO. Success was relatively limited until Costas discovered a system with an enormous epoxidation efficiency (TON > 1000) using only 1.2 equivalents of H_2O_2 in the presence of a large excess (14-fold) of acetic acid.[73] In spite of the chirality of the ligand, no enantioselectivity results were reported. However, the remarkable activity spurred the optimisation of the system by several groups, who were able to combine values of TON > 1000 with catalyst loads of only 0.1 mol%. Bryliakov's group discovered that enantioselectivity increased with the bulkiness of the carboxylic acid additive used and *ee* values >90% were reported for the epoxidation of chalcones or chromenes, as summarised in Scheme 7.26.[74] In this system, ligand **24**

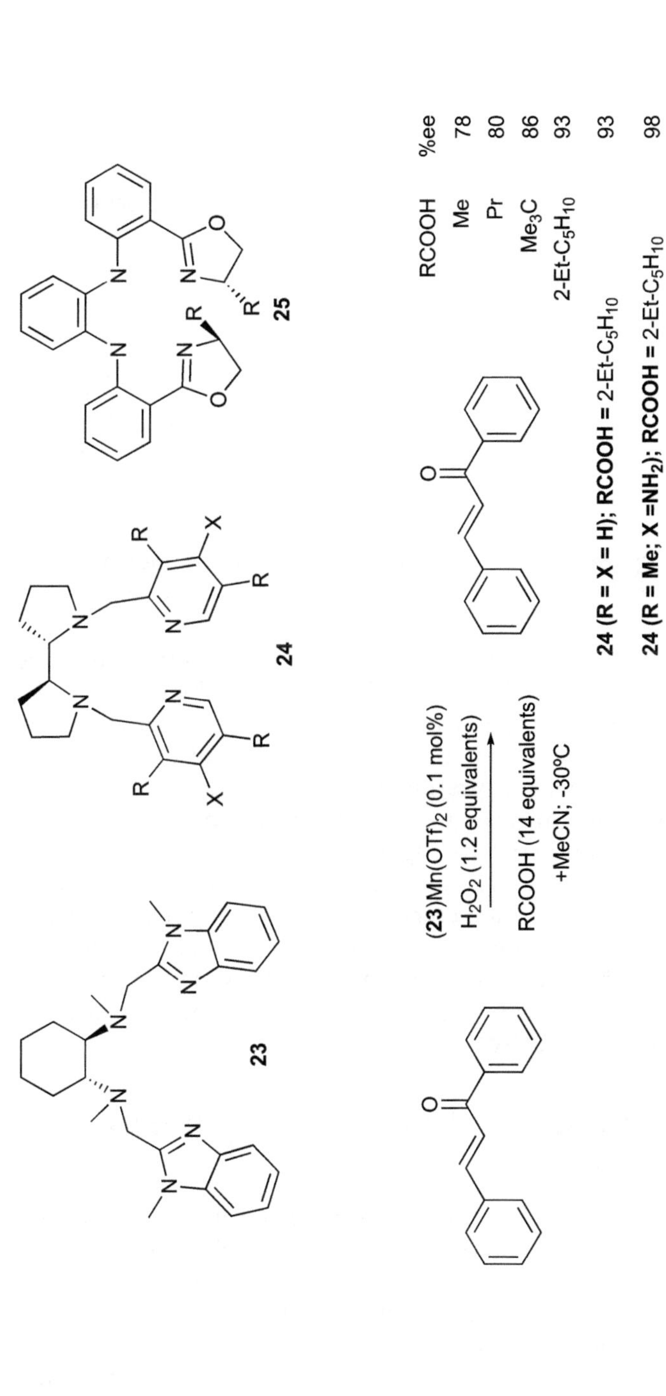

Scheme 7.26 Examples of aminopyridine ligands that perform olefin epoxidation with H_2O_2 in the presence of Mn^{II} and Fe^{II} ions.[5]

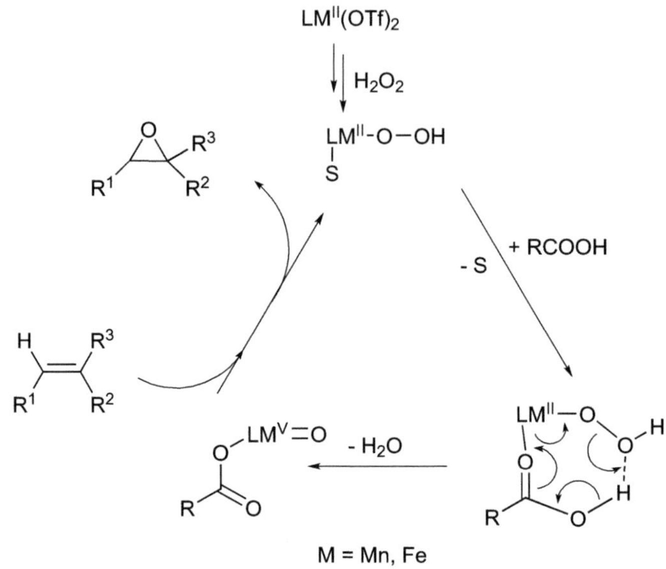

Scheme 7.27 Epoxidation mechanism with M(aminopyridine) ($M = Fe^{II}$ and Mn^{II}) complexes.[5]

performed at the same level as **23** but ligand **25** performed even better.[75] As shown in the scheme, for 2-Et-hexanoic, different substituents at the pyridine rings in ligand **26** affect the % *ee* obtained through electronic effects.

The dependence of selectivity on the bulkiness of the carboxylic additive indicates that the acid is present in the catalytically active species. The corresponding reaction mechanism is proposed in Scheme 7.27 as derived from mechanistic and spectroscopic studies.[74]

From the green chemistry point of view, these systems are advantageous due to their use of H_2O_2, high efficacy (high TON), low catalyst loads, oxidant economy and high *ee* values but are limited in their substrate range that is optimal for chalcones and chromenes. This limitation is even more important for the Fe^{II} complexes where high selectivity is no longer accompanied by high activity.[5]

7.3 Transition Metal-catalysed Asymmetric Sulfoxidation

Although sulfur is an element with a large number of available oxidation states, this chapter considers only the catalytic formation of organic sulfoxides, R^1R^2SO, that are formally generated by addition of one O atom to one of the two lone pairs on the S atom of a sulfide, R^1R^2S. From the stereochemical point of view, the unshared lone pair functions like a bond. Therefore, a deferentially substituted sulfoxide like R^1R^2SO will have two enantiomers, as shown in Scheme 7.1.

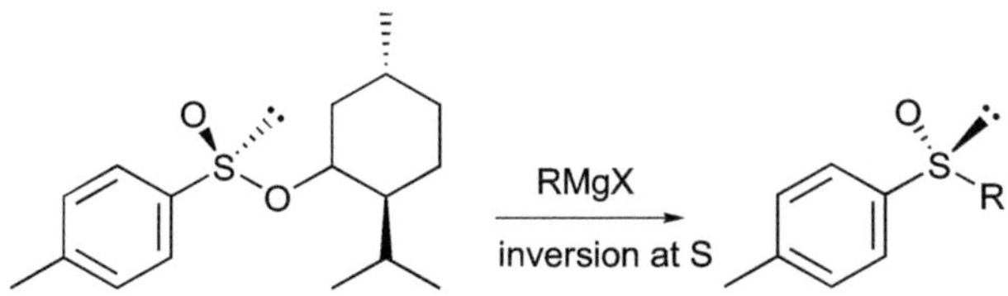

Scheme 7.28 The Andersen stoichiometric synthesis of chiral sulfoxides.

Sulfoxides play important roles in biology and the simplest example, Me_2SO (DMSO), almost became an anti-inflammatory drug with effects similar to those of aspirin. In cell culture research, DMSO is still the solvent of choice to dissolve and test water-insoluble molecules since up to 10% DMSO–water mixtures are still non-toxic to cells *in vitro*. Many drugs containing the sulfoxide function are made as racemates. However, it was often found that the biological or therapeutic activity of one the enantiomers was higher, sometimes much higher than that of the other. Therefore, a pressing need for ways of preparing enantiomerically pure sulfoxides emerged in the pharmaceutical industry. The sulfoxide function is also an important tool in the armamentarium of asymmetric organic synthesis since it can play a role in directing stereochemical outcomes of other intramolecular transformations, as in domino reactions. These will not be discussed here but can be found in extensive literature reviews.[76,77]

Since the natural "chiral pool" does not provide useful chiral sulfoxide-containing building blocks, chiral resolution either by biological or chemical methods is rather limited. This leaves enantioselective synthesis as the most general way to achieve the preparation of enantioenriched or enantiopure sulfoxides. The Andersen method,[78] introduced in 1962 and exemplified in Scheme 7.28, is effective and highly sophisticated, in terms of stereochemical control yet is not catalytic and will not be treated here.

In the following sections, we will concentrate on asymmetric oxygen transfer from TOs to sulfides mediated by chiral transition metal catalysts. Such systems surpassed the importance of biological and organic oxidation processes.

7.3.1 Asymmetric Sulfoxidation with Sharpless-type Catalysts

Inspired by the success of Sharpless epoxidation catalysis, Kagan's group reported the first catalytic asymmetric synthesis of sulfoxides. Modifying the Sharpless protocol by addition of H_2O, they found that a mixture of $Ti(O^iPr)_4$–DET–H_2O–TBHP = 1 : 2 : 1 : 1.1 is able to oxidise a variety of prochiral sulfides in CH_2Cl_2 at *ca.* −20 °C, to produce the corresponding sulfoxides with *ee* values between 75 and 90% for alkyl aryl sulfoxides and 50–70% for dialkyl sulfoxides (Scheme 7.29).[79] The enantiomer that was favoured could be predicted according to the mnemonic proposed by Jørgensen (Scheme 7.29).[80]

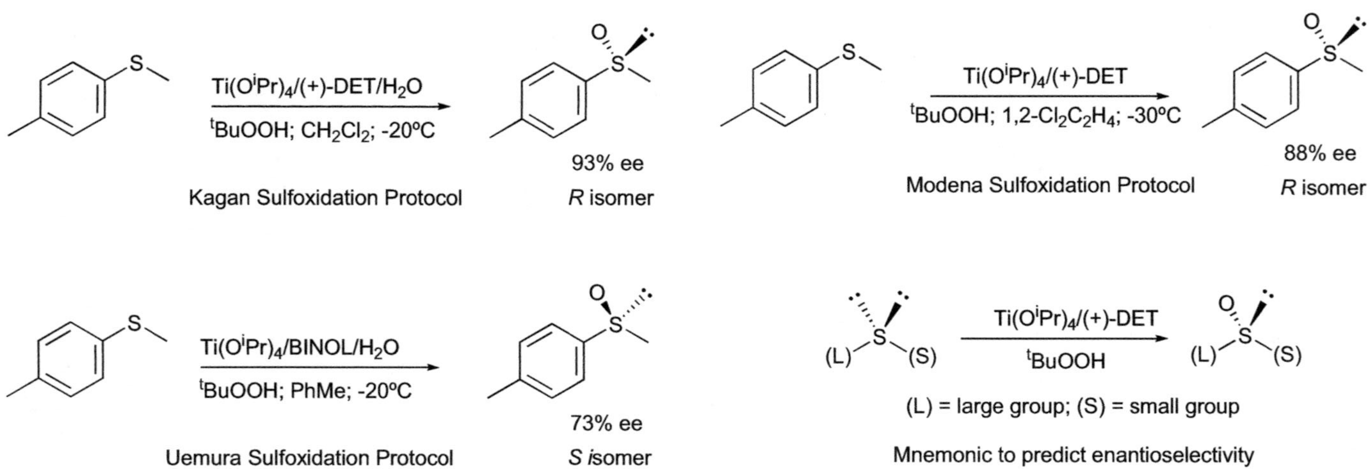

Scheme 7.29 Examples of asymmetric sulfide oxidation (ASO) and mnemonic predicting favoured enantiomer.

In the case of the archetypal model *p*-tolyl methyl sulfide, the (*R*) sulfoxide is obtained after 4 h with 90% *ee* and 90% yield relative to $Ti(^iPrO)_4$. No over-oxidation to the sulfone is observed in this and most other cases. Increasing the bulkiness of methyl to benzyl under the same reaction conditions gives the sulfoxide with 7% *ee* and 41% yield in a slow reaction. Modifications of the *para* and *ortho* positions of the aryl substituent in Ar*MeS do not change the *ee* values of the MeAr*SO product. However, Me(alkyl)S gives lower *ee* values, *ca.* 50–70%. A recent study using a chiral Ti–hydrobenzoin catalyst also confirms the need for tuning the catalytic reaction conditions and even the order of addition of the reagents to maximise the enantioselectivity in the sulfoxidation of aryl benzyl sulfides.[81]

The addition of water to the Sharpless-type mixture is a non-obvious input if we consider that it poisons the SAE process. In the case of R_2S, oxidation enantioselectivity is enhanced if H_2O is added but only within certain limits. At one equivalent of H_2O, *ee* values reach a maximum, but dry conditions or too much H_2O significantly diminish selectivity. The role of water is similar to that of modifying the dimeric Sharpless catalyst in favour of a monomeric Ti species (see below). In fact, in the same year, Modena reported an anhydrous protocol (the Modena protocol) that also used $Ti(O^iPr)_4$, DET and TBHP with dichloroethane instead of dichloromethane as the solvent (Scheme 7.29).[82] This shows that the water is not essential for the epoxidation although it may favour the formation of a rather active species.

Improvements and optimisations included the use of cumyl hydroperoxide (CHP) instead of TBHP leading the ASO of several substrates to *ee* values >99%, and the replacement of water with ^iPrOH in the mixture $Ti(O^iPr)_4$–(R,R)-DET–^iPrOH–TBHP (0.1 : 2 : 1 : 1.1) that becomes catalytic in Ti.

A number of chiral diols (*e.g.* BINOL (1,1′-bi-2-naphthol) and several of its derivatives, mandelic acid, camphene diols) have been used in attempts to replace DET. The modification of the original Kagan's protocol (Scheme 7.29) that replaces tartrate with BINOL and CH_2Cl_2 with toluene is now known as the Uemura protocol,[83] and has produced very good results in terms of enantioselectivity.

It is clear that small variations in any of the reaction parameters (hydroperoxide, solvent, water, chiral diols) can change the results in terms of yield and selectivity, sometimes in drastic ways. For instance, use of the F_8-BINOL (fluorinated at the distal C6 rings) can improve reactivity compared to BINOL as well as improve and invert enantioselectivity under similar reaction conditions.[84]

The Scettri group's work on furyl hydroperoxides vividly illustrates how small changes in the ROOH structure or reaction conditions can affect the outcome of the asymmetric sulfoxidation of a given sulfide substrate.[85] In some cases, the *ee* values increase with reaction time due to convergent kinetic resolution. This process means that over time the minor enantiomer

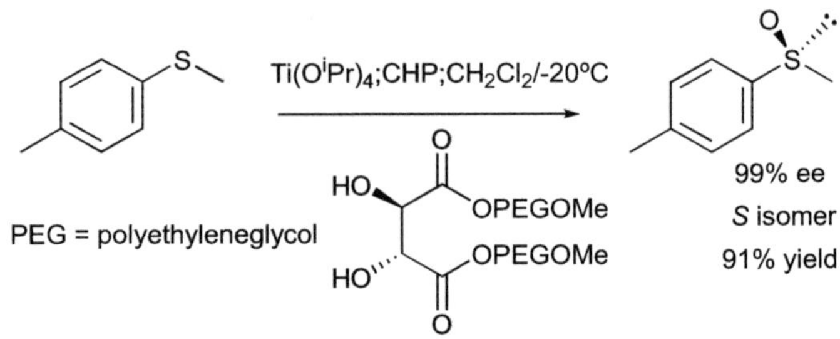

Scheme 7.30 Improved catalysts based on pegylated tartrate ligands.[86]

is selectively oxidised into sulfone, enriching the solution in the major enantiomer. Of course, the yield of sulfoxide decreases.

Attempts at improving the sustainability of this technology by facilitating catalyst recycling as well as product separation include methods of supporting and immobilising the catalysts. Particularly successful was the use of Ti-BINOL oligomers or the support of the tartrate ligand in a PEG polymer,[86] which yielded the remarkable results in Scheme 7.30.

In terms of pharmaceutical applications, this Ti-based ASO technology is very well established. The family of proton pump inhibitors (PPI) initiated with the anti-ulcer omeprazole, the once top-selling drug in the world and sold as a racemate, provides a remarkable example (Figure 7.7).

In these compounds, the sulfoxide function is usually installed in the last synthetic step. Kagan's protocol with a few modifications provides the basis for the multi-ton scale industrial synthesis of esomeprazole (Astra-Zeneca), the (*S*)-enantiomer of omeprazole. As seen in Scheme 7.31, the ASO of its sulfide precursor (pyrmethazole) was developed optimising Kagan's protocol until it became very efficient. The main modifications were the use of CHP instead of TBHP and the addition of the very bulky amine N(^{i}Pr)$_2$Et). This type of reagent, absent in the original Kagan's protocol, turned out to be crucial for improved yield and selectivity.[87]

Although enantiospecific protocols were developed for several other PPIs, they are all commercialised as racemates, except for esomeprazole where the pure enantiomer has a clinical advantage over the racemate.[88] Conditions similar to those for the esomeprazole industrial synthesis, namely in terms of solvent and added base, were later used for the ASO of other sulfides previously inaccessible through Kagan's protocol. The synthesis of the hypolipidemic candidate RP 73163 (now abandoned) is one such example developed at Rhone Poulenc Rorer. Many other examples can be found in reviews and specialized textbooks.[86,89–94]

The robustness of the esomeprazole process and its similarity to the original Kagan's protocol prompted a very thorough study of the mechanism of the ASO of pyrmethazole.[95] This extremely elegant study used the same conditions of the industrial process and established that unsubstituted imidazole methyl sulfide (**A** in Scheme 7.32) is a reliable model

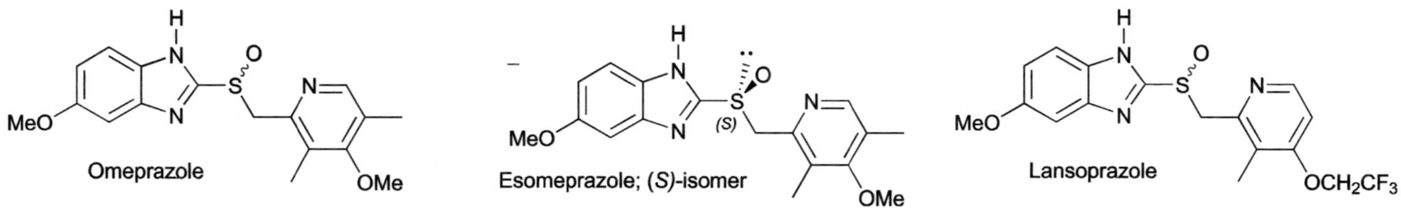

Figure 7.7 Some important proton pump inhibitors (PPI).

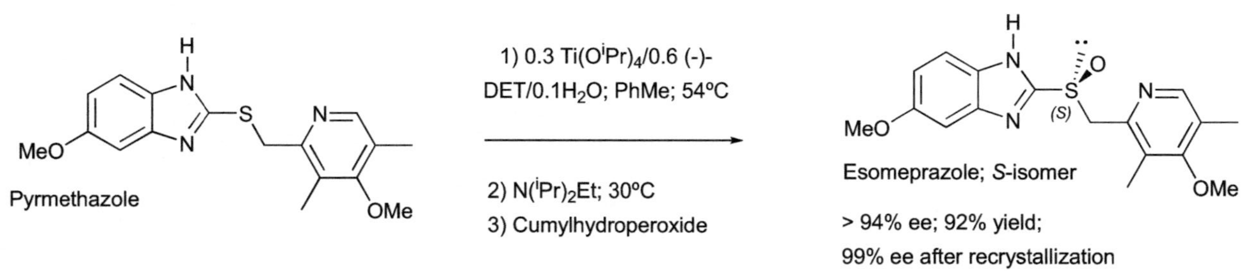

Scheme 7.31 Synthesis of esomeprazole.

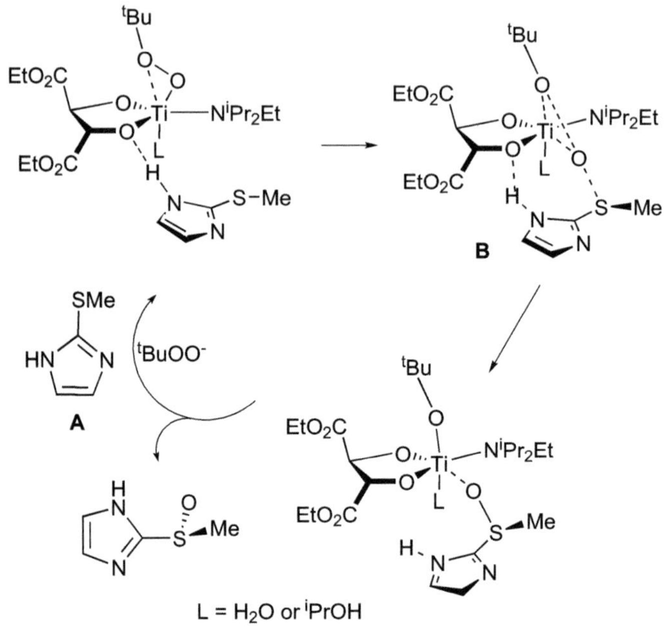

Scheme 7.32 Proposed mechanism for the industrial production of esomeprazole.[95]

for the sulfoxidation of pyrmethazole. It then showed that addition of H_2O to the Ti(O^{i}Pr)$_4$–DET mixture (1 : 2 in $CHCl_3$) leads to the disappearance of dimeric species similar to those in the Sharpless mechanism of epoxidation, leaving a species like [Ti(DET)$_2$L$_2$] (L = H_2O or ^{i}PrOH) as the pre-catalyst. Then, the role of N(^{i}Pr)$_2$Et was shown to be that of coordinating and stabilising the catalyst as a [Ti(DET)L$_3$(N(^{i}Pr)$_2$Et)] species. TBHP then replaces two of the labile solvent ligands forming the active species **B**, [Ti(DET)L(N(^{i}Pr)$_2$Et)(η^2-^{t}BuOO)]. The interaction of this species with imidazole methyl sulfide was modelled by DFT (Density Functional Theory) calculations leading to the last steps of the mechanism. One important remark is that the O–O–S interaction is essentially linear, as already proposed by Jørgensen,[80] and does not involve Ti–S bonds. Given the conditions tested, this is most likely the mechanism of the real esomeprazole synthesis.

It is important to realise that in ASO reactions conditions are often incompatible with some functionalities present in a given sulfide, precluding the sulfoxidation in the last reaction step. One example from Otsuka Pharmaceutical Co. is the synthesis of a platelet adhesion inhibitor (OPC-29030), where the sulfoxide was installed on an intermediate through an air and moisture compatible variation of the Uemura protocol with mandelic acid instead of DET as the chiral auxiliary (Scheme 7.33).[96]

This kind of assembling sequence may be useful in the cases where the chiral sulfoxide decisively influences the enantioselectivity of the later reaction steps, as will be discussed below.

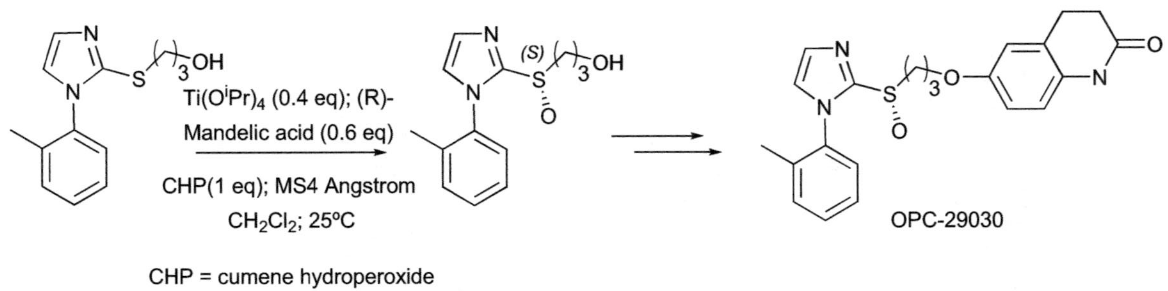

Scheme 7.33 Example of ASO on a pharmaceutical intermediate.

7.3.2 Asymmetric Sulfoxidation with Jacobsen–Katsuki-type Catalysts

The amazing ability of the $[Mn^{III}(salen)]^{+}$ complexes in AOE led to the testing of Jacobsen's catalyst in asymmetric sulfoxidation, yet with disappointing results. On the contrary, some first- and second-generation Katsuki catalysts, **28** and **29** in Figure 7.8,[97] oxidised alkyl aryl sulfides almost quantitatively with *ee* values >90% using PhIO as the TO. These values are really very high and no complementary kinetic resolution is necessary.

Using simple Fe^{III} salts in the presence of the salen ligand of Jacobsen's catalyst, a broad range of sulfoxides has been obtained with high *ee* values, using TBHP as the oxidant in water. Such systems had never been explored before but hold promise in terms of their green potential.[98]

7.3.3 Asymmetric Sulfoxidation with M(bis-hydroxamic) Catalysts

Although many Mo complexes have shown high activity in catalytic sulphoxidation, Yamamoto's group was the first to obtain excellent *ee* values in the sulfoxidation of aryl methyl sulfides using the chiral BHA ligand shown in Scheme 7.24 with $R = 4\text{-}^{i}PrPh$, $MoO_2(acac)_2$ and TritylHP or CHP as the oxidant.[99] This result proves once more the high capacity of this type of ligands to induce chirality at the metal sphere.[67]

7.3.4 Catalytic ASO Processes Using Environmentally Sustainable O_2 as TO

Mukaiyama (1995) was the first to report ASO with O_2 as the TO using the same system mentioned above for olefin epoxidation.[62] Excess of auxiliary pivaldehyde remained as a serious problem in spite of some good

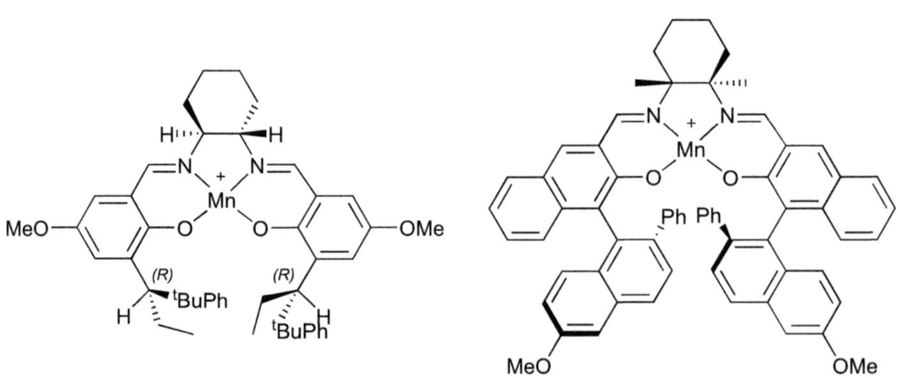

28 **29**

Figure 7.8 Examples of Katsuki-type catalysts for ASO.

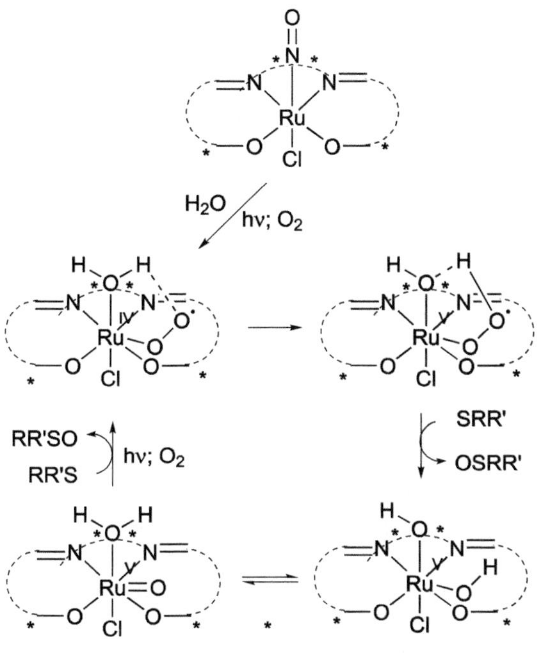

Scheme 7.34 Mechanism proposed for the sulfoxidation (or epoxidation) catalysed by a Ru(salen) complex with O_2 as the TO.[63]

enantioselectivities (>90%) obtained for alkyl aryl sulfides. Again, the breakthrough came from Katsuki's group and the [Ru(NO)(salen*)Cl] complex, which was able to oxidise aryl methyl and other sulfides with O_2 under light irradiation. This complex is represented as **19** in Figure 7.6 with R = Me and Ar = Ph. This amazing oxidation mechanism is proposed to take place according to Scheme 7.34. It involves one H_2O molecule, which is first deprotonated and then reprotonated along the two successive oxidation steps. Without any doubt, this reaction and mechanism type is a great example of a sustainable oxidation process that solely involves O_2 as the oxidant and only produces the oxidised molecule since the other by-product, water, is recycled in the process. If this can be optimised and mastered, it will certainly become a major source of oxidative processes. This is represented for sulfoxidation but the process is analogous for epoxidation, just replacing the sulfide by the epoxide.[63]

7.3.5 Catalytic ASO Processes Using Environmentally Sustainable H_2O_2 as TO

Asymmetric sulfoxidation with vanadium complexes gained an important expression when Bolm's group introduced the vanadyl complexes derived from tridentate Schiff bases and $VO(acac)_2$ (Scheme 7.35).[100] These ASO

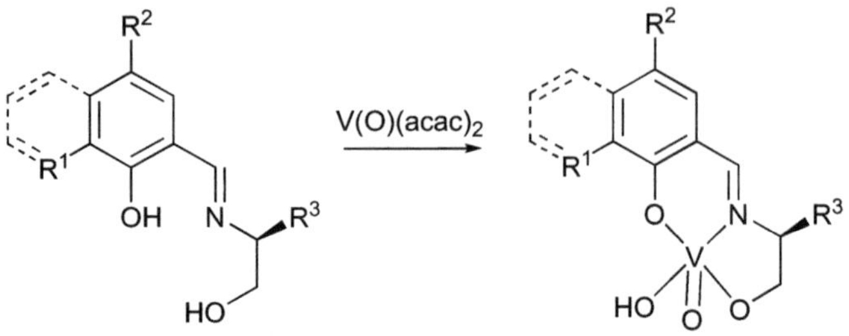

30: R^1, R^2 = I; R^3 = ^{t}Bu

31: R^1= fused C6 ring; R^2 = Br; R^3 = ^{t}Bu

Scheme 7.35 The Bolm-type tridentate Schiff bases as versatile ligands for ASO.

catalysts work with 30% aqueous H_2O_2 and their selectivity can be tuned by changes in R^1, R^2 and R^3 up to 85% *ee*. The ligands can be readily made from enantiopure 1,2-amino alcohols and the catalysts can be used in very low amounts (even down to 0.01 mol%) and the reactions do not need an inert atmosphere or moisture control. Changing the chirality of the amino alcohols allows the independent preparation of both enantiomers.

Using the oxidation of tBuSStBu as a model, it was found that R^2 does not affect the enantioselectivity and both R^1 and R^3 are crucial because they are close to the reaction centre. However, it was found that R^2 groups have an electronic influence on the reactivity. Tuning all these substituents results in the design of the optimum ligand being substrate-specific. A solid array was designed for screening the best set of ligand substituents for ASO of aryl alkyl sulfides with this system. From that screening, it was concluded that ligands **30** and **31** (Scheme 7.35) are the most suited for this type of reaction. Type **30** ligands in combination with several metals other than V have also been used in other areas of oxidative chemistry.[100]

Notwithstanding, a huge variety of substituents affecting the R^3 and R^1 positions have been devised, many with more than one chiral centre. Although in some cases very high values of *ee* are obtained, ligand/substrate specificity is often encountered as well as unexpected disappointing results. Substituents on the aryl sulfides affect the selectivity to a lesser extent than the ligand substituents. In practice, these ASO reactions often involve forming the Schiff base catalyst *in situ* and then adding the substrate and H_2O_2. The use of low temperatures (*ca.* -20 °C) usually improves selectivity and yield, and the best solvents are chloroform and dichloromethane. The rate of addition of H_2O_2 influences the outcome of the reaction in terms of selectivity. Fast addition of H_2O_2 leads to a high concentration of $[V(O)(O_2)_2]^-$. This side product is a very powerful oxidation catalyst. However, it is not chiral and only produces racemic sulfoxides, thus lowering the overall selectivity of the reaction.

The preparation of pure and well-characterised molecular catalysts from the reaction of the Schiff base and the $VO(acac)_2$ has also been reported.[101]

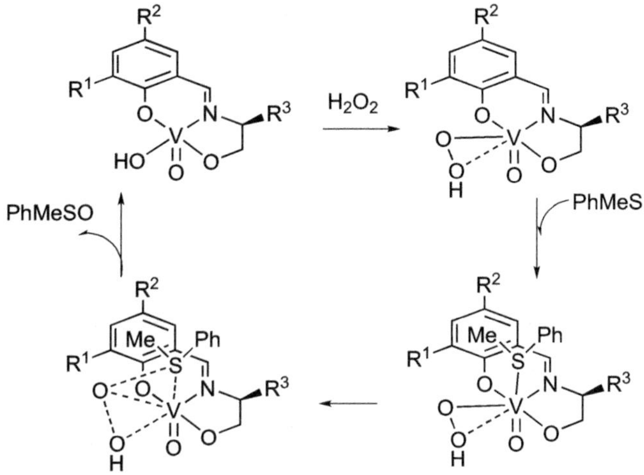

Scheme 7.36 Mechanism proposed for the Bolm-type ASO reaction.[101]

Such catalysts produce better enantioselectivity relative to their counterparts performed *in situ*, probably because they exclude competition from non-selective $[V(O)(O_2)_2]^-$, an effect already mentioned above for V(HA) epoxidation catalysis (Section 7.2.4).

One of the most consistent proposals for the mechanism of these reactions is depicted in Scheme 7.36. It agrees with the fact that the marked O atoms of $H_2{}^{18}O_2$ end up in the final sulfoxide and in the OH group attached to V, and also with kinetic measurements indicating that the coordination of SR_2 is the rate-determining step.

Iron complexes, with a cheap highly abundant metal, have not yet produced extensive results. However, ligand **30** (Scheme 7.35) mixed *in situ* with $Fe(acac)_3$, the substrate and then aqueous H_2O_2 in CH_2Cl_2 at room temperature under air lead to high *ee* values of sulindac, a non-steroidal anti-inflammatory drug. The process is reported to bring interesting environmental gains over earlier protocols.

7.3.6 M(salen), M(salan) and M(salalen) Sulfoxidation with H_2O_2 as TO

The result of the catalytic experiments using salan tetradentate ligands has been reported for several metal ions. In Scheme 7.37 the complex formed by reaction of the salan ligand with $VO(acac)_2$ oxidises sulfides with very high enantioselectivities using H_2O_2 as the TO. The mechanism proposed is substantially different from that in Scheme 7.36 and is more in line with the mechanisms proposed for the Mn(salen) systems.[102]

Taking advantage of the high stereoselection power of their second-generation salen ligands, Katsuki's group prepared an iron complex with a salan ligand (**32** in Figure 7.9), which allowed the aqueous synthesis of a

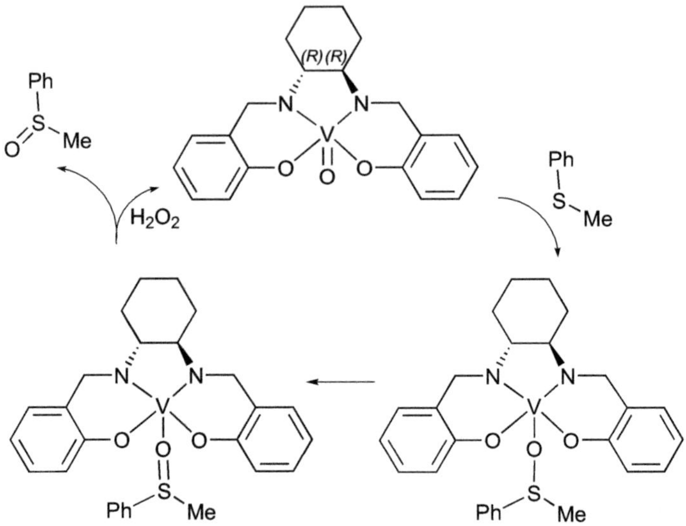

Scheme 7.37 Mechanism proposed for the VO^{2+}(salen)-catalysed ASO reaction.[102]

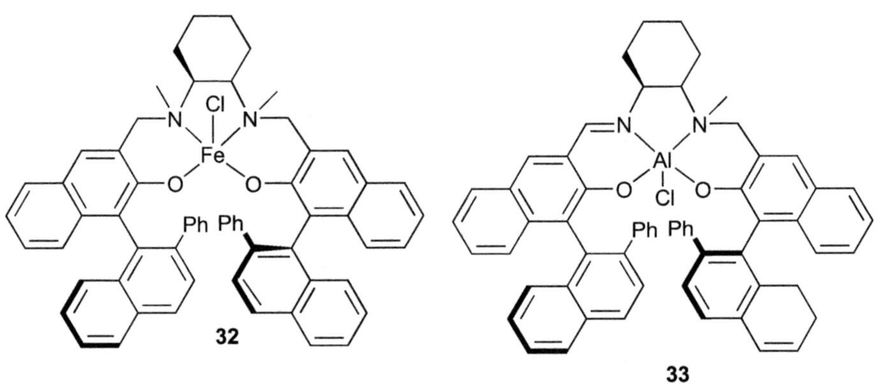

Figure 7.9 One salan (**32**) and one salalen (**33**) complex with excellent performances in ASO catalysis with H_2O_2.

wide number of dialkyl and aryl alkyl sulfoxides in very high enantiomeric excesses using H_2O_2 as TO.[103]

Ti complexes like the dimer **22** in Figure 7.6 also performed better than most other Ti species with Bolm-type Schiff bases in terms of enantioselectivity in sulfoxidation with H_2O_2. It is easy to assign this result to the extremely powerful enantioselection capacity imposed on metal complexes by second-generation Katsuki-type ligands. A most remarkable example of this power is the salalen complex of Al^{III} (**33** in Figure 7.9) an excellent catalyst for ASO, attaining *ee* values close to 99% for aryl methyl sulfides with H_2O_2 as the oxidant. Such high *ee* values result from the fact that the already

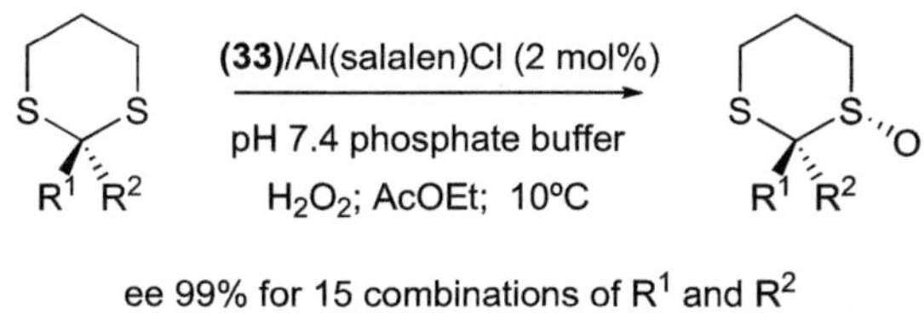

Scheme 7.38 General scheme for the ASO of cyclic dithioacetals with H_2O_2 and the Al(salalen) catalyst.

high enantioselectivity of sulfoxidation synergises with the faster oxidation of the less abundant enantiomer to the corresponding sulfone. This kinetic resolution process leads to very high ee values. This Al(salalen) complex at concentrations as low as 0.002–0.01 mol% allows ASO of aryl methyl sulfides with H_2O_2 under solvent-free or highly concentrated solutions with ee values often close to 99%.[104] Such reaction characteristics facilitate product separation and save on the use of solvents. Adding the fact that the TO is aqueous H_2O_2, and the reaction takes place in MeOH, the process is quite "green" and has already been implemented in large scale. The catalytically active species proposed is a η^2-hydroperoxo ligand formed and operating in a cycle very similar to that of the V complex in Scheme 7.36.

Beyond the aryl methyl sulfides, the Al(salalen)Cl complex **33** catalyses sulfoxidation of cyclic dithioacetals with extremely high levels of selectivity for 15 examples with R^1 and R^2 as H, aryls or alkyls in Scheme 7.38.[105] On the contrary, acyclic dithioacetals do not undergo high enantioselective SO with these catalysts.

7.4 Conclusion

More than three decades of intensive research and development have shaped alkene epoxidation and sulfoxidation as extremely useful and reliable methods for the synthesis of chiral synthons, intermediates and products in organic synthesis and the pharmaceutical industry. Almost quantitative enantiomeric excesses and very high yields have been consistently achieved for several substrate types. The Sharpless asymmetric allylic alcohol epoxidation and Jacobsen's epoxidation catalyst for unfunctionalized alkenes are standard tools in alkene epoxidation. Kagan's, Modena's and Bolm's catalysts are highly effective sulfoxidation technologies. Present development targets are focused on achieving even higher catalytic efficiency while decreasing environmental footprint through the use of less toxic catalysts, solvents and waste-free TOs.

Accordingly, recent research has been mainly directed towards the discovery and development of catalytic systems obeying the highest possible number of "green chemistry" rules. The metals involved in the catalytic delivery of active "O" to the alkene and/or thioether substrates are not a main question since first-row transition metals have efficiently taken up this role. In any case, Fe-based catalysts would be highly welcome and a few have already appeared.[103] Halogenated solvents like dichloromethane and chloroform have been hard to displace and in some cases seem almost irreplaceable. The ideal solvent, water, will most likely have to be complemented with non-aqueous solvents in order to bring lipophilic alkenes and thioethers in contact with the catalysts. However, alternative solutions may emerge from greener solvent systems, like ionic liquids or supercritical fluids.[91] So far, catalyst heterogenization has not provided outstanding results and further research is required. Eventually, microfluidic technologies may provide effective solutions to these problems. Nonetheless, the very high selectivity often achieved combined with very high yields has done much in terms of atom economy, waste decrease and simplification of purification procedures when compared to classic stoichiometric oxidation processes. The efforts to use green oxidants like O_2 and H_2O_2 are at the forefront of the research on these and other oxidation processes. In the case of sulphoxidation, the use of H_2O_2 as the TO is already largely dominant. In the case of epoxidation, it seems that the use of Ti–salan[64] and Mn–aminopyridine complexes[73] can utilize H_2O_2 with very high catalyst efficiency. Recently, the range of alkenes susceptible to selective epoxidation has been extended through the use of the BHA technologies, which mostly use alkyl hydroperoxides but have shown potential for using H_2O_2. As a new family of "privileged ligands", BHAs may be advantageous due to high flexibility in terms of coordination sphere design and tuning.[67]

The activation and control of the oxidative power of molecular oxygen, O_2, is still in its infancy, in spite of the amazing results produced by Ru complexes of second-generation Katsuki-type salen ligands.[64] Exploring and expanding the control over the exquisite mechanism of these catalysts in both epoxidation and sulfoxidation poses a very high research target that could lead to waste-free, fully sustainable oxidation processes. Their very existence proves the endeavor is worth pursuing.

References

1. S. Huber, M. Cokoja and F. E. Kühn, *J. Organomet. Chem.*, 2014, **751**, 25.
2. W. Adam, R. Curci and J. O. Edwards, *Acc. Chem. Res.*, 1989, **22**, 205.
3. S. E. Denmark and Z. C. Wu, *J. Org. Chem.*, 1997, **62**, 8964.
4. D. Limnios and C. G. Kokotos, *J. Org. Chem.*, 2014, **79**, 4270.
5. K. Bryliakov, *Environmentally Sustainable Catalytic Asymmetric Oxidations*, CRC Press, Boca Raton, 2015.
6. R. Irie, T. Uchida and K. Matsumoto, *Chem. Lett.*, 2015, **44**, 1268.
7. M. Tokunaga, J. F. Larrow, F. Kakiuchi and E. N. Jacobsen, *Science*, 1997, **277**, 936.
8. T. Katsuki and K. B. Sharpless, *J. Am. Chem. Soc.*, 1980, **102**, 5974.
9. S. S. Woodard, M. G. Finn and K. B. Sharpless, *J. Am. Chem. Soc.*, 1991, **113**, 106.
10. M. G. Finn and K. B. Sharpless, *J. Am. Chem. Soc.*, 1991, **113**, 113.

11. H. S. Rzepa, *Chemistry with a twist*, http://www.ch.imperial.ac.uk/rzepa/blog/?p=8588.
12. Y. Gao, R. M. Hanson, J. M. Klunder, S. Y. Ko, H. Masamune and K. B. Sharpless, *J. Am. Chem. Soc.*, 1987, **109**, 5765.
13. L. Canali, J. K. Karjalainen, D. C. Sherrington and O. Hormi, *Chem. Commun.*, 1997, 123–124.
14. R. Ballesteros, M. Fajardo, I. Sierra and I. del Hierro, *J. Mol. Catal. A: Chem.*, 2009, **310**, 83.
15. R. van Grieken, R. A. Garcia, G. Calleja and J. Iglesias, *Catal. Commun.*, 2007, **8**, 655.
16. E. J. Corey and D. C. Ha, *Tetrahedron Lett.*, 1988, **29**, 3171.
17. A. G. Myers, R. Glatthar, M. Hammond, P. M. Harrington, E. Y. Kuo, J. Liang, S. E. Schaus, Y. Wu and J.-N. Xiang, *J. Am. Chem. Soc.*, 2002, **124**, 5380.
18. A. G. Myers, M. Hammond, Y. S. Wu, J. N. Xiang, P. M. Harrington and E. Y. Kuo, *J. Am. Chem. Soc.*, 1996, **118**, 10006.
19. K. K. Krishnan, A. M. Thomas, K. S. Sindhu and G. Anilkumar, *Tetrahedron*, 2016, **72**, 1.
20. S. Y. Ko, A. Lee, S. Masamune, L. A. Reed, K. B. Sharpless and F. J. Walker, *Science*, 1983, **220**, 949.
21. S. Y. Ko, A. Lee, S. Masamune, L. A. Reed, K. B. Sharpless and F. J. Walker, *Tetrahedron*, 1990, **46**, 245.
22. T. Katsuki and V. Martin, *Asymmetric Epoxidation of Allylic Alcohols: The Katsuki–Sharpless Epoxidation Reaction*, John Wiley & Sons, Inc., Hoboken, NJ, USA, 1996.
23. A. Riera and M. Moreno, *Molecules*, 2010, **15**, 1041.
24. B. E. Rossiter, T. Katsuki and K. B. Sharpless, *J. Am. Chem. Soc.*, 1981, **103**, 464.
25. T. Kimura, N. Yamamoto, Y. Suzuki, K. Kawano, Y. Norimine, K. Ito, S. Nagato, Y. Iimura and M. Yonaga, *J. Org. Chem.*, 2002, **67**, 6228.
26. K. E. Henegar and M. Cebula, *Org. Process Res. Dev.*, 2007, **11**, 354.
27. G. Beck, *Synlett*, 2002, 837.
28. L. Jianfeng; P. Zhenkun, *Chin. Pat.*, CN20121355753, 2012.
29. W. Zhang, J. L. Loebach, S. R. Wilson and E. N. Jacobsen, *J. Am. Chem. Soc.*, 1990, **112**, 2801.
30. R. Irie, K. Noda, Y. Ito, N. Matsumoto and T. Katsuki, *Tetrahedron Lett.*, 1990, **31**, 7345.
31. T. Fukuda, R. Irie and T. Katsuki, *Synlett*, 1995, 197.
32. N. S. Finney, P. J. Pospisil, S. Chang, M. Palucki, R. G. Konsler, K. B. Hansen and E. N. Jacobsen, *Angew. Chem., Int. Ed. Engl.*, 1997, **36**, 1720.
33. E. N. Jacobsen, W. Zhang, A. R. Muci, J. R. Ecker and L. Deng, *J. Am. Chem. Soc.*, 1991, **113**, 7063.
34. T. L. Siddall, N. Miyaura, J. C. Huffman and J. K. Kochi, *J. Chem. Soc., Chem. Commun.*, 1983, 1185.

35. E. G. Samsel, K. Srinivasan and J. K. Kochi, *J. Am. Chem. Soc.*, 1985, **107**, 7606.
36. K. Srinivasan, P. Michaud and J. K. Kochi, *J. Am. Chem. Soc.*, 1986, **108**, 2309.
37. D. Feichtinger and D. A. Plattner, *Angew. Chem., Int. Ed. Engl.*, 1997, **36**, 1718.
38. T. Linker, *Angew. Chem., Int. Ed. Engl.*, 1997, **36**, 2060.
39. J. Dao, D. Benoit and C. J. Hawker, *J. Polym. Sci., Part A-1: Polym. Chem.*, 1998, **36**, 2161.
40. L. Cavallo and H. Jacobsen, *Angew. Chem., Int. Ed.*, 2000, **39**, 589.
41. Y. G. Abashkin and S. K. Burt, *Org. Lett.*, 2004, **6**, 59.
42. L. Kürti, M. M. Blewett and E. J. Corey, *Org. Lett.*, 2009, **11**, 4592.
43. T. Katsuki, *J. Mol. Catal. A: Chem.*, 1996, **113**, 87.
44. E. M. McGarrigle and D. G. Gilheany, *Chem. Rev.*, 2005, **105**, 1563.
45. H. Jacobsen and L. Cavallo, *Chem. – Eur. J.*, 2001, **7**, 800.
46. T. Hamada, T. Fukuda, H. Imanishi and T. Katsuki, *Tetrahedron*, 1996, **52**, 515.
47. N. S. Finney, P. J. Pospisil, S. Chang, M. Palucki, R. G. Konsler, K. B. Hansen and E. N. Jacobsen, *Angew. Chem., Int. Ed. Engl.*, 1997, **36**, 1720.
48. L. Cavallo and H. Jacobsen, *J. Org. Chem.*, 2003, **68**, 6202.
49. S. Liao and B. List, *Angew. Chem., Int. Ed.*, 2010, **49**, 628.
50. W. Zhang and E. N. Jacobsen, *J. Org. Chem.*, 1991, **56**, 2296.
51. M. Palucki, P. J. Pospisil, W. Zhang and E. N. Jacobsen, *J. Am. Chem. Soc.*, 1994, **116**, 9333.
52. I. Arends and R. A. Sheldon, *Appl. Catal., A*, 2001, **212**, 175.
53. D. Feichtinger and D. A. Plattner, *Chem. – Eur. J.*, 2001, **7**, 591.
54. J. P. Collman, L. Zeng and J. I. Brauman, *Inorg. Chem.*, 2004, **43**, 2672.
55. W. Adam, K. J. Roschmann, C. R. Saha-Möller and D. Seebach, *J. Am. Chem. Soc.*, 2002, **124**, 5068.
56. H. Fu, G. C. Look, W. Zhang, E. N. Jacobsen and C. H. Wong, *J. Org. Chem.*, 1991, **56**, 6497.
57. W. Zhang, N. H. Lee and E. N. Jacobsen, *J. Am. Chem. Soc.*, 1994, **116**, 425.
58. S. B. Chang, J. M. Galvin and E. N. Jacobsen, *J. Am. Chem. Soc.*, 1994, **116**, 6937.
59. E. N. Jacobsen, W. Zhang and M. L. Guler, *J. Am. Chem. Soc.*, 1991, **113**, 6703.
60. H. Nishikori, C. Ohta and T. Katsuki, *Synlett*, 2000, 1557.
61. J. F. Larrow, E. N. Jacobsen, Y. Gao, Y. P. Hong, X. Y. Nie and C. M. Zepp, *J. Org. Chem.*, 1994, **59**, 1939.
62. T. Yamada, K. Imagawa, T. Nagata and T. Mukaiyama, *Chem. Lett.*, 1992, 2231.
63. H. Tanaka, H. Nishikawa, T. Uchida and T. Katsuki, *J. Am. Chem. Soc.*, 2010, **132**, 12034.

64. S. Koya, Y. Nishioka, H. Mizoguchi, T. Uchida and T. Katsuki, *Angew. Chem., Int. Ed.*, 2012, **51**, 8243.
65. K. Matsumoto, Y. Sawada, B. Saito, K. Sakai and T. Katsuki, *Angew. Chem., Int. Ed.*, 2005, **44**, 4935.
66. Y. Sawada, K. Matsumoto, S. Kondo, H. Watanabe, T. Ozawa, K. Suzuki, B. Saito and T. Katsuki, *Angew. Chem., Int. Ed.*, 2006, **45**, 3478.
67. Z. Li and H. Yamamoto, *Acc. Chem. Res.*, 2013, **46**, 506.
68. A. U. Barlan, A. Basak and H. Yamamoto, *Angew. Chem., Int. Ed.*, 2006, **45**, 5849.
69. J. A. Brito, B. Royo and M. Gómez, *Catal. Sci. Technol.*, 2011, **1**, 1109.
70. C. Wang and H. Yamamoto, *J. Am. Chem. Soc.*, 2014, **136**, 1222.
71. L. Luo and H. Yamamoto, *Org. Biomol. Chem.*, 2015, **13**, 10466.
72. S. J. Mehrman, A. F. Abdel-Magid, C. A. Maryanoff and B. P. Medaer, in *Organometallics in Process Chemistry*, Springer, Berlin, Heidelberg, p. 153.
73. I. Garcia Bosch, X. Ribas and M. Costas, *Adv. Synth. Catal.*, 2009, **351**, 348.
74. O. Y. Lyakin, R. V. Ottenbacher, K. P. Bryliakov and E. P. Talsi, *ACS Catal.*, 2012, **2**, 1196.
75. W. Dai, J. Li, B. Chen, G. Li, Y. Lv, L. Wang and S. Gao, *Org. Lett.*, 2013, **15**, 5658.
76. I. Fernandez and N. Khiar, *Chem. Rev.*, 2003, **103**, 3651.
77. H. Pellissier, *Tetrahedron*, 2006, **62**, 1619.
78. K. K. Andersen, *Tetrahedron Lett.*, 1962, **3**, 93.
79. P. Pitchen, E. Dunach, M. N. Deshmukh and H. B. Kagan, *J. Am. Chem. Soc.*, 1984, **106**, 8188.
80. K. A. Jørgensen, *J. Chem. Soc., Perkin Trans. 2*, 1994, 117.
81. M. A. M. Capozzi, C. Centrone, G. Fracchiolla, F. Naso and C. Cardellicchio, *Eur. J. Org. Chem.*, 2011, 4327.
82. F. Di Furia, G. Modena and R. Seraglia, *Synthesis*, 1984, 325.
83. N. Komatsu, Y. Nishibayashi, T. Sugita and S. Uemura, *Tetrahedron Lett.*, 1992, **33**, 5391.
84. L. J. P. Martyn, S. Pandiaraju and A. K. Yudin, *J. Organomet. Chem.*, 2000, **603**, 98.
85. A. Massa, F. R. Siniscalchi, V. Bugatti, A. Lattanzi and A. Scettri, *Tetrahedron: Asymmetry*, 2002, **13**, 1277.
86. G. E. O'Mahony, P. Kelly, S. E. Lawrence and A. R. Maguire, *Archivoc*, 2011, 1.
87. H. Cotton, T. Elebring, M. Larsson, L. Li and H. Sörensen, *Tetrahedron*, 2000, **11**, 3819.
88. T. Andersson and L. Weidolf, *Clin. Drug Invest.*, 2008, **28**, 263.
89. H. B. Kagan, *Asymmetric Synthesis of Chiral Sulfoxides in Organosulfur Chemistry in Asymmetric Synthesis*, ed. T. Toru and C. Bolm, Wiley-VCH Verlag GmbH & Co. KGaA, Weinheim, Germany, 2008, ch. 1, p. 1.
90. E. Wojaczyńska and J. Wojaczyński, *Chem. Rev.*, 2010, **110**, 4303.

91. B. Zhang, M.-D. Zhou, M. Cokoja, J. Mink, S.-L. Zang and F. E. Kühn, *RSC Adv.*, 2012, **2**, 8416.
92. H. Srour, P. Le Maux, S. Chevance and G. Simonneaux, *Coord. Chem. Rev.*, 2013, **257**, 3030.
93. J. Legros, J. R. Dehli and C. Bolm, *Adv. Synth. Catal.*, 2005, **347**, 19.
94. P. Pitchen, *Asymmetric Synthesis of Sulfoxides: Two case Studies, in Chirality in Industry II: Developments in the Commercial Manufacture and Applications of Optically Active Compounds*, ed. A. N. Collins, G. Sheldrake and J. Crosby, J. Wiley and Sons,Chichester, 1997, ch. 19, p. 381.
95. M. Seenivasaperumal, H.-J. Federsel and K. J. Szabó, *Adv. Synth. Catal.*, 2009, **351**, 903.
96. M. Matsugi, N. Fukuda, Y. Muguruma, T. Yamaguchi, J. Minamikawa and S. Otsuka, *Tetrahedron*, 2001, **57**, 2739.
97. C. Kokubo and T. Katsuki, *Tetrahedron*, 1996, **52**, 13895.
98. D. Chakraborty, P. Malik and V. K. Goda, *Appl. Organomet. Chem.*, 2012, **26**, 21.
99. A. Basak, A. U. Barlan and H. Yamamoto, *Tetrahedron: Asymmetry*, 2006, **17**, 508.
100. C. Bolm and F. Bienewald, *Angew. Chem., Int. Ed. Engl.*, 1995, **34**, 2640.
101. Q. Zeng, H. Wang, W. Weng, W. Lin, Y. Gao, X. Huang and Y. Zhao, *New J. Chem.*, 2005, **29**, 1125.
102. J. Sun, C. Zhu, Z. Dai, M. Yang, Y. Pan and H. Hu, *J. Org. Chem.*, 2004, **69**, 8500.
103. H. Egami and T. Katsuki, *J. Am. Chem. Soc.*, 2007, **129**, 8940.
104. K. Matsumoto, T. Yamaguchi, J. Fujisaki, B. Saito and T. Katsuki, *Chem. – Asian J.*, 2008, **3**, 351.
105. J. Fujisaki, K. Matsumoto, K. Matsumoto and T. Katsuki, *J. Am. Chem. Soc.*, 2011, **133**, 56.

CHAPTER 8

C–C Bond Formation in the Sustainable Synthesis of Pharmaceuticals

L. M. D. R. S. MARTINS,* A. M. F. PHILLIPS AND A. J. L. POMBEIRO

Universidade de Lisboa, Centro de Química Estrutural, Instituto Superior Técnico, Av. Rovisco Pais, 1049-001 Lisboa, Portugal
*Email: luisamargaridamartins@tecnico.ulisboa.pt

8.1 Introduction

The development of eco-benign and clean synthetic methodologies for pharmaceuticals and other fine chemicals is an important goal of current research in chemistry. Catalysis had a significant impact in the design of innovative processes, new technologies and environmentally compatible synthetic routes to replace stoichiometric, multistep methods using toxic and/or hazardous reagents. Metal-catalysed reactions are particularly attractive in this context as they have the potential to provide atom-efficient routes for the synthesis of highly functionalized molecules. In particular, metal-catalysed cross-coupling reactions for the creation of carbon–carbon bonds are of central importance in organic chemistry.[1–4]

In 2010, the Nobel Prize in Chemistry was awarded[5,6] to Professors Richard F. Heck (University of Delaware), Akira Suzuki (University of Hokkaido) and Ei-ichi Negishi (Purdue University) for their achievements in

Green Chemistry Series No. 54
Sustainable Synthesis of Pharmaceuticals: Using Transition Metal Complexes as Catalysts
Edited by Mariette M. Pereira and Mário J. F. Calvete

Published by the Royal Society of Chemistry, www.rsc.org

Pd-catalysed C–C bond-forming reactions. Their discoveries had a great impact on the development of new drugs and materials, and are currently used in many industrial chemical processes for the synthesis of pharmaceuticals and other biologically active compounds.[7–9]

Palladium-catalysed C–C bond-forming cross-coupling occurs between an aryl/vinyl halide or pseudo-halide and an organometallic reagent (Figure 8.1a, left side). There are three basic steps in the catalytic cycle: oxidative addition, transmetallation and reductive elimination: first, the oxidative addition of the organic halide/pseudo-halide R–X to the palladium(0) complex $[PdL_n]$ generates a palladium(II) $[PdL_n(R)(X)]$ intermediate; then, an organometallic reagent M–R′ transfers the nucleophilic R′ to palladium, while M forms a "bond" with X (transmetallation); in the third step (reductive elimination), R and R′ couple and leave the Pd catalyst.

As depicted in Figure 8.1, since the first report by Heck regarding carbon–carbon bond formation using Pd catalysis,[1] Pd-catalysed cross-couplings assumed different designations as researchers found new possibilities to couple different reagents. In 1975, Sonogashira reported the mild coupling of acetylene gas and aryl halides using a combination of sub-stoichiometric Cu and Pd co-catalysts.[10] In 1977 Negishi introduced organozinc compounds as nucleophiles for palladium-catalysed cross-coupling, which gave superior yields compared to other organometallic compounds.[11,12] Moreover, those highly selective organozinc compounds in palladium-catalysed cross-coupling allowed for the presence of a wide range of functional groups.

In 1979 Suzuki and co-workers reported[13,14] that organoboron compounds can be used in Pd-catalysed cross-coupling with vinyl, aryl or alkyl halides, in the presence of a base. The base activates organoboron reagents as boronate intermediates to facilitate the transmetallation. The weak nucleophilicity of organoboron compounds, along with their stability and the fact that they operate under very mild conditions, made this reaction very practical for the pharmaceutical industry.[7,8]

Palladium-catalysed cross couplings using an organotin compound as the nucleophile were developed by Stille in 1978.[15,16] Organotin compounds are stable and allow mild reaction conditions and therefore the Stille reaction could be used as an alternative to the Negishi or Suzuki reactions for substrates with sensitive functional groups. However, the toxic character of organotin compounds has limited their industrial application. Later, Hiyama reported[17] efficient cross-coupling reactions of aryl silanes with aryl, alkenyl, and alkyl halides or pseudo-halides using a fluoride or a base as the silane activating agent, and many other possibilities followed.[1–4,7–9]

The Heck–Mizoroki reaction (often called the Heck reaction)[18–20] does not involve a transmetallation step (Figure 8.1b, right side). Instead, the alkene substrate undergoes a migratory insertion with the Pd(II) species, followed by a β-hydride elimination to give the product.[21] The presence of a base is required to induce the reductive elimination of HX. The nature of

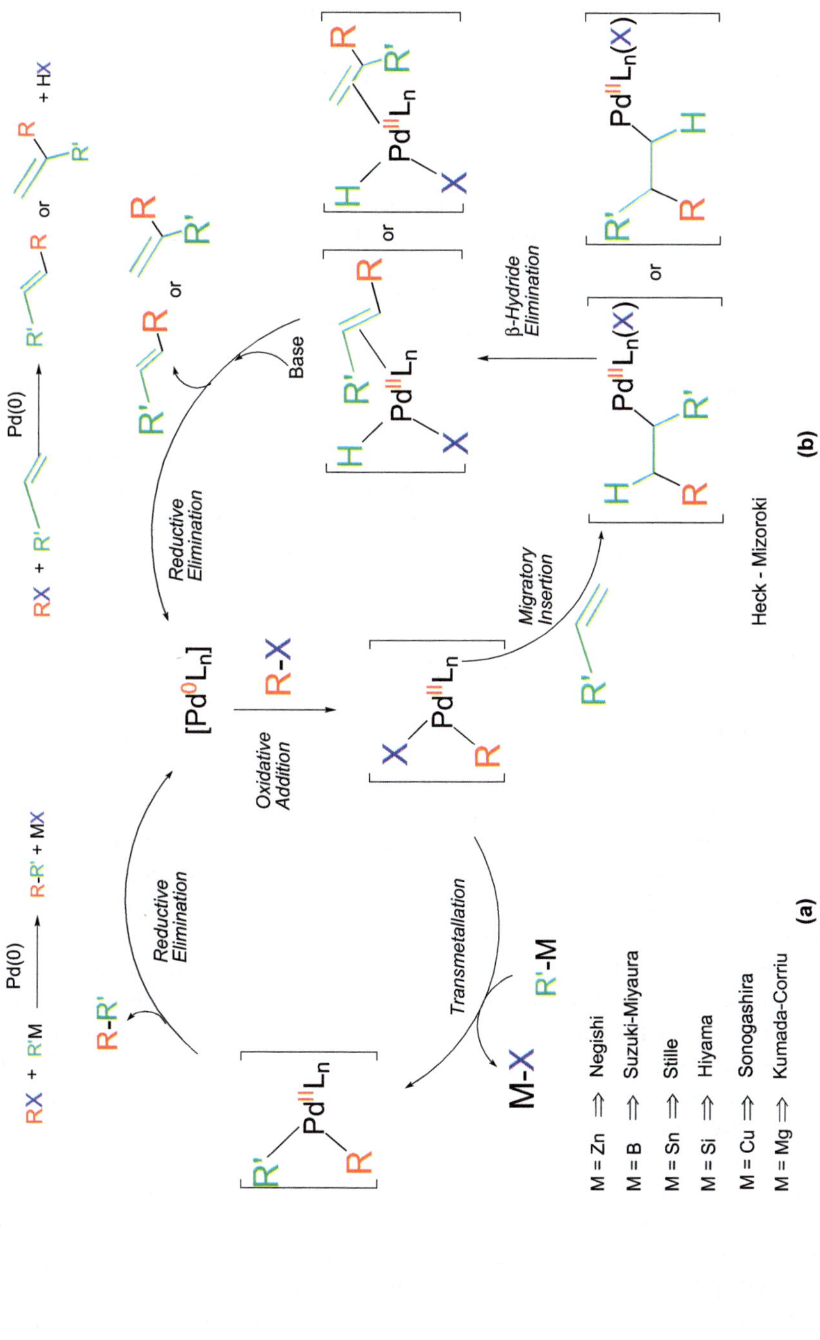

Figure 8.1 General mechanism of cross-coupling reactions and the Heck–Mizoroki reaction.

substituents on the olefin can influence the regioselectivity of the coupled products, determining the formation of linear or branched ones. More recently, the substrate range has been expanded to include triflates, carbonyl and sulfonyl chlorides, diazonium salts, iodonium salts and chlorides.[22–26]

In the years that followed, Pd-catalysed cross-coupling reactions were extended to C–heteroatom bond formation, namely C–N bond formation by coupling free amines with aryl halides (Buchwald–Hartwig amination)[27,28] and C–O and C–S bond formation.[1–4]

During the last few decades, the introduction of novel ligands, pre-catalysts, and reaction designs, as well as the more in-depth understanding of reaction mechanisms, have significantly contributed to the improvement of cross-coupling reactions. In fact, substrates considered initially difficult or improbable can now be used successfully. One example is the successful coupling of organolithiums with aryl halides developed by Feringa, one of the 2016 Nobel Laureates in Chemistry.[29] Nowadays, several palladium catalysts are available for the synthesis of important intermediates in the manufacture of pharmaceuticals, agrochemicals, dyes and other industrial products.

In the following sections, applications of transition-metal-catalysed C–C coupling reactions with significance for the manufacture of active pharmaceutical ingredients (API) or other drug components in the pharmaceutical industry are presented. They include not only the above-mentioned types of reactions, but also others with interest for those purposes.

8.2 C–C Coupling Reactions

8.2.1 Suzuki–Miyaura Coupling

The most representative coupling for the synthesis of pharmaceuticals is the Suzuki–Miyaura reaction, first reported in 1979,[13,14] which is (Section 8.1) the Pd-catalysed cross-coupling of an organoboron reagent and an organohalide or sulfonate (Figure 8.1a). Although the reaction can be very sensitive to the presence of oxygen, the mild reaction conditions, wide functional group compatibility and commercial availability of boronic acids make Suzuki–Miyaura coupling routinely used in the pharmaceutical industry for the preparation of highly functionalized molecules. Besides palladium, nickel can sometimes be employed as a catalyst.[30] Some examples of pharmaceuticals (or candidates) where either homogeneous or heterogeneous catalysed Suzuki–Miyaura couplings are included in their synthetic routes are presented below.

One of the first examples of industrial-scale Suzuki–Miyaura coupling in pharmaceuticals is the SmithKline Beecham Pharmaceuticals synthesis of SB-245570 for treatment of depression by coupling of boronic acid **1** and aryl bromide **2** (Figure 8.2a).[31,32] Inexpensive and readily available Pd supported on charcoal (Pd/C), used without additional ligands, was found to be the most efficient catalyst, providing **3** in excellent yield and purity (Pd contamination

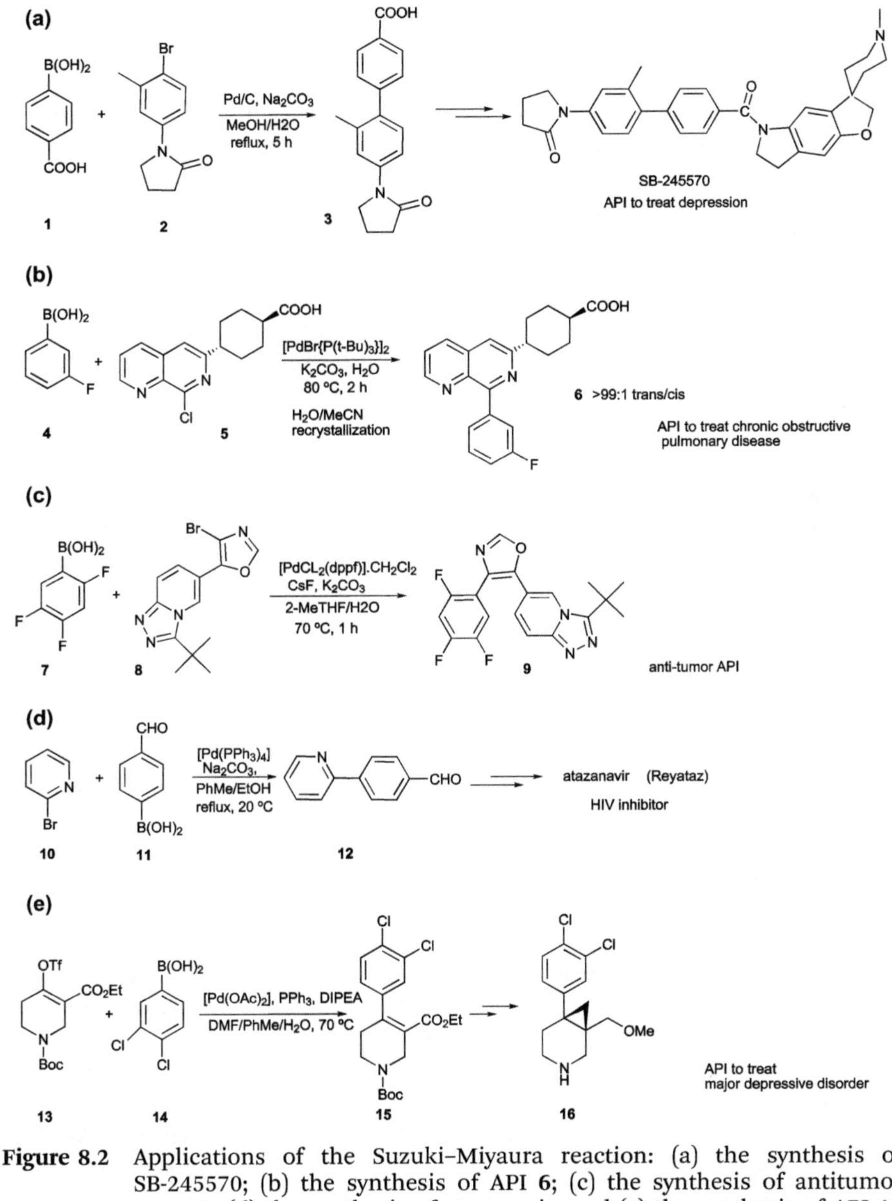

Figure 8.2 Applications of the Suzuki–Miyaura reaction: (a) the synthesis of SB-245570; (b) the synthesis of API **6**; (c) the synthesis of antitumor agent **9**; (d) the synthesis of atazanavir; and (e) the synthesis of API **16** to treat major depressive disorder.

minimized). Metallic Pd distributed only at the carbon surface proved to be more efficient than when distributed both in the pores and at the surface.

Jiang and co-workers at Novartis described the synthesis of the phosphodiesterase-4 inhibitor **6** for treatment of asthma and chronic obstructive pulmonary disease[33] where the last step of the synthesis is the Suzuki–Miyaura coupling of boronic acid **4** and chloride **5** (Figure 8.2b). Pd(I) dimer [PdBrP(*t*-Bu)$_3$]$_2$ was the chosen catalyst, affording inhibitor **6**, after recrystallization, as a >99 : 1 *trans–cis* mixture (with 1 ppm of Pd).

Researchers at Pfizer used[34] a Suzuki–Miyaura coupling of commercially available 2,4,5-trifluorophenylboronic acid **7** with bromide **8**, in the presence of [PdCl$_2$(dppf)$_3$].CH$_2$Cl$_2$ (dppf = 1,10-bis(diphenylphosphino)ferrocene), for the synthesis of a potent and selective inhibitor of the stress-activated kinase p38α (**9**), isolated with excellent purity (99.74% by HPLC) and Pd and Fe levels both below 10 ppm (Figure 8.2c). Potentially **9** has the same therapeutic properties as the currently commercialized therapies.

At Bristol-Myers Squibb, Xu and co-workers included the Suzuki C–C coupling of 2-bromopyridine **10** and boronic acid **11** to the synthesis of **12** (trade name Reyataz), a strong human immunodeficiency virus protease inhibitor (Figure 8.2d).[35] The coupling required low [Pd(PPh$_3$)$_4$] loading and **12** was isolated with acceptable levels of residual Pd (5–50 ppm).

Researchers at GlaxoSmithKline implemented the Suzuki–Miyaura coupling of triflate **13** and boronic acid **14** to **15** in the pilot plant-scale synthesis of **16**, a potent serotonin, noradrenaline, and dopamine reuptake inhibitor to treat major depressive disorder (Figure 8.2e).[36]

8.2.2 Negishi and Stille Couplings

Pd- or Ni-catalysed[11] Negishi coupling of organozinc reagents is not employed for the large-scale synthesis of pharmaceuticals as extensively as the Suzuki reaction,[7,9] mainly owing to its operational complexity, the generation of substantial zinc waste and the need of a balance between reactivity and functional group compatibility (due to the higher ionic character of the C–Zn bond).

Researchers at Abbott used the Negishi coupling of aryl zinc **17** and aryl bromide **18** in the presence of [PdCl$_2$(PPh$_3$)$_2$] in the synthesis a nonsteroidal ligand for the glucocorticoid receptor (Figure 8.3a).[37] The intermediate **19** precipitated, was easily isolated by filtration after reaction completion, and converted subsequently into the nonsteroidal ligand **20**.

Liu and co-workers at Hunan University in China synthesized the retinoid adapalene (trade name Differin) for topical treatment of psoriasis, by the ZnCl$_2$-mediated coupling of aryl bromide **21** and the Grignard reagent **22** in the presence of [PdCl$_2$(PPh$_3$)$_2$] (Figure 8.3b).[38] This method provided **23** in 86% yield and with residual Pd/Zn/Mg <20 ppm.

The Stille reaction[15] (Section 8.1) has been scarcely used on account of the toxicity exhibited by the organotin reagents and the difficult elimination of tin by-products from drug intermediates and APIs. Nevertheless, organotin reagents are usually available, stable, and compatible with several functional groups. For example, researchers at Pfizer included the Stille coupling of imidazolyl stannane (**24**) and iodothienopyridine (**25**) to **26** in the synthesis of vascular endothelial growth factor receptor (VEGFR) kinase inhibitor **27** (Figure 8.3c).[39]

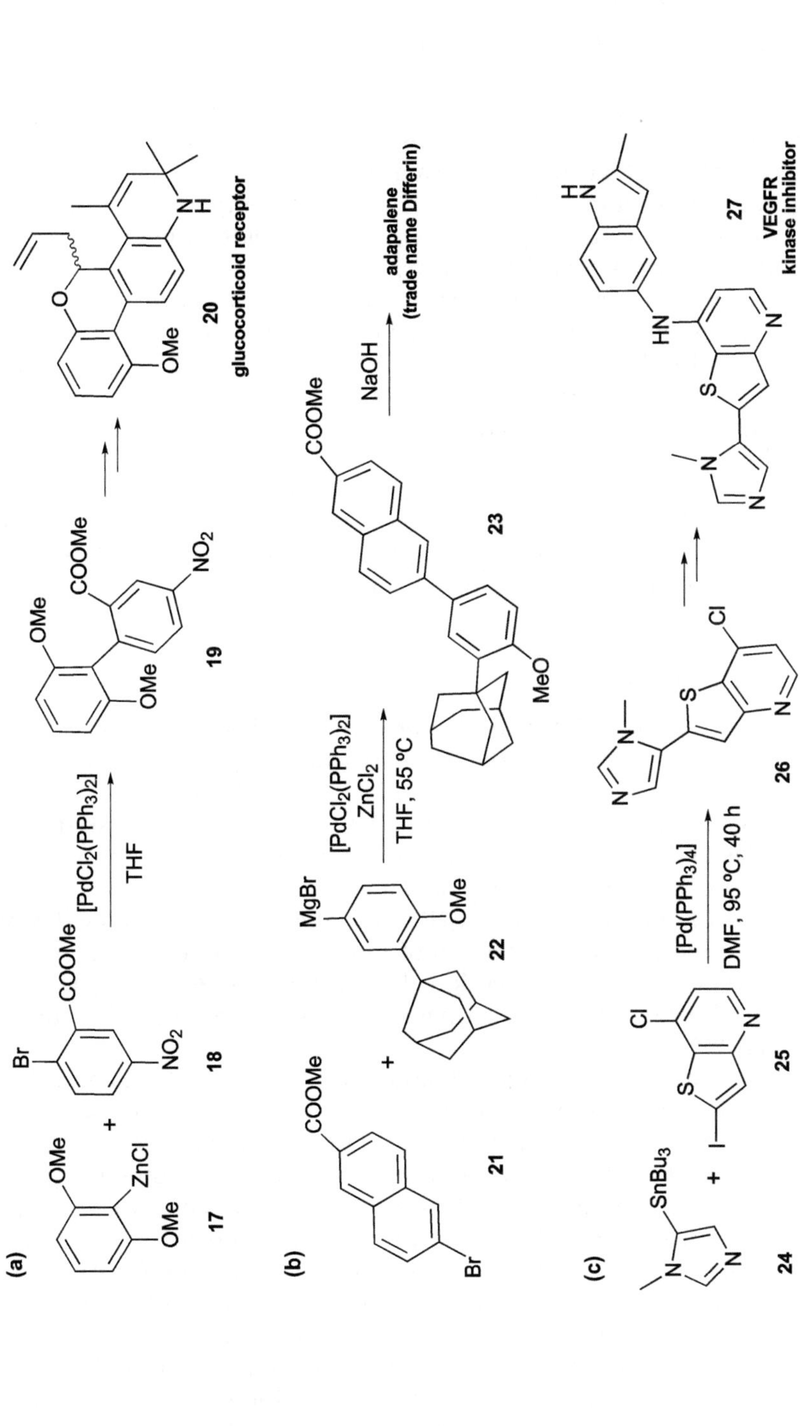

Figure 8.3 Applications of the Negishi and Stille reactions: (a) the synthesis of ligand **20** for glucocorticoid receptor; (b) the synthesis of adapalene; and (c) the synthesis of VEGFR kinase inhibitor **27**.

8.2.3 Sonogashira Coupling

The Sonogashira reaction, also known as the Sonogashira-Hagihara reaction, is the most useful method to prepare conjugated enynes and arenynes by coupling of sp^2-hybridized carbons and terminal alkynes (Section 8.1),[40–42] resulting in important intermediates for the synthesis of pharmaceuticals. Cu/Pd is the typical catalytic system, although Sonogashira coupling could be promoted by other metal/ligand complexes.[43] Advantages of this coupling include high efficiency and yields, functional group compatibility and its practical simplicity.

Thomas and co-workers at Abbott Laboratories used Sonogashira coupling of iodide **28** with 3-butyn-2-ol (**29**) in the synthesis of a 5-lipoxygenase inhibitor.[44] The reaction was carried out using low loadings of Pd and Cu, with PPh_3 additive, at room temperature to afford **30** in quantitative yield (Figure 8.4a), which was subsequently converted into the inhibitor fenleuton (**31**).

The Sonogashira reaction of iodoresorcinol **32** and propyne gas (**33**) to give **34**, was part of the method developed by Berliner and co-workers at Pfizer (Figure 8.4b) to obtain 4-hydroxy-2-methylbenzofuran (**35**), an important intermediate of various pharmaceutical compounds.[45]

8.2.4 Heck Coupling

The versatile coupling of organic halides with olefins in the presence of catalytic palladium and base, the Heck reaction (Figure 8.1b), is widely applied for the synthesis of pharmaceuticals[7,46] and for other industrial applications. In large-scale applications, the most common substrates are aryl bromides and iodides. The Heck reaction of (nitro and ester) activated methyl-2-chloro-5-nitrobenzoate (**36**) and 4-fluorostyrene **37**, with equimolar amounts of $PdCl_2$, $P(OEt)_3$ and $[n\text{-}Bu_4N]Br$ in *N,N*-dimethylacetamide (DMAc) at 90 °C was used for the preparation of intermediate **38** by Robinson and co-workers at AstraZeneca to develop the synthesis of **39** (90% yield), an anti-proliferative agent for the treatment of breast cancer and other tumors (Figure 8.5a).[47]

The Heck reaction of iodide **40** and 2-vinylpyridine **41** catalysed by $Pd(OAc)_2$ in the presence of $P(o\text{-tolyl})_3$, LiBr, and proton sponge **42** as the base in 2-methylpyrrolidinone (NMP) at 110 °C for 28 h was used for the preparation of axitinib **43**, an inhibitor of vascular endothelial growth factor for the treatment of cancer (Figure 8.5b), as described by Flahive and co-workers at Pfizer.[48] In addition to the very well-known intramolecular Heck-type coupling of **44** that provides morphine skeleton **45** (Figure 8.5c),[49] several methods have been reported for morphine (**46**) synthesis. Researchers at GlaxoSmithKline reported the intramolecular Heck reaction on iodo alkene **47** to the intermediate **48** leading to the chiral tetrahydroquinoline carboxylic acid *N*-methyl-D-glucamine salt **49**, a glycine antagonist for the treatment of nicotine craving (Figure 8.5d).[50]

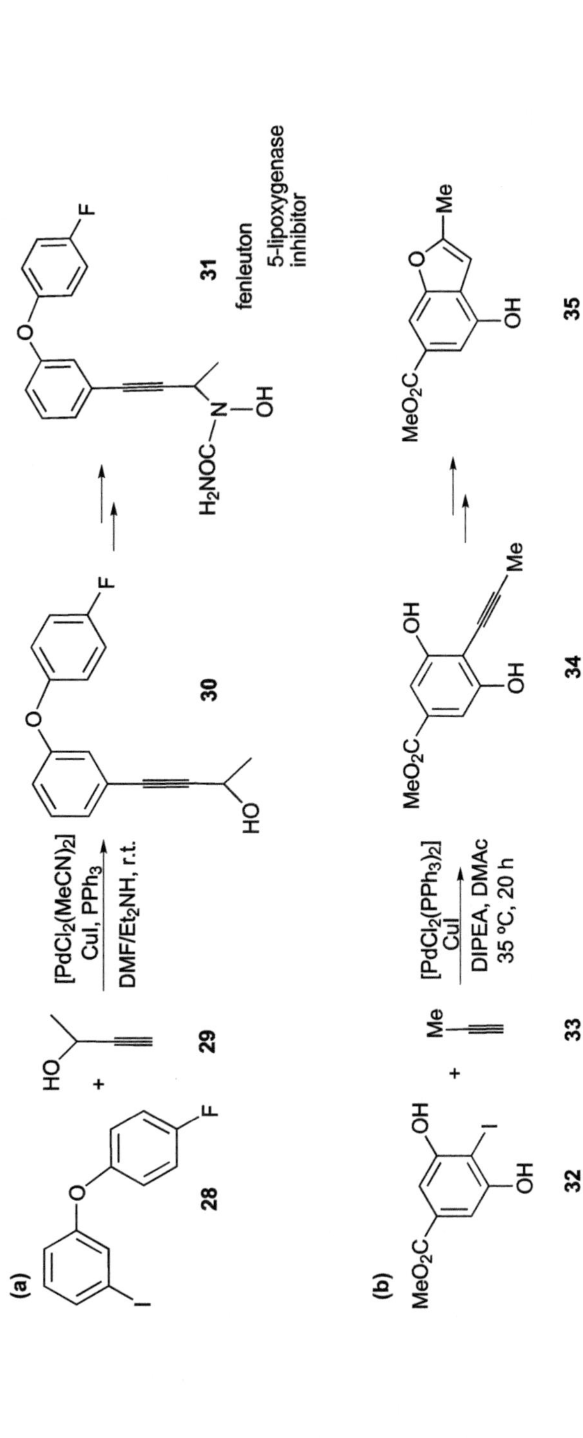

Figure 8.4 Applications of the Sonogashira reaction: (a) the synthesis of fenleuton and (b) the synthesis of 4-hydroxy-2-methylbenzofuran.

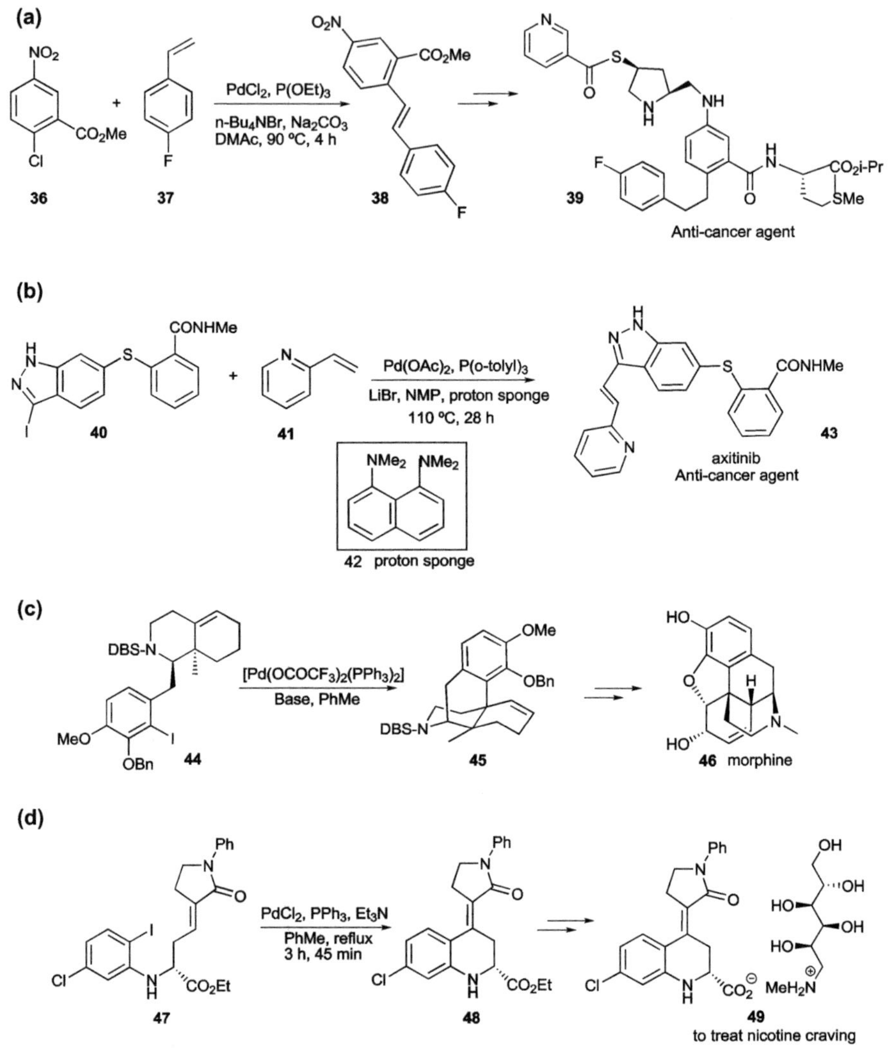

Figure 8.5 Applications of the Heck reaction: (a) the synthesis of anticancer agent **39**; (b) the synthesis of axitinib; (c) the synthesis of morphine; and (d) the synthesis of API **49**.

8.2.5 Decarboxylative C–C Coupling

In the presence of a catalyst system consisting of palladium(II) bromide, copper(II) oxide and 1,10-phenanthroline, 1-bromo(4-dimethoxymethyl)-benzene (**50**) is coupled with 2-cyanocarboxylic acid (**51**) for the construction of biaryl moiety **52** (Figure 8.6),[51] an intermediate in the preparation of valsartan (**53**), an angiotensin II receptor antagonist mainly used for treatment of high blood pressure. This representative example of efficient

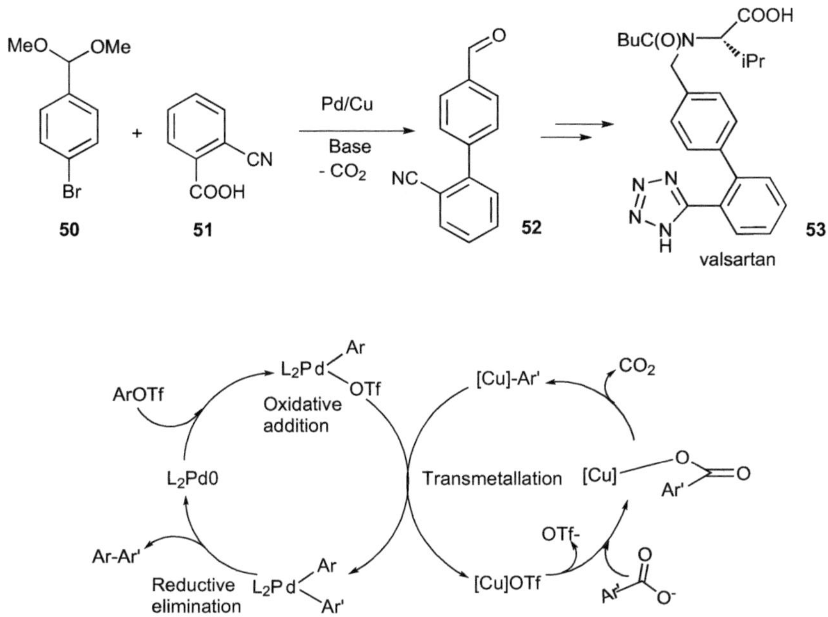

Figure 8.6 Application of the decarboxylative C–C coupling: synthesis of valsartan.

decarboxylative C–C coupling is a concise route to **53** with substantial economic and ecological advantages.

8.2.6 The Kumada–Corriu Coupling

Also known as Kumada coupling or Kumada–Tamao–Corriu coupling, the reaction is named after its discoverers.[52–54] Originally it was described as the Ni-catalysed coupling of a Grignard reagent with a vinyl or aryl halide, but now it has a broader scope (Figure 8.7a). It follows a typical coupling mechanism with oxidative addition, transmetallation and reductive elimination steps (Figure 8.1). Pd is often preferred to Ni since it is less toxic and exhibits higher activity and selectivity. Fe is receiving increasing attention because for catalysis cheap, readily available, nontoxic iron salts are used as pre-catalysts, but the scope is limited.[55] The ligands are important. Tamao and Kumada introduced phosphine ligands to modulate reactivity[56] and their results subsequently had an important impact in the development of cross-coupling methodology as a whole.[52–54]

One limitation of this reaction, from here onwards denoted as KC coupling, is that coupling partners bearing base-sensitive functionalities are not tolerated due to the nature of the organomagnesium reagents. Since they are also highly reactive, they have limited scope in the pharmaceutical industry, but there are examples. One of these is a process to synthesize ST1535, a highly selective adenosine A_{2A} receptor ligand antagonist to treat Parkinson's disease, reported by researchers at the University of Urbino and Sigma-Tau.[57]

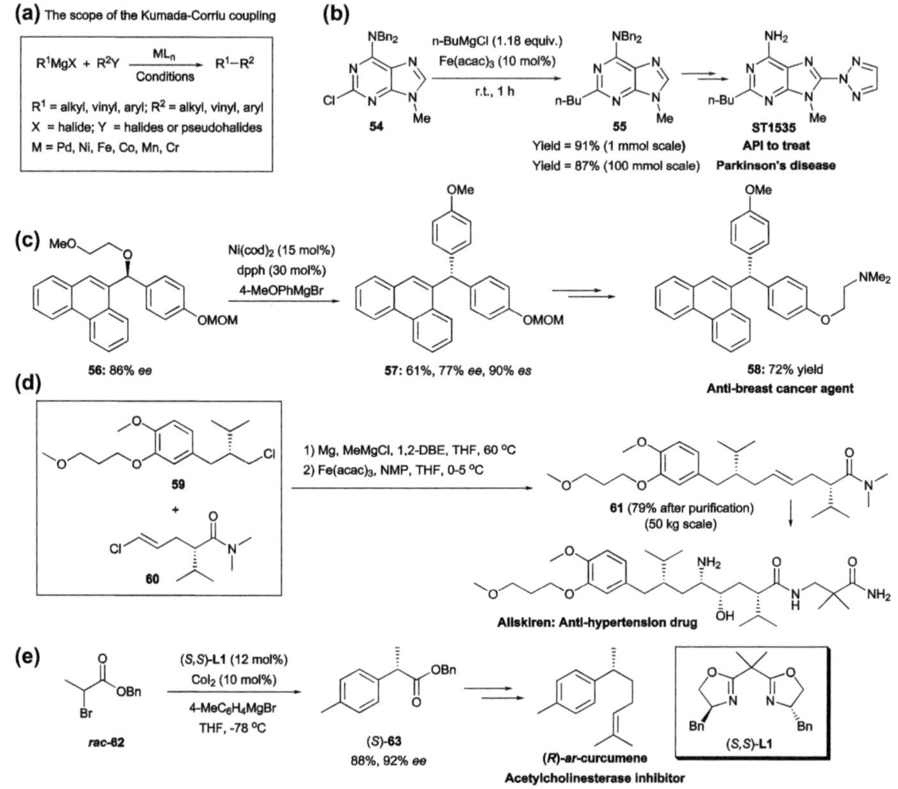

Figure 8.7 The Kumada–Corriu cross-coupling: (a) the reaction and its scope; (b) the synthesis of KC1535; (c) the synthesis of anti-breast cancer agent **61**; (d) the synthesis of Aliskiren; and (e) the synthesis of *ar*-curcumene.

An Fe-catalysed KC coupling introduced alkyl chains at the less reactive 2-position of 2-chloropurine intermediate **54**, generating key intermediate **55**. Fe is known to perform well in the presence of nitrogenated molecules, whereas Pd catalysts may show low reactivity when they bind to the heteroatoms in the substrate and the product, are present in larger excess and become deactivated. There were no competitive reactions (Figure 8.7b).

The first stereospecific KC coupling was developed recently at the University of California[58] and applied to the production of an anti-breast cancer agent. In a difficult to achieve Ni-catalysed process, which proceeded with inversion of configuration at the benzylic centre, enantio-enriched benzylic ether **56** was converted into optically active triarylmethane **57**. The high enantio-specificity of the reaction was possible particularly when 1,6-bis(diphenylphosphino)hexane (dpph) was used as the ligand. Intermediate **57** was subsequently converted into anti-breast cancer agent **58**, a tamoxifen analogue (Figure 8.7c). Homo-coupling products were not detected.

Researchers from Dr. Reddy's Laboratories (Telangana, India),[59] developed a KC cross-coupling to prepare an intermediate of aliskiren, a direct renin inhibitor to treat hypertension. How problems related to the initiation and batch-to-batch reproducibility encountered during process scale-up were solved was recently reported. The initiation, *i.e.* the oxidative addition step in which the organic halide (**59**) is transferred to one Mg reactive site, is the rate-limiting step. The Grignard dissolves away from the metal site as it forms, leaving the metal exposed for further reactions. This step usually requires heat and it is also highly exothermic, but it must proceed in a controlled manner. In this work, Mg powder performed better than Mg turnings, 1,2-dibromoethane (DBE) was used as the initiator, and total exclusion of moisture and air was necessary because the new Grignard was unstable, decomposed, and side products were obtained otherwise. MeMgCl was added to the THF solution containing halide and DBE to react with any moisture still present in the solvent and in the reactor and to weaken the MgO layer. The extra measures improved the results substantially, and **59** could be coupled with vinyl chloride **60** to afford **61** in very high yields, which was subsequently used to produce aliskiren (Figure 8.7d).

A report from the China Agricultural University (Beijing) described the first example of an enantioselective cobalt-catalysed KC coupling.[60] It involved substitution at a C(sp^3)-centre (on **62**), a difficult transformation to achieve enantioselectively. A catalyst-controlled enantioselective KC coupling was used, which yielded **63** in very high *ee*s, thus establishing the stereochemistry of the chiral centre of the aromatic sesquiterpene (*R*)-(+)-*ar*-curcumene, an essential oil of the rhizomes of *Curcuma aromatica* Salisb. It displays acetylcholinesterase inhibitory activity, besides being the main component of a sex-attracting pheromone produced by the red shoulder stink bug *Thyanta pallidovirens* (Figure 8.7e).

8.2.7 The α-Arylation of Enolates

Nucleophilic aromatic substitution is a difficult transformation to achieve. In 1997 Miura, Buchwald, Hartwig, and Muratake, independently developed Pd-catalysed processes for the α-arylation of enolates derived *in situ* from simple non-activated ketones in the presence of base under mild conditions.[61–63] These methods worked well with a wide range of substrates and nowadays the reaction has a much broader scope (Figure 8.8a). Pd is the most common catalyst and the reaction follows the usual coupling mechanism: oxidative addition, transmetallation, and reductive elimination. The reaction is very useful for the enantioselective synthesis of chiral quaternary carbon centres. There have been applications in total synthesis, as illustrated in Figure 8.8.

Enolate arylation is a key synthetic step in the enantioselective synthesis of the calcitonin gene-related peptide antagonist BMS-846372, being developed at Bristol-Myers Squibb (USA) as an API for the treatment of migraine (Figure 8.8b).[64] Aiming at a large-scale application, Pd catalysis was chosen for this step, in which the relative stereochemistry of the molecule is established. The reaction takes place under substrate control, with the

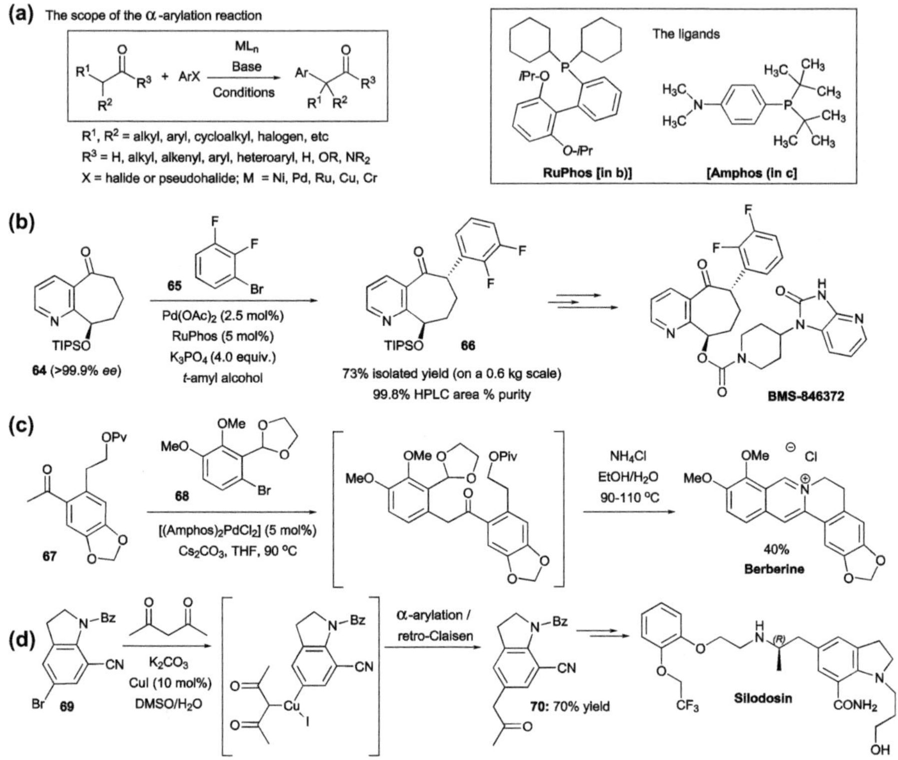

Figure 8.8 The Pd-catalysed α-arylation of ketone enolates and applications in API synthesis.

stereochemistry established in the previous step dictating the outcome. A bulky alcohol protecting group (TIPS) was used in **64** to direct the arylation with **65** to the desired face, but the maximum diastereoselectivity (*dr*) obtained was only 6:1 in favour of **66** as a result of epimerization. The *dr* could, however, be raised through recrystallization. RuPhos worked well as the ligand.

In 2014 researchers at the University of Oxford and GlaxoSmithKline developed a novel route to obtain the isoquinoline core structure of berberine and other members of the protoberberine family of alkaloids.[65] These secondary plant metabolites possess significant biological activities as a result of their ability to bind or intercalate DNA. In a "one-pot" procedure, catalysed by [(Amphos)$_2$PdCl$_2$], consisting of sequential ketone α-arylation-aromatization-cyclization, these alkaloids could be obtained in higher yields than previously recorded (Figure 8.8c), using ketone **67** and aryl halide **68** as reaction partners.

Researchers at the University of Milan and Dipharma Francis[66] reported a novel route to obtain enantiopure silodosin, the API of a medication for the symptomatic treatment of benign prostatic hyperplasia. This first application of a CuI-promoted ketone α-arylation in total synthesis could be performed on a gram scale when pentane 2,4-dione was reacted with aryl bromide **69** to produce key intermediate **70** (Figure 8.8d). It was also the shortest route reported so far for this API.

8.2.8 The Hayashi–Miyaura Reaction

The rhodium-catalysed 1,4-conjugate addition of aryl and 1-alkenyl boronic acids to enones, the Hayashi–Miyaura reaction, was first reported in 1997.[67,68] The reaction, its scope and mechanism, are shown in Figure 8.9.[69] This Michael-type addition is an alternative to the corresponding reactions of Grignard, cuprate and organozinc reagents, with the added advantage that organoboronic acids are stable in protic and aqueous media and many are commercially available in a variety of forms. It has large functional group tolerance and the reagents do not add in a 1,2-manner; however, protodeboronation may compete. In enantioselective versions, there is no competitive non-catalysed addition of the organoboron reagents owing to their low intrinsic nucleophilicity.

The Hayashi–Miyaura reaction is attractive for large-scale applications due to the ready availability of reagents of low toxicity (which ultimately degrade to boric acid) that are easy to handle due to their low reactivity, as well as the mild reaction conditions used.

One example is the preparation, starting with **71** (Figure 8.10a), of a chiral ester needed for the synthesis of an API the AstraZeneca group was developing.[70] The initial bench-top method was a Cu-mediated 1,4-addition of a Grignard to unsaturated ester **72** with ephedrine as the chiral auxiliary. However, the scale-up presented problems: the Cu-contaminated waste, the reagent (dibutylboron triflate) was expensive and ephedrine was subject to controlled-substances legislation in the U.K. The Hayashi–Miyaura reaction, still unknown at the manufacturing level, was attempted, but the usual conditions had to be modified (Figure 8.10a). It was found that cheap 3,5-difluorophenyl boronic acid (**73**) could be used as a reagent, but equimolar amounts of base were needed. *Iso*-PrOH replaced water as the solvent, eliminating the previously observed aggregation of the base, while also suppressing protodeboronation. Under these conditions, the boron reagent could be reduced to new lower levels and chiral ester (*R*)-**74** was obtained in high yield. The use of Smopex-234/oxone as a rhodium scavenger resulted in a final rhodium impurity content of <30 ppm.

Recently, Wu and co-workers from the National Taiwan Normal University and ScinoPharm developed an enantioselective route to (*S*)-SKF 383,[71] the enantiomer of a benzazepine which is a potent D1/D5 dopamine receptor partial agonist with potential to treat central nervous system diseases. The chiral centre was established by rhodium-catalysed addition of phenylboronic acid to nitrostyrene derivative **75**. The product (**76**) was obtained only when KHF_2 was used as the additive, a success attributed to the formation of an intermediate organotrifluoroborate, which transmetallates

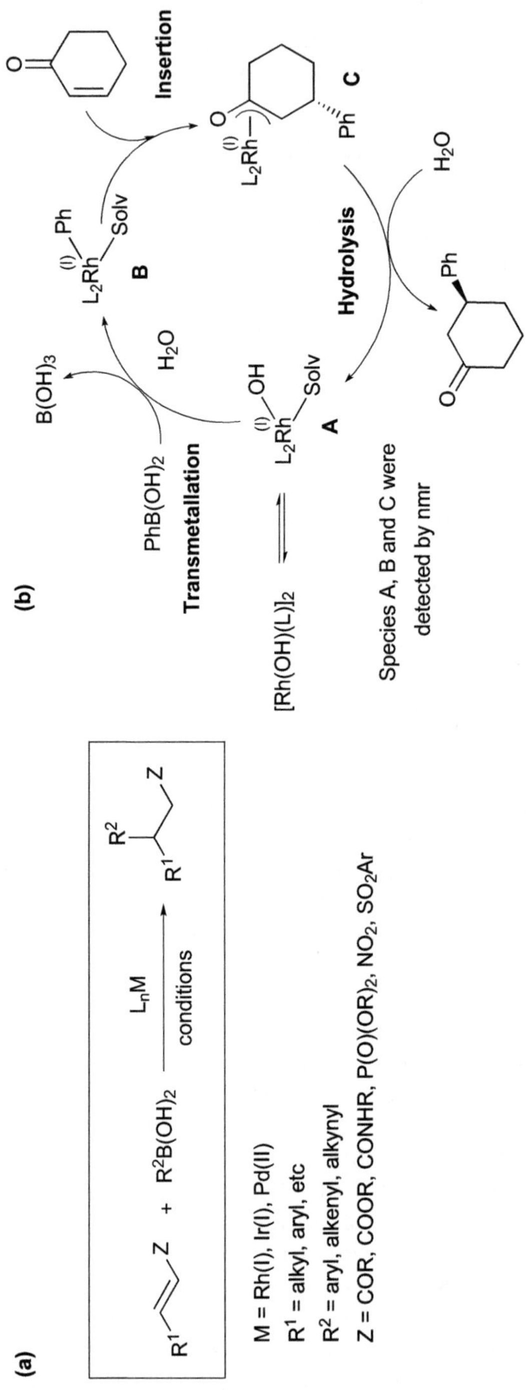

Figure 8.9 (a) The Hayashi–Miyaura reaction and (b) the mechanism proposed by Hayashi.

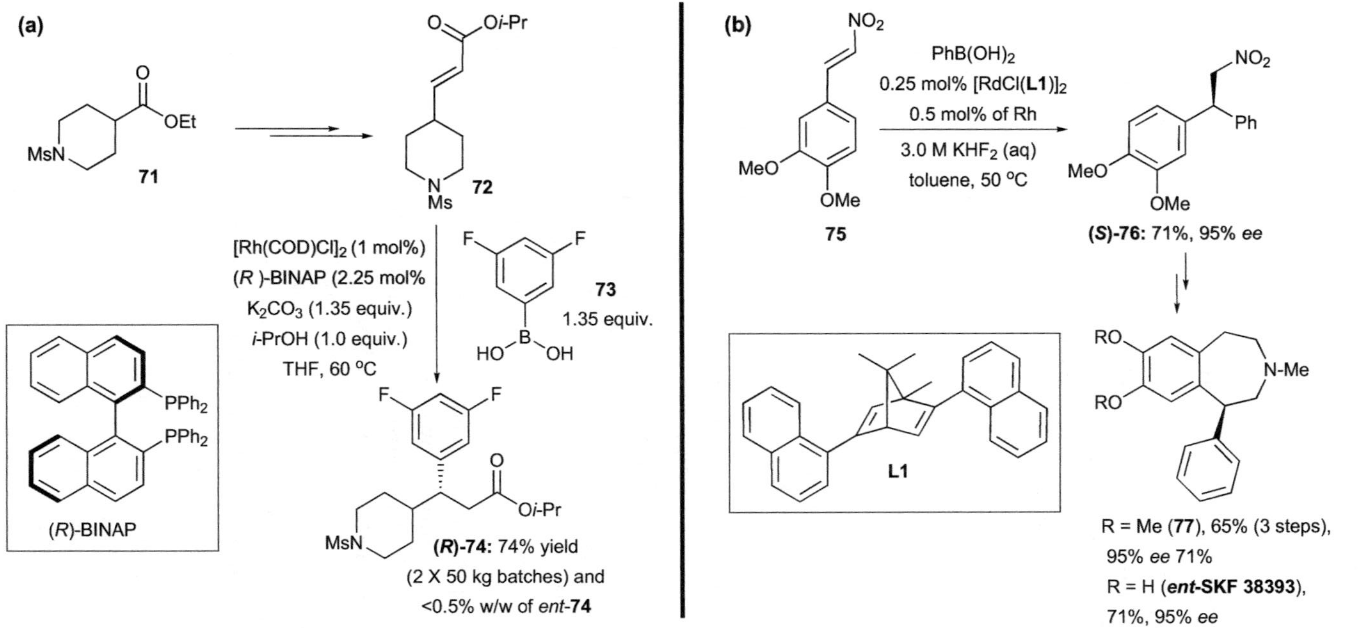

Figure 8.10 Applications of the Hayashi–Miyaura reaction: (a) AstraZeneca's API intermediate **83** and (b) ScinoPharm's *ent*-SKF 38393.

with rhodium, facilitating the reaction (Figure 8.10b). Temperatures and rhodium concentrations lower than those previously reported could be used, an advantage for manufacturing processes. Methyl-substituted benzazepine **77** or ent-38393 could be obtained from **76** in high yields.

8.2.9 Tsuji–Trost Allylation

This reaction, also known as Tsuji–Trost allylic substitution or alkylation, concerns the transition-metal-catalysed alkylation of allylic compounds[72–74] first reported by Tsuji and co-workers in 1965.[75] The initial stoichiometric version was later converted into a catalytic one.[72–74] Trost reported the first asymmetric version.[76] Pd is the most used metal and the catalytically active species is Pd(0).[72–74] The reaction proceeds *via* an intermediate π-allyl metal complex, to which a carbon, soft nitrogen, oxygen or sulfur nucleophile adds. The reaction and mechanism[77] are shown in Figure 8.11.

The reaction is regioselective and the nucleophile adds to the least substituted end of an unsymmetrical system regardless of the initial position of the leaving group. The nature of the ancillary ligand and the nucleophile may influence regioselectivity. The Tsuji–Trost reaction is also stereoselective and with soft nucleophiles double inversion (leading to retention of configuration) is observed, but with hard nucleophiles only inversion occurs.

A recent application comes from the Scripps Research Institute in the synthesis of (−)-neothiobinupharidine, a dimeric thiaspirane nuphar alkaloid with anticancer properties that is found in the freshwater plant *Nuphar lutea*.[78] An enantioselective tandem Buchwald conjugate reduction-modified Tsuji–Trost allylation of cyclopentenone **78** (Figure 8.12a) establishes the first chiral centre. The Cu-catalysed reduction produces an enolate trapped by a π-allyl palladium species generated *in situ* from allyl methyl carbonate and methyllithium. *Anti*-functionalized ketone **79** was obtained with very high stereoselectivity and minimal amounts of by-products. However, since the product was volatile, the yields were modest. Later in the sequence, a Grubbs ring-closing metathesis (RCM) cyclization was used to produce an intermediate quinolizidine.

Researchers from Université Pierre et Marie Curie and the CNRS in Paris have developed approaches to obtain γ-lactones and γ-lactams through C3–C4 ring closure using Pd-catalysed Tsuji–Trost reactions.[79] Although not the most commonly used strategy, this Pd-catalysed allylic alkylation uses mild conditions and is compatible with several functional groups. It provided access to picropodophyllin and derivatives (Figure 8.12b). This substance is an insulin-like growth factor I receptor kinase inhibitor, crucial for the growth and survival of cancer cells. It is a C2 stereoisomer of podophyllotoxin, a substance with potent anticancer activity, but cannot be used for therapy because it causes several adverse side effects. One of the applications of picropodophyllin is in kinase phosphatase biology research.

Although there are two possibilities for cyclization of an allylpalladium intermediate like **IV** (5-exo and 7-endo), incorporation of a bulky silyl group

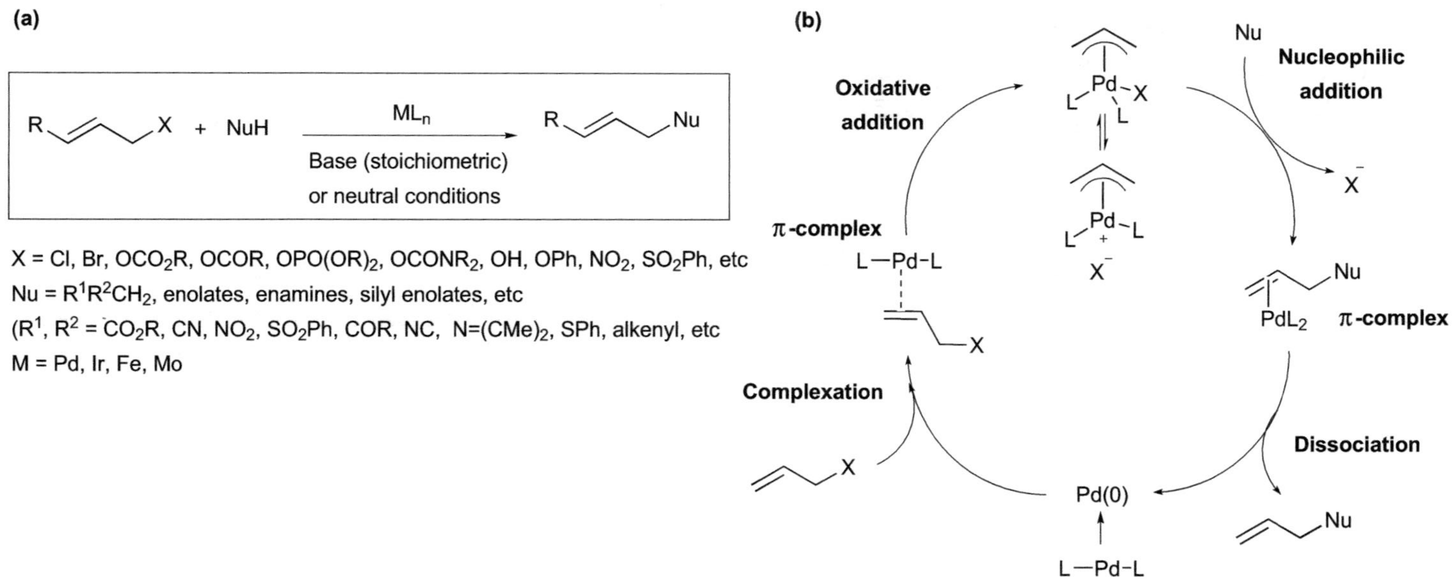

Figure 8.11 The Tsuji–Trost reaction (a) and its mechanism (b).

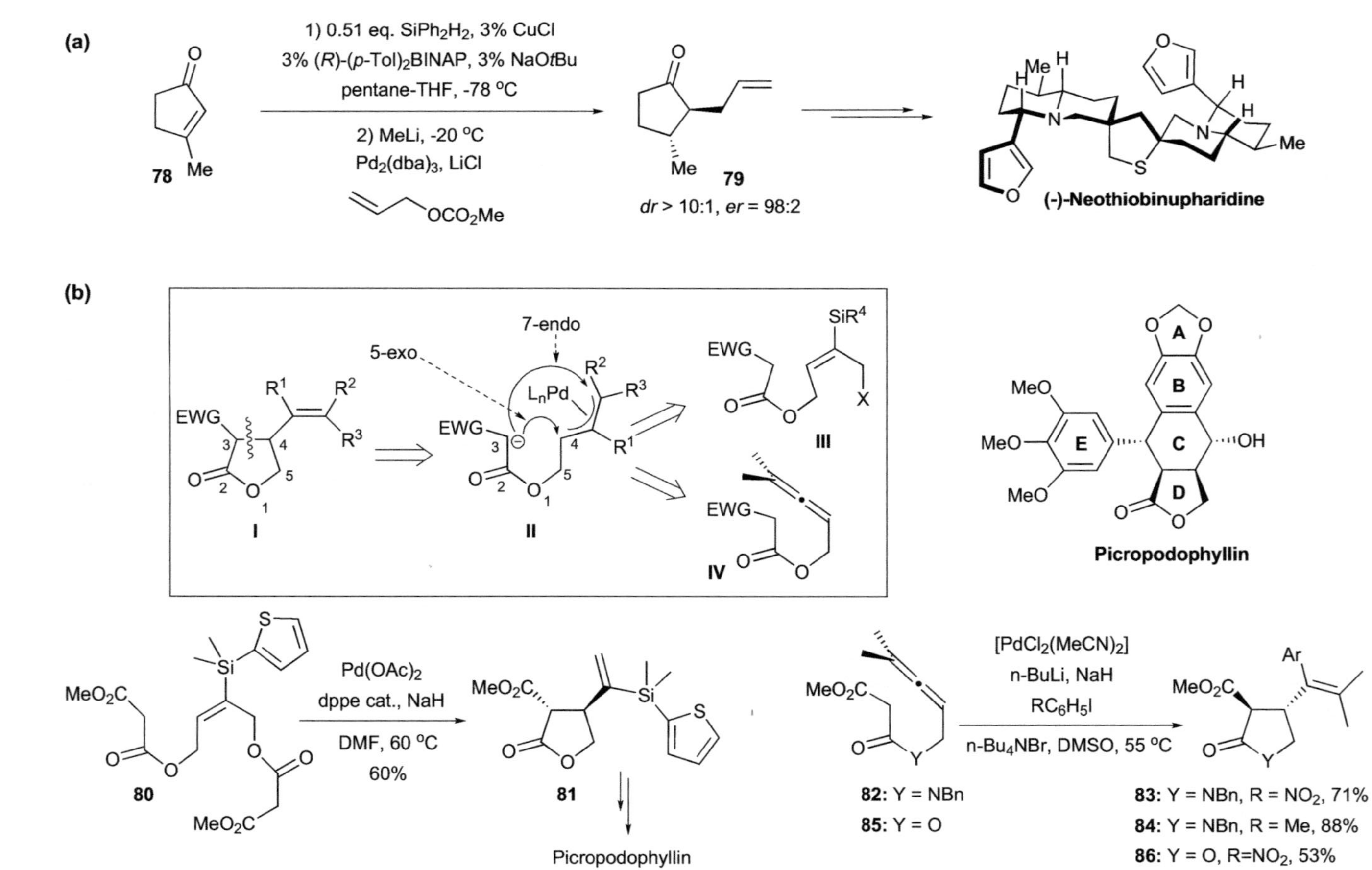

Figure 8.12 Applications of the Tsuji–Trost allylation.

on one of the vinylic carbon atoms, as in **80**, forces the palladium intermediate to adopt a conformation in which only the 5-exo ring closure is possible and a single product is obtained in the reaction, *trans*-γ-lactone (**81**). The product is subsequently subjected to Pd-catalysed Hiyama coupling to install rings C and D, and further functionalization provided picropodophyllin. The Pd-catalysed reactions of *N*-α-allenyl malonamides (**82**) proceed *via* analogous intermediates to afford α-styryl-γ-lactams **83** or **84**. An analogous strategy was used recently by the Poli group, and Merino and collaborators at the University of Zaragoza, to convert *O*-α-allenyl malonates (**85**) into γ-lactones (**86**), although the reaction conditions had to be modified in order to obtain higher yields.[80]

8.2.10 Aromatic Cyanation

Cyanation is an important reaction since the resulting cyanides, although useful in their own right, may be intermediates for a wide range of functional groups.[81,82] Takagi discovered a way to perform cyanation using palladium catalysis, which opened the way for many new developments. The reaction, its scope[81,82] and general mechanism[83] are shown in Figure 8.13. Catalysis avoids the use of stoichiometric amounts of metal cyanide reagents and the concomitant metal waste production. Since HCN is very toxic, other cyanide sources are used, mostly cyanide salts. $K_4[Fe(CN)_6]$, introduced by Beller in 2004, is a nontoxic source, also used in the food and wine industries.[84] Pd is the most used catalyst and co-catalysts, which facilitate palladium reduction, are also used. High concentrations of cyanide may cause catalyst inhibition and low turnover numbers (TONs) owing to the formation of inactive palladium species, *i.e.* $[HPd(CN)_3]^-$, $[ArPd(CN)_3]^-$ and $[Pd(CN)_4]_2$.[85] To overcome the problem and maintain an adequate CN^- concentration, the solubility of the cyanide source may be adjusted by judicious choice of solvents, by continuous dosage of cyanide to the reaction mixture, *e.g.* with acetone cyanohydrin, or using two-phase systems.

Several pharmaceuticals contain cyanated APIs, but transition-metal-catalysed processes are seldom used. The synthesis of the antidepressant citalopram, a selective serotonin reuptake inhibitor and one of the world's top-selling drugs, is an example. It was developed by H. Lundbeck Laboratories in Denmark in 1989 as Celexa.[86] Later it was found that the antidepressant activity was caused mainly by the (*S*)-enantiomer, escitalopram. Before the patent expired, Lundbeck and Forest performed a racemic switch and began marketing Lexapro (EU), also known as Cipralex (USA), whose API is escitalopram oxalate. In the manufacturing process,[87] esterification of diol **87** with (+)-di-*p*-toluoyl-D-tartaric acid diol (DTTA) provides a way to obtain pure **88** with the desired stereochemistry through fractional crystallization of the resulting diastereoisomers and further processing. The cyano group is introduced in benzofuran intermediate **88** with Pd catalysis at temperatures lower than would be needed for direct nucleophilic displacements (Figure 8.14a).

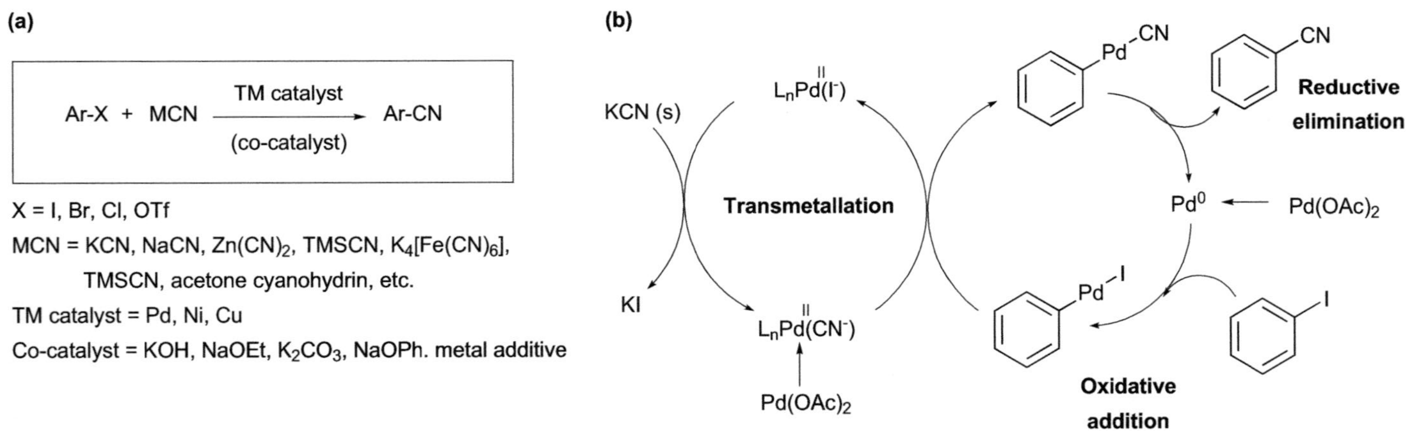

Figure 8.13 (a) The Pd-catalysed cyanation of aryl iodides and (b) the mechanism proposed by Takagi and co-workers.

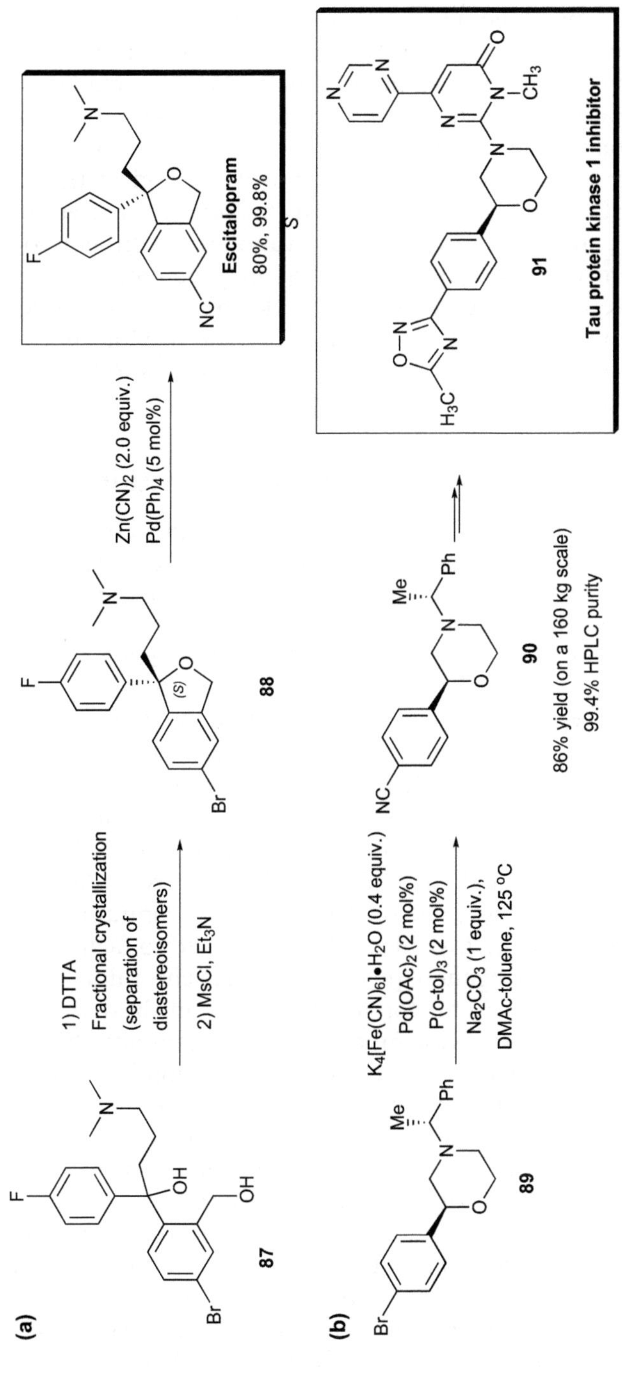

Figure 8.14 Transition-metal catalysed cyanation reactions: (a) Lundbeck and Forest's escitalopram synthesis and (b) Mitsubishi Tanabe's API for the treatment of Alzheimer's disease.

Mitsubishi Tanabe of Osaka is developing a tau protein kinase 1 inhibitor to treat Alzheimer's disease and related disorders.[88] The ways found to overcome problems encountered during scale-up have been reported. Pd-catalysed cyanation of **89** with $K_4[Fe(CN)_6]$ was used to prepare the API's oxadiazolephenyl unit *via* key intermediate **90**. The reaction was slow and stalled irreversibly near the end, and the results were not reproducible, suggesting catalyst deactivation. There was also homo-dimerization. $K_4[Fe(CN)_6]$ may cause catalyst deactivation because it gives rise to very high cyanide concentrations in which inactive palladium species are formed. Heterogeneous conditions with insoluble cyanoferrate, difficult to reproduce in large-scale processes, are usually used. Some problems were solved using dimethyl acetate (DMAc)–toluene dilution, keeping oxygen concentrations low ($<1\%$) and optimizing the cyanoferrate particle size for impeller stirring. The use of $P(o\text{-Tol})_3$ as the ligand sped up the reaction and further improved the results in the preparation of inhibitor **91** (Figure 8.14b).

8.2.11 Nozaki–Hiyama–Kishi Coupling Reaction

The Nozaki–Hiyama–Kishi (NHK) reaction,[89] *i.e.* the chromium-catalysed coupling reaction between a carbonyl compound and a halide to produce an alcohol, was reported in 1977 by Nozaki and Hiyama[90] and extended to a version catalytic in Ni by these researchers and Kishi in 1986.[89] A nucleophilic organochromium(III) intermediate is produced that adds to the carbonyl compound. The reaction has broad scope because of the low basicity of the organochromium reagents (Figure 8.15a,b). It is highly chemoselective for aldehydes, but the need for an excess of highly toxic chromium is a disadvantage. Generally, face selectivity is poor, but the reactions of allyl halides are highly stereoselective, yielding *anti* homoallyl alcohols regardless of the original configuration of the allyl halides, and alkenyl halides or triflates tend to completely retain their double-bond geometry.[89] Fürstner and Shi developed a version using catalytic amounts of chromium in 1996.[91] For catalyst regeneration, trimethylchlorosilane was used to facilitate chromium release, and nontoxic manganese powder to reduce chromium (Figure 8.15c).

The NHK reaction is useful for making medium-sized rings. The mild reaction conditions and the wide functional group compatibility also make it useful for large-scale applications. The synthesis of eribulin, the most complex synthetic drug in use, with 19 stereogenic centres, demonstrates the power of this reaction. Eribulin is marketed by Eisai of Japan as Halaven for the treatment of patients with metastatic breast cancer. It was recently approved for use in many countries. It works by interacting with tubulin and suppressing microtubule assembly during mitosis. Eribulin is a fully synthetic analogue of a marine natural product, halichondrin B, found in *Halichondria* sponges, first synthesized by Kishi in 1992.[92,93] New improvements were recently made to the manufacturing process by Eisai researchers.[94,95]

Sequential asymmetric Cr–Ni-catalysed couplings and Williamson ether cyclization are used in two key synthetic steps: the first joins subunit

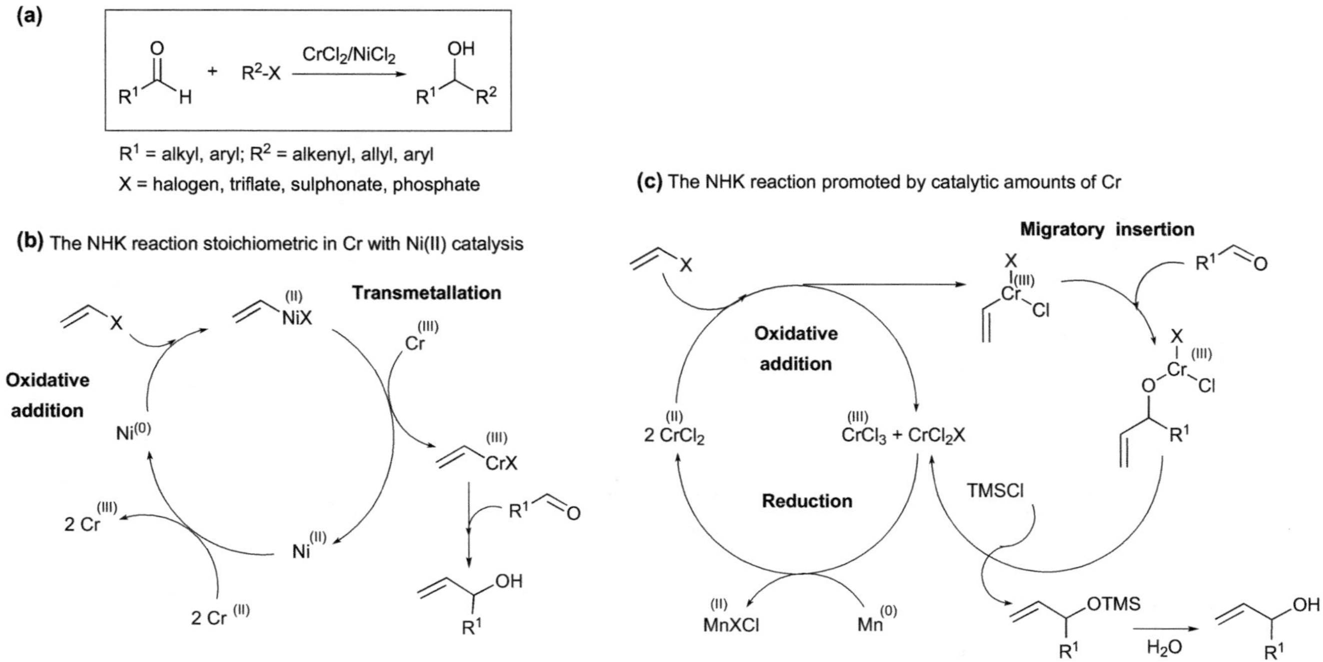

Figure 8.15 (a) The scope and the mechanisms of the NHK reaction: (b) catalytic in Ni with Ni–Cr catalysis and (c) catalytic in chromium.

C14–C19 (**92**) to C20–C26 (**93**) (Figure 8.16).[94] Vinyl bromide-aldehyde coupling with chiral ligand **CLA** establishes the C20 chiral centre. The coupled product (**94**) is treated with silica gel in 2-propanol to induce cyclization to the C20–C17 tetrahydrofuran ring in **95** which is then converted in a few steps to the C14–C26 subunit (**96**). A mixture of diastereoisomers is produced at this stage. The C26–C27 triflate-aldehyde coupling obtained through the reaction between **96** and **97** is performed in the presence of *ent*-CLA and the new stereogenic centre is created with 20 : 1 *dr*. Subsequent cyclic etherification was achieved with KHMDS at low temperature to afford the C23–C27 pyran ring in **98**. The conversion of **96** into **98**, *i.e.* the coupling of C14–C26 to C27–C35, is another example of a sequential asymmetric Cr–Ni-catalysed couplings–Williamson ether cyclization. Further structural modifications, chromatography and crystallization provided the C14–C35 fragment as a white crystalline solid. This was the first report of Ni–Cr coupling on a kilogram scale in fixed equipment in a pilot plant.

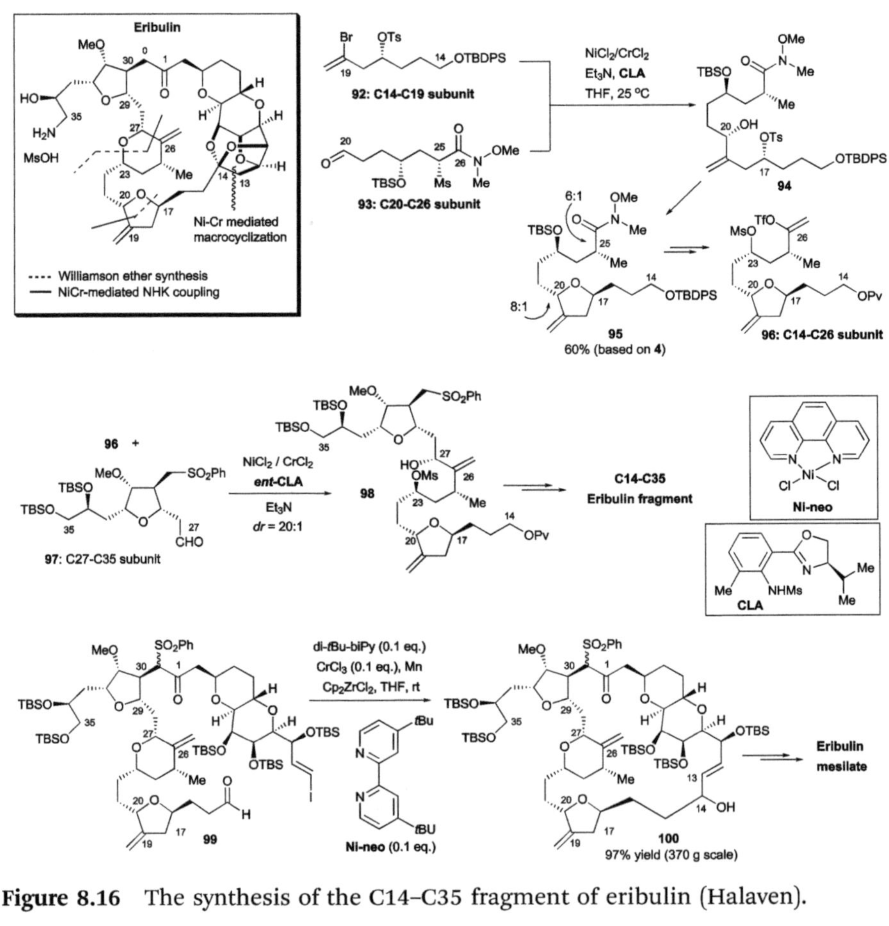

Figure 8.16 The synthesis of the C14–C35 fragment of eribulin (Halaven).

For the final stages of the synthesis, an intramolecular NHK macrocyclization catalytic in Cr was developed recently for **99**. Mn(0) is used to regenerate the catalyst and provide **100**.[95] Instead of TMSCl, $CpZrCl_2$ is used to facilitate the release of chromium by transmetallation with the chromate.

The large-scale synthesis of an anticancer marine product, discodermolide, at Novartis and the University of Cambridge includes another application of NHK coupling.[96] Paterson's one-pot NHK/Peterson elimination strategy was used to obtain the terminal diene (C21–C24 fragment) of the 24-carbon polyketide. $CrCl_2$ alone was used in less than stoichiometric proportions, to promote the coupling reaction of aldehyde **101** with allylic bromide **102** (Figure 8.17a). The initially obtained mixture of diastereomeric β-hydroxysilanes **103** underwent a Peterson *syn*-elimination upon basic treatment in MeOH–water to provide the desired (*Z*)-olefin **104**.

A synthetic strategy to obtain (+)-amphirionin-4, another bioactive polyketide of natural origin, came from Purdue University.[97] This substance was first isolated from an *Amphidinium* KCA09051 strain in the benthic sea sand, Iriomote Island. It has exceptionally potent proliferation-promoting activity (95% promotion) on murine bone marrow stomal ST-2 cells making it an interesting target for medical applications. A traditional Ni–Cr-catalysed NHK coupling between vinyl iodide **105** and aldehyde **106** was used to construct the C8 stereocentre (Figure 8.17b). This reaction took place at the end of the procedure, just prior to C8-alcohol deprotection, and the product was obtained in good yield (65%).

8.2.12 C–H Activation

C–H activation is in a broad sense a way to functionalize directly a C–H bond converting it into a carbon–carbon or carbon–heteroatom bond (Figure 8.18). The functionalization of organic molecules has been pursued by chemists for many years,[98–102] but only recently have new methods emerged to couple C–H covalent bonds directly, without any pre-functionalization, using transition-metal catalysis, which allow for the rapid construction of complex molecules under mild conditions.

For many years, the major targets were converting light alkanes to alcohols and linear alkanes to α-olefins,[103] aiming to obtain fuels, detergents, polymers. In the last few years the number of publications on C–H functionalization has grown exponentially with the development of new methods that allow a reaction to take place at a non-activated C–H bond instead of it occurring at another functional group present in the molecule. This is the main challenge in this area, but there are others: (i) the need to differentiate between different C–H bonds, even when the other bonds are more acidic or weaker; (ii) the strength of the C–H bonds since they are amongst the strongest chemical bonds; (iii) how to achieve mono functionalization; (iv) the problem, if C–X bonds are formed, that the new bond may be more reactive than the C–H bond. Directing groups are often included in starting materials to enhance site selectivity.

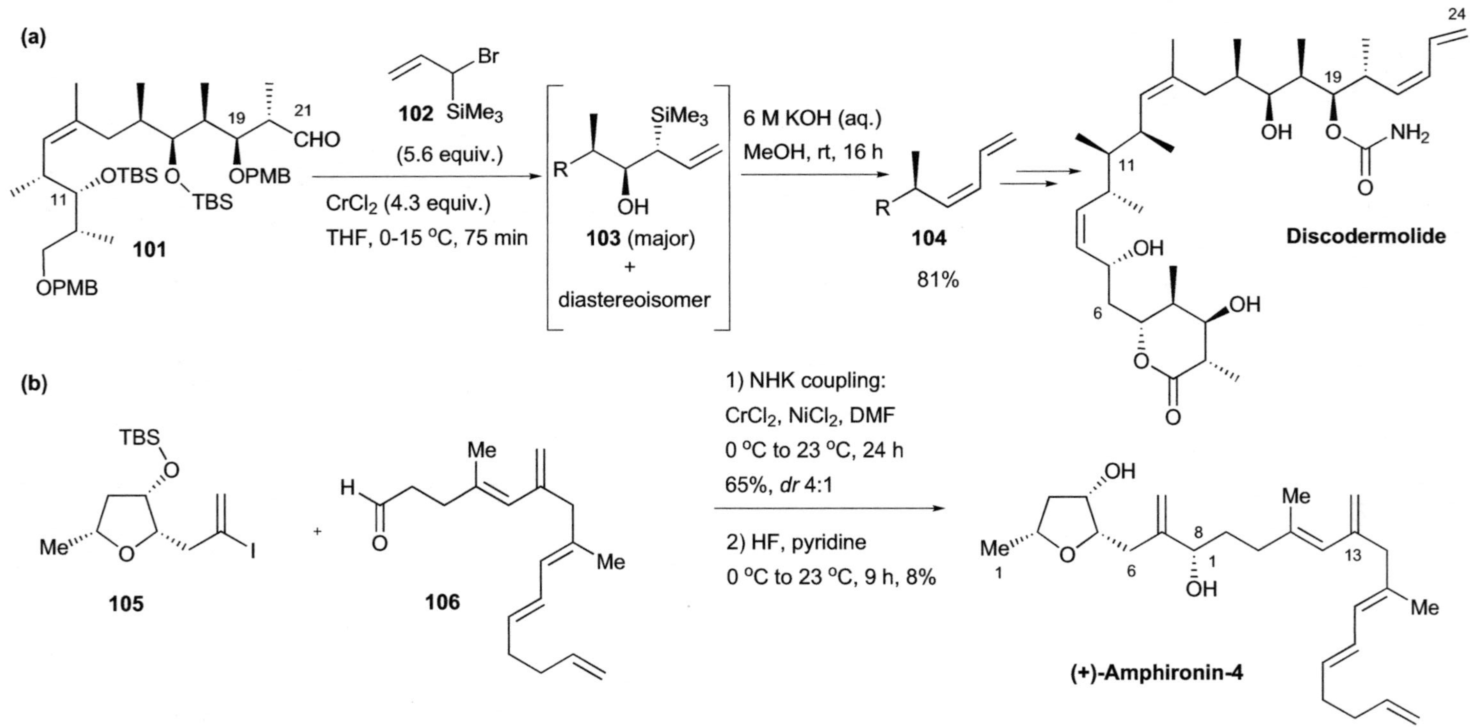

Figure 8.17 Applications of the Nozaki–Hiyama–Kishi coupling reaction in the total synthesis of natural products: the syntheses of (a) discodermolide and (b) (+)-amphirionin.

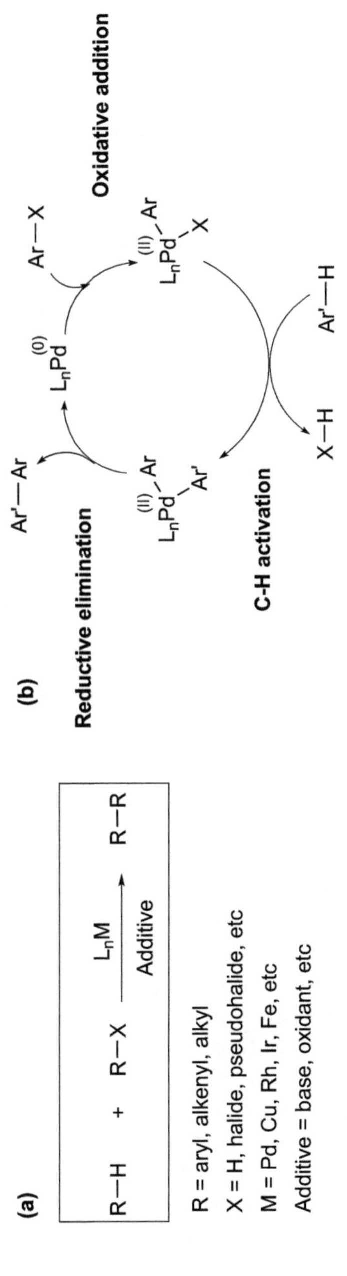

Figure 8.18 C–H activation/C–C coupling: the reaction and mechanism for Pd(0)/Pd(II) catalysis.

Other challenges arise when C–H activation is combined with C–C coupling, *e.g.* in Pd catalysis. C–C coupling is promoted by Pd(0) and C–H coupling by Pd(II), and the ligands required for one process may not be compatible with the other. If another organometallic reagent is used, the Pd(II) species may prefer to react with it rather than with the inactive C–H bond; an oxidant is also then required to convert Pd(II) to Pd(0). Oxidants are also required in cross-dehydrogenative coupling processes.

Despite the challenges, this chemistry gives rise to shorter synthetic routes and higher yields, as well as being cheaper and eventually greener since fewer steps are needed. It has been adopted by the pharmaceutical industry, even for late-stage functionalization in complex molecules. One example is a novel serine palmitoyltransferase (SPT) inhibitor being developed by Pfizer.[104] The new API has potential to treat heart disease since it is known that inhibition of SPT may raise high-density lipoprotein (HDL) cholesterol and high HDL cholesterol levels are associated with a lower the risk of health problems related to the cardiovascular system. In the medicinal chemistry route, the key intermediate was first obtained in five steps from 4-fluoro-3-nitrobenzoic acid. A shorter three-step process was subsequently developed, starting from aminobenzoate **107**, which after further transformations into α-chloroacetanilide **108** could provide the key intermediate oxindole **109** *via* Pd-catalysed $C_{sp}{}^2$–H/$C_{sp}{}^3$–Cl coupling using Buchwald's C–H activation methodology (Figure 8.19a). Nitrogen deprotection and further derivatization provided the SPT inhibitor **110**.

Another example is the synthesis of 2*H*-pyranonaphthoquinones, developed at the China Pharmaceutical University and the National Engineering and Research Centre for Target Drugs (Nanjing, China).[105] These chemical entities, *i.e.*, the lapachones, often have important biological activities: anti-inflammatory, antitumor, antimicrobial. They are present in very small amounts in many natural products, hence there is interest in methods to obtain them synthetically. The procedure involved a Pd(II)-catalysed C–H bond activation/C–C coupling/intramolecular Tsuji–Trost reaction cascade (Figure 8.19b). When diversely substituted 2-hydroxy-1,4-naphthoquinones (**111**) and *tert*-butyl-(2-methylbut-3-en-2-yl) carbonate (**112**) were mixed in a reactor with [Pd(OAc)$_2$] and an oxidant, the double cascade took place readily under mild conditions to produce ring C of lapachone **113**. An oxidant was needed since although Pd(II) is required for C–H activation, the Tsuji–Trost allylation is catalysed by Pd(0). The Tsuji–Trost reaction in this case involved C–O bond formation, with concomitant ring closure.

Researchers from Sumitomo Dainippon Pharma in Osaka reported in 2016 procedures to synthesize an API for the treatment of patients with Alzheimer's disease, now undergoing clinical trials.[106] The new compound is a phosphodiesterase type 4 (PDE4) inhibitor, *i.e.* it blocks the degrading action of phosphodiesterase 4, which should lead to nerve regeneration and memory improvement. Although the process used initially, starting from 4-fluoro-3-nitrobenzoic acid, provided the product on a 14 kg scale, it was desired to find a shorter synthetic route and avoid the use of 2-aminophenol,

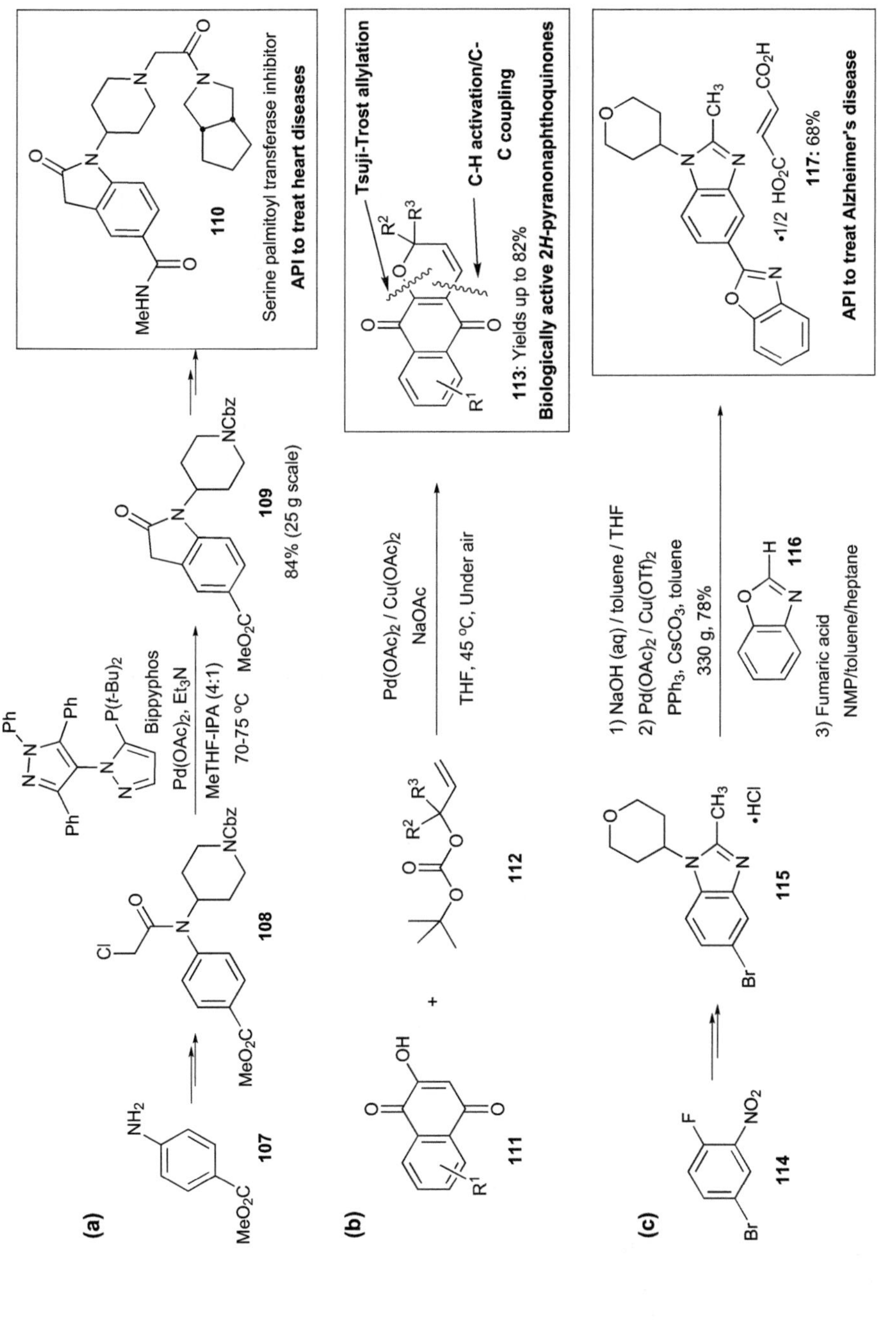

Figure 8.19 Applications of tandem C–H activation/C–C coupling in total synthesis.

one of the reagents used, which is a known mutagenic. For the second-generation synthesis, a new starting material (**114**) was used to prepare benzoimidazole **115**, which could be coupled directly to the second structural component of the drug, benzoxazole **116** *via* a Pd–Cu-catalysed C–Br/C–H coupling using Miura's catalytic system (Figure 8.19c).[106] In this way phosphodiesterase inhibitor **117** could be obtained in high yields.

8.3 Conclusion

Current manufacture of pharmaceuticals (APIs or other drug components) includes transition-metal-catalysed C–C cross-coupling reactions as key steps of the synthetic processes in view of their mildness, functional group compatibility and the high turnover of used catalysts. Moreover, modern protocols allow the purge of the metal catalyst, providing high-purity compounds.

The continued development of optimized catalysts, ligands, additives and reactions conditions (*e.g.*, temperature, solvent, pH) that led already to the above achievements will certainly expand the versatility of such transition-metal-catalysed transformations.

Abbreviations

acac	acetylacetonate
API	Active Pharmaceutical Ingredient
CLA	*N*-[2-(4-isopropyl-4,5-dihydro-oxazol-2-yl)-6-methyl-phenyl]-methanesulfonamide
COD	1,5-cyclooctadiene
1,2-DBE	1,2-dibromoethane
DIPEA	diisopropylethylamine
DMAc	*N,N*-dimethylacetatamide
DMF	*N,N*-dimethylformamide
DMSO	dimethylsulfoxide
dppe	1,2-bis(diphenylphosphino)ethane
dppf	1,10-bis(diphenylphosphino)ferrocene
dr	diastereomeric ratio
DTTA	(+)-di-*p*-toluoyl-D-tartaric acid
ee	enantiomeric excess
IPA	isopropanol
KC coupling	Kumada–Corriu coupling
NHK coupling	Nozaki–Hiyama–Kishi coupling
NMP	2-methylpyrrolidinone
OTf	triflate
RCM	ring-closing metathesis
ROM	ring-opening metathesis
r.t.	room temperature
RuPhos	2-dicyclohexylphosphino-2′,6′-diisopropoxybiphenyl

THF	tetrahydrofuran
Tol	toluene
TMS	trimethylsilyl
TON	turnover number

Acknowledgements

The support from FCT (Fundação para a Ciência e a Tecnologia, Portugal) (UID/QUI/00100/2013, PTDC/QEQ-ERQ/1648/2014 and PTDC/QEQ-QIN/3967/2014 projects) is acknowledged.

References

1. *New Trends in Cross-Coupling: Theory and Applications*, ed. T. Colacot, RSC Catalysis Series, RSC Publishing, Cambridge, 2015.
2. *Metal-Catalysed Cross-Coupling Reactions and More*, ed. A. de Meijere, S. Bräse and M. Oestreich, Wiley-VCH Verlag GmbH & Co. KGaA, Weinheim, 2014.
3. *Metal-Catalysed Cross-Coupling Reactions*, ed. A. de Meijere and F. Diederich, 2nd edn, Wiley, New York, 2004.
4. *Palladium-Catalysed Coupling Reactions: Practical Aspects and Future Developments*, ed. Á. Molnár, Wiley-VCH Verlag GmbH & Co. KGaA, Weinheim, 2013.
5. A. Suzuki, *Angew. Chem., Int. Ed.*, 2011, **50**, 6722.
6. E. Negishi, *Angew. Chem., Int. Ed.*, 2011, **50**, 6738.
7. J. Magano and J. R. Dunetz, *Chem. Rev.*, 2011, **111**, 2177.
8. X.-F. Wu, H. Neumann and M. Beller, *Chem. Soc. Rev.*, 2011, **40**, 4986.
9. V. B. Phapale and D. J. Cardenas, *Chem. Soc. Rev.*, 2009, **38**, 1598.
10. K. Sonogashira, Y. Tohda and N. Hagihara, *Tetrahedron Lett.*, 1975, **16**, 4467.
11. E.-I. Negishi, A. O. King and N. Okukado, *J. Org. Chem.*, 1977, **42**, 1821.
12. A. O. King, N. Okukado and E.-I. Negishi, *Chem. Commun.*, 1977, 683.
13. N. Miyaura, K. Yamada and A. Suzuki, *Tetrahedron Lett.*, 1979, **20**, 3437.
14. N. Miyaura and A. Suzuki, *J. Chem. Soc., Chem. Commun.*, 1979, 866.
15. D. Milstein and J. K. Stille, *J. Am. Chem. Soc.*, 1978, **100**, 3636.
16. D. Milstein and J. K. Stille, *J. Am. Chem. Soc.*, 1979, **101**, 4992.
17. Y. Hatanaka, S. Fukushima and T. Hiyama, *Chem. Lett.*, 1989, 1711.
18. R. F. Heck, *J. Am. Chem. Soc.*, 1968, **90**, 5538.
19. R. F. Heck, *J. Am. Chem. Soc.*, 1969, **91**, 6707.
20. R. F. Heck and J. P. Nolley, *J. Org. Chem.*, 1972, **37**, 2320.
21. Y. Fujiwara, I. Moritani, S. Danno, R. Asano and S. Teranishi, *J. Am. Chem. Soc.*, 1969, **91**, 7166.
22. N. J. Whitcombe, K. K. Hii and S. E. Gibson, *Tetrahedron*, 2001, **57**, 7449.
23. V. Farina, *Adv. Synth. Catal.*, 2004, **346**, 1553.
24. A. F. Littke and G. C. Fu, *Org. Synth.*, 2005, **81**, 63.

25. B. Basu and T. Frejd, *Acta Chem. Scand.*, 1996, **50**, 316.
26. A. F. Littke and G. C. Fu, *Angew. Chem., Int. Ed.*, 2002, **41**, 4176.
27. A. S. Guram, R. A. Rennels and S. L. Buchwald, *Angew. Chem., Int. Ed.*, 1995, **34**, 134.
28. J. Louie and J. F. Hartwig, *Tetrahedron Lett.*, 1995, **36**, 3609.
29. M. Giannerini, V. Hornillos, C. Vila, M. Fañanás-Mastral and B. L. Feringa, *Angew. Chem., Int. Ed.*, 2013, **52**, 13329.
30. D. Zim and A. L. Monteiro, *Tetrahedron Lett.*, 2002, **43**, 4009.
31. D. S. Ennis, J. McManus, W. Wood-Kaczmar, J. Richardson, G. E. Smith and A. Carstairs, *Org. Process Res. Dev.*, 1999, **3**, 248.
32. N. Miyaura, T. Yanagi and A. Suzuki, *Synth. Commun.*, 1981, **11**, 513.
33. X. Jiang, G. T. Lee, E. B. Villhauer, K. Prasad and M. Prashad, *Org. Process Res. Dev.*, 2010, **14**, 883.
34. B. Li, R. A. Buzon and Z. Zhang, *Org. Process Res. Dev.*, 2007, **11**, 951.
35. Z. Xu, J. Singh, M. D. Schwinden, B. Zheng, T. P. Kissick, B. Patel, M. J. Humora, F. Quiroz, L. Dong, D.-M. Hsieh, J. E. Heikes, M. Pudipeddi, M. D. Lindrud, S. K. Srivastava, D. R. Kronenthal and R. H. Mueller, *Org. Process Res. Dev.*, 2002, **6**, 323.
36. V. I. Elitzin, K. A. Harvey, H. Kim, M. Salmons, M. J. Sharp, E. A. Tabet and M. A. Toczko, *Org. Process Res. Dev.*, 2010, **14**, 912.
37. Y.-Y. Ku, T. Grieme, P. Raje, P. Sharma, H. E. Morton, M. Rozema and S. A. King, *J. Org. Chem.*, 2003, **68**, 3238.
38. Z. Liu and J. Xiang, *Org. Process Res. Dev.*, 2006, **10**, 285.
39. J. A. Ragan, J. W. Raggon, P. D. Hill, B. P. Jones, R. E. McDermott, M. J. Munchhof, M. A. Marx, J. M. Casavant, B. A. Cooper, J. L. Doty and Y. Lu, *Org. Process Res. Dev.*, 2003, **7**, 676.
40. H. Doucet and J.-C. Hierso, *Angew. Chem., Int. Ed.*, 2007, **46**, 834.
41. M. M. Heravi and S. Sadjadi, *Tetrahedron*, 2009, **65**, 7761.
42. R. Chinchilla and C. Najera, *Chem. Rev.*, 2007, **107**, 874.
43. H. Plenio, *Angew. Chem., Int. Ed.*, 2008, **47**, 6954.
44. A. V. Thomas, H. H. Patel, L. A. Reif, S. R. Chemburkar, D. P. Sawick, B. Shelat, M. K. Balmer and R. R. Patel, *Org. Process Res. Dev.*, 1997, **1**, 294.
45. M. A. Berliner, E. M. Cordi, J. R. Dunetz and K. E. Price, *Org. Process Res. Dev.*, 2010, **14**, 180.
46. M. Prashad, *Top. Organomet. Chem.*, 2004, **6**, 181.
47. S. Abbas, L. Ferris, A. K. Norton, L. Powell, G. E. Robinson, P. Siedlecki, R. J. Southworth, A. Stark and E. G. Williams, *Org. Process Res. Dev.*, 2008, **12**, 202.
48. E. J. Flahive, B. L. Ewanicki, N. W. Sach, S. A. O'Neill-Slawecki, N. S. Stankovic, S. Yu, S. M. Guinness and J. Dunn, *Org. Process Res. Dev.*, 2008, **12**, 637.
49. C. Y. Hong, N. Kado and L. E. Overman, *J. Am. Chem. Soc.*, 1993, **115**, 11028.
50. A. Banks, G. F. Breen, D. Caine, J. S. Carey, C. Drake, M. A. Forth, A. Gladwin, S. Guelfi, J. F. Hayes, P. Maragni, D. O. Morgan, P. Oxley,

A. Perboni, M. E. Popkin, F. Rawlinson and G. Roux, *Org. Process Res. Dev.*, 2009, **13**, 1130.
51. L. J. Goossen and B. Melzer, *J. Org. Chem.*, 2007, **72**, 7473.
52. C. C. C. Johansson Seechurn, M. O. Kitching, T. J. Colacot and V. Snieckus, *Angew. Chem., Int. Ed.*, 2012, **51**, 5062.
53. M. M. Heravi and P. Hajiabbasi, *Monatsh. Chem.*, 2012, **143**, 1575.
54. J. Terao and N. Kambe, *Acc. Chem. Res.*, 2008, **41**, 1545.
55. I. Bauer and H.-J. Knölker, *Chem. Rev.*, 2015, **115**, 3170.
56. M. Kumada, *Pure Appl. Chem.*, 1980, **52**, 669.
57. F. Bartoccini, G. Piersanti, S. Armaroli, A. Cerri and W. Cabri, *Tetrahedron Lett.*, 2014, **55**, 1376.
58. B. L. H. Taylor, M. R. Harris and E. R. Jarvo, *Angew. Chem., Int. Ed.*, 2012, **51**, 7790.
59. S. Gangula, U. K. Neelam, S. R. Baddam, V. H. Dahanukar and R. Bandichhor, *Org. Process Res. Dev.*, 2015, **19**, 470.
60. L. Wu, J.-C. Zhong, S.-K. Liu, F.-P. Liu, Z.-D. Gao, M. Wang and Q.-H. Bian, *Tetrahedron: Asymmetry*, 2016, **27**, 78.
61. C. C. C. Johansson and T. J. Colacot, *Angew. Chem., Int. Ed.*, 2010, **49**, 676.
62. D. Prim, S. Marque, A. Gaucher and J.-M. Campagne, in *Organic Reactions*, ed. S. E. Denmark, Wiley, Hoboken, NJ, 2012, vol. 76, ch. 2, p. 49.
63. A. J. Burcke and C. S. Marques, *Catalytic Arylation Methods: From the Academic Lab to Industrial Processes*, Wiley-VCH, 2015, ch. 8, p. 376.
64. L. V. Desai, M. B. Hay, D. K. Leahy, C. Wei, D. Fanfair, T. Rosner and Y. Hsiao, *Tetrahedron*, 2013, **69**, 5677.
65. A. E. Gatland, B. S. Pilgrim, P. A. Procopiou and T. J. Donohoe, *Angew. Chem., Int. Ed.*, 2014, **53**, 14555.
66. F. Calogero, P. Allegrini, E. Attolino and D. Passarella, *Eur. J. Org. Chem.*, 2015, 6011.
67. G. Berthon, and T. Hayashi, in *Catalytic Asymmetric Conjugate Reactions*, ed. A. Córdova, Wiley-VCH, Weinheim, 2010, p. 1.
68. H. J. Edwards, J. D. Hargrave, S. D. Penrose and C. G. Frost, *Chem. Soc. Rev.*, 2010, **39**, 2093.
69. T. Hayashi, M. Takahashi, Y. Takaya and M. Ogasawara, *J. Am. Chem. Soc.*, 2002, **124**, 5052.
70. S. Brock, D. R. J. Hose, J. D. Moseley, A. J. Parker, I. Patel and A. J. Williams, *Org. Process Res. Dev.*, 2008, **12**, 496.
71. K.-C. Huang, B. Gopula, T.-S. Kuo, C.-W. Chiang, P.-Y. Wu, J. P. Henschke and H.-L. Wu, *Org. Lett.*, 2013, **15**, 5730.
72. L. Kürti and B. Czakó, *Strategic Applications of Named Reactions in Organic Synthesis*, Elsevier, San Diego, California, 2005, p. 458.
73. Z. Wang, *Comprehensive Organic Name Reactions and Reagents*, John Wiley & Sons, 2010, 2812.
74. B. M. Trost, *Org. Process Res. Dev.*, 2012, **16**, 185.
75. J. Tsuji, H. Takahashi and M. Morikawa, *Tetrahedron Lett.*, 1965, **49**, 4387.

76. B. M. Trost and P. E. Strege, *J. Am. Chem. Soc.*, 1977, **99**, 1649.
77. B. Aakermark, S. Hansson, B. Krakenberger, A. Vitagliano and K. Zetterberg, Organometallics, 1984, **3**, 679.
78. D. J. Jansen and R. A. Shenvi, *J. Am. Chem. Soc.*, 2013, **135**, 1209.
79. M. Vitalal, G. Prestat, D. Lopes, D. Madec, C. Kammerer, G. Poli and L. Girnita, *J. Org. Chem.*, 2008, **73**, 5795.
80. C. Kammerer-Pentier, A. D. Martinez, J. Oble, G. Prestat, P. Merino and G. Poli, *J. Organomet. Chem.*, 2012, **714**, 53.
81. X.-F. Wu, H. Neumann and M. Beller, *Chem. Soc. Rev.*, 2011, **40**, 4986.
82. J. Kim, H. J. Kim and S. Chang, *Angew. Chem., Int. Ed.*, 2012, **51**, 11948.
83. K. Takagi, T. Okamoto, Y. Sakakibara, A. Ohno, S. Oka and N. Hayama, *Bull. Chem. Soc. Jpn.*, 1976, **49**, 3177.
84. M. Sundermeier, A. Zapf and M. Beller, *Angew. Chem., Int. Ed.*, 2003, **42**, 1661.
85. S. Erhardt, V. V. Grushin, A. H. Kilpatrick, S. A. Macgregor, W. J. Marshall and D. C. Roe, *J. Am. Chem. Soc.*, 2008, **130**, 4828.
86. K. P. Bøgesø and C. Sánchez, *Analogue-Based Drug Discovery III*, ed. J. Fischer, C. R. Ganellin and D. P. Rotella, Wiley-VCH, Weinheim, 1st edn, 2013, ch. 11, pp. 269–294.
87. S. M. Bech, O. Nielsen, H. Petersen, H. Ahmadian, H. Pedersen, P. Brosen, F. Geiser, J. Lee, G. Cox, O. Dapremont, C. Suteu, S. P. Assenza, S. Hariharan and U. Nair, *WO Pat.*, 2003006449, 2003.
88. M. Utsugi, H. Ozawa, E. Toyofuku and M. Hatsuda, *Org. Process Res. Dev.*, 2014, **18**, 693.
89. For reviews on the NHK reaction see (a) G. C. Hargaden and P. J. Guiry, *Adv. Synth. Catal.*, 2007, **349**, 2407; (b) Z. Wang, *Comprehensive Organic Name Reactions and Reagents*, John Wiley & Sons, Hoboken, 2010, vol. 3, pp. 2076–2080; (c) L. Kürti and B. Czakó, *Strategic Applications of Named Reactions in Organic Synthesis*, Elsevier, San Diego, California, 2005, pp. 318–319.
90. Y. Okude, S. Hirano, T. Hiyama and H. Nozaki, *J. Am. Chem. Soc.*, 1977, **99**, 3179.
91. A. Fürstner and N. Shi, *J. Am. Chem. Soc.*, 1996, **118**, 2533.
92. G. R. Pettit, C. L. Herald, M. R. Boyd, J. E. Leed, C. Dufresne, D. L. Doubek, J. M. Schmidt, R. L. Cerny, J. N. A. Hooper and K. C. Rützler, *J. Med. Chem.*, 1990, **34**, 3339.
93. T. D. Aicher, K. R. Buszek, F. G. Fang, C. J. Forsyth, S. H. Jung, Y. Kishi, M. C. Matelich, P. M. Scola, D. M. Spero and S. K. Koon, *J. Am. Chem. Soc.*, 1992, **114**, 3162.
94. B. C. Austad, F. Benayoud, T. L. Calkins, S. Campagna, C. E. Chase, H.-W. Choi, W. Christ, R. Costanzo, J. Cutter, A. Endo, F. G. Fang, Y. Hu, B. M. Lewis, M. D. Lewis, S. McKenna, T. A. Noland, J. D. Orr, M. Pesant, M. J. Schnaderbeck, G. D. Wilkie, T. Abe, N. Asai, Y. Asai, A. Kayano, Y. Kimoto, Y. Komatsu, M. Kubota, H. Kuroda, M. Mizuno, T. Nakamura, T. Omae, N. Ozeki, T. Suzuki, T. Takigawa, T. Watanabe and K. Yoshizawa, *Synlett*, 2013, **24**, 327.

95. T. Fukuyama, H. Chiba, T. Takigawa, Y. Komatsu, A. Kayano and K. Tagami, *Org. Process Res. Dev.*, 2016, **20**, 100.
96. A. Bach, G.-P. Chen, W. Chen, P. Geng, G. T. Lee, E. Loeser, J. McKenna, F. R. Kinder, Jr., K. Konigsberger, K. Prasad, T. M. Ramsey, N. Reel, O. Repic, L. Rogers, W.-C. Shieh, R.-M. Wang, L. Waykole, S. Xue, G. Florence and I. Paterson, *Org. Process Res. Dev.*, 2004, **8**, 113.
97. A. K. Ghosh and P. R. Nyalapatla, *Org. Lett.*, 2016, **18**, 2296.
98. J. F. Hartwig, *J. Am. Chem. Soc.*, 2016, **138**, 2.
99. H. M. L. Davies and D. Morton, *J. Org. Chem.*, 2016, **81**, 343.
100. G. B. Shul'pin, *Catalysts*, 2016, **6**, 50.
101. X. Chen, K. M. Engle, D.-H. Wang and J.-Q. Yu, *Angew. Chem., Int. Ed.*, 2009, **48**, 5094.
102. O. Daugulis, *Top. Curr. Chem.*, 2010, **292**, 57.
103. A. J. L. Pombeiro, in *Advances in Organometallic Chemistry and Catalysis: The Silver/Gold Jubilee ICOMC Celebratory Book*, ed. A. J. L. Pombeiro, J. Wiley & Sons, 2013, ch. 2, pp. 15–25.
104. E. J. Kiser, J. Magano, R. J. Shine and M. H. Chen, *Org. Process Res. Dev.*, 2012, **16**, 255.
105. J. Bian, X. Qian, N. Wang, T. Mu, X. Li, H. Sun, L. Zhang, Q. You and X. Zhang, *Org. Lett.*, 2015, **17**, 3410.
106. K. Kuroda, S. Tsuyumine and T. Kodama, *Org. Process Res. Dev.*, 2016, **20**, 1053.

CHAPTER 9

Metal-catalysed Metathesis Reactions for Greener Synthon/Drug Synthesis

E. N. DOS SANTOS,* A. V. GRANATO AND A. G. SANTOS

Universidade Federal de Minas Gerais, Departamento de Química-ICEx, Av. Antônio Carlos 6627, Belo Horizonte 31270-901, Brazil
*Email: nicolau@ufmg.br

9.1 Introduction

Olefin metathesis is a catalysed reaction that involves the breaking and re-generation of carbon–carbon double bonds (C=C) in a way that the ylidene ($=CR^1R^2$) fragments are exchanged between different molecules. The word metathesis comes from the Greek for "change position". A simple example is the metathesis between ethylene and but-2-ene resulting in propene, shown in Figure 9.1. Formally, the fragment $=CH_2$ in ethylene changes position with $=CHCH_3$ in the but-2-ene molecule. This reaction is currently used to supply the growing worldwide demand for propene and is operated at a million ton scale.[1]

More recently, this reaction acquired an important status in green chemistry through the introduction of bio-refineries based on plant oils. Linear olefins, usually obtained from petroleum, are obtained from renewable sources along with new bio-renewable feedstock for polymers, lubricants, surfactants, *etc.* It is worth mentioning that these bio-refineries operate under mild reaction conditions, which increases their

Green Chemistry Series No. 54
Sustainable Synthesis of Pharmaceuticals: Using Transition Metal Complexes as Catalysts
Edited by Mariette M. Pereira and Mário J. F. Calvete

Published by the Royal Society of Chemistry, www.rsc.org

cost-effectiveness and lowers their carbon footprint.[2] Furthermore, metathesis can easily integrate tandem sequences, in which the catalyst or its derivatives catalyse additional reactions, making the synthetic routes even more efficient.[3]

Owing to the utmost industrial relevance of olefin metathesis, the 2005 Nobel Prize in Chemistry was awarded to R. Grubbs (Caltech), R. Schrock (MIT), and I. Chauvin (IFP) for their contribution to the field.[4] Besides large-scale petrochemical applications, its importance to fine chemicals and pharmaceuticals is increasing and will be the focus of this chapter.

This reaction offers more possibilities than the example in Figure 9.1 or the definition suggests at first glance. Metathesis is classified into certain sub-types, which facilitates appreciating its synthetic potential. In Figure 9.2, cross metathesis (CM), ring-closing metathesis (RCM), ring-opening metathesis (ROM), ring-opening metathesis polymerization (ROMP), and acyclic diene polymerization metathesis (ADMET) are generically represented. Although not immediately obvious, all these transformations would fit into the definition of olefin metathesis and can be accomplished with metathesis catalysts. When two olefin molecules of the same type react to produce a different one, the reaction is named self-metathesis (SM). For example, the reverse reaction of Figure 9.1, *i.e.*, the transformation of two molecules of propene into ethylene and but-2-ene, is classified as SM.

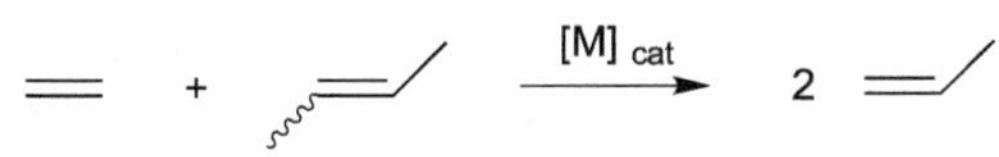

Figure 9.1 Metathesis of ethylene and but-2-ene.

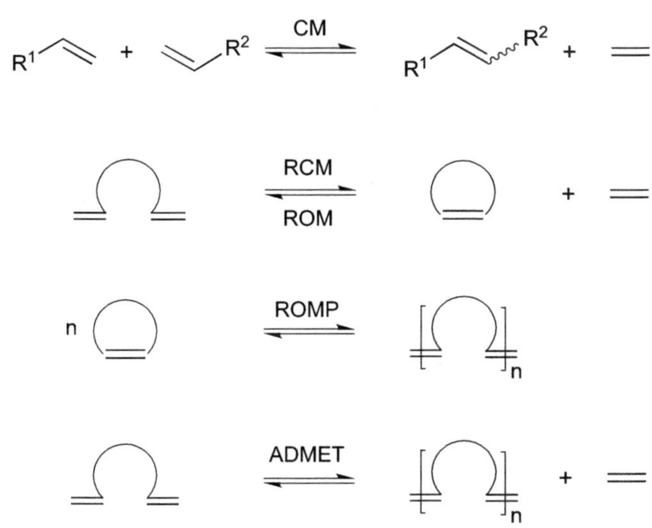

Figure 9.2 Types of metathesis reactions.

While ROMP and ADMET are employed in the specialty polymer industry, important applications of ROM are found for fine chemicals. RCM is particularly important for pharmaceuticals and CM is widely employed both in the commodities and fine chemicals industries.[1,5]

9.2 Mechanistic Aspects

Metathesis is a reaction catalysed by a metal–carbene (metal–ylidene) species and can be classified as organometallic catalysis. The catalytic site can be generically represented by the formula $L_nM{=}CR^1R^2$, in which M is a transition metal centre in a compound (either a Werner complex or a metal oxide), $=CR^1R^2$ is an ylidene ligand (a Schrock-type carbene), and L_n represents the other ligands in the coordination sphere of the metal. The most useful metals for metathesis are Mo, W, Re, and Ru, as detailed in the next section. Chauvin and Hérrison[6] played a decisive role in the elucidation of the mechanism currently accepted for metathesis, which is depicted in Figure 9.3.

The catalytically active species is either formed *in situ* through the combination of appropriate precursors or by a simple ligand loss in a well-defined transition metal complex containing an ylidene moiety. In both cases, a metal compound containing an ylidene ligand next to a vacant coordination site is formed (**A**), and catalysis takes place. The C=C of one reactant coordinates the vacant site to form **B** (step a), the fragments ylidene and C=C react in the coordination sphere of the metal to form the metallacyclobutane **C** (step b), which is the key intermediate. This intermediate can be split in two manners: either reverting step b or at the dashed line, following step c. The latter manner leads to the formation of **D** with a new

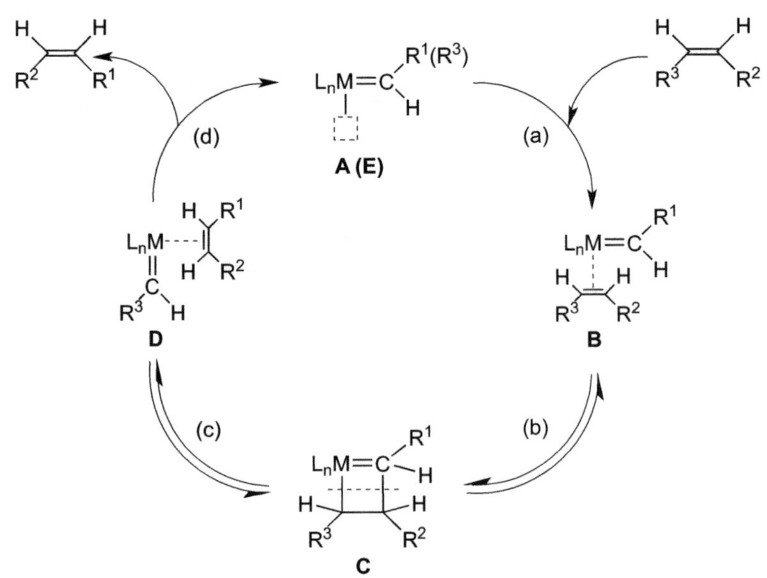

Figure 9.3 General mechanism for metathesis.

C=C and a new ylidene moiety. The dissociation of C=C from the co-ordination sphere of the metal (step d) leads to the formation of a new olefin and regenerates the catalytic species. Rigorously speaking, the cycle is not closed in Figure 9.3 because the metal–ylidene species formed in step **d** has a different substituent (R^3) to the original one (R^1), but **E** will make an equivalent cycle reacting with another molecule containing a C=C, and eventually **A** will be formed again. At this point, it must be clear that the catalyst promotes the exchange of $=CR^1R^2$ fragments in compounds containing C=C double bonds. To predict the possible products, it is useful to imagine cutting the C=C in half and making all possible combinations with the resulting fragments.

It is important to point out that there is no significant change in the enthalpy of the reaction when both reactants and products are olefins. No elementary step in the catalytic cycle has a high energy barrier and thus the reaction tends to reach thermodynamic equilibrium. Strategies such as the use of an excess of one reactant or the withdrawal of one of the products are particularly useful to increase yields. In the example of Figure 9.1, to get higher conversion of but-2-ene, a high partial pressure of ethylene is employed and the unreacted ethylene is recovered by distillation and then recycled.

9.3 Catalysts for Metathesis

The heart of the metathesis reaction is the catalyst. Although there is no such a thing as "the best metathesis catalyst", the choice of the appropriate catalyst is determinant for the success of a particular transformation. The molecular environment of the catalytic site (ligands) is critical to selectivity, as well as to stability. At first, metathesis catalysts were limited to metal oxides of various early transition metals, such as W, Mo, and Re, supported on solids such as alumina or silica (heterogeneous catalysis) and were activated at quite high temperatures in the presence of the reactants. This type of catalyst is still in use in various large-scale industrial processes, but they are quite sensitive to impurities and have low tolerance to functional groups, which limits their use to petrochemical processes involving highly purified olefins.[7] Furthermore, some combinations of soluble early transition metal chlorides and oxochlorides with activating organometallic compounds from group 13 or 14 rendered molecular catalyst systems that were active at temperatures below 100 °C in solution (homogeneous catalysis). Again, these systems presented low tolerance to functional groups and low catalyst stability.[1]

Inspired by the early findings in olefin metathesis and the mechanistic elucidation by Chauvin, the groups of Schrock (MIT)[8] and Grubbs (Caltech)[9] started searching for stable metal–carbene complexes capable of catalysing the metathesis reaction. As they were introduced, such catalysts started to be referred to as "well-defined", as opposed to the above-mentioned "ill-defined" combinations. They presented higher tolerance to functional groups and impurities, and worked at lower temperatures. Although their synthesis involved rather complex organometallic manipulations, once synthesised, they could be quite easily employed in organic synthesis. The introduction of the well-defined molecular catalysts contributed decisively to the inclusion of metathesis in the toolbox of modern organic chemistry. The well-defined catalysts are coordination complexes (Werner complexes), *i.e.*, they have a transition metal central atom surrounded by molecules or ions that donate electrons (ligands), and are organometallics because they have at least one metal–carbon bond. Presently, hundreds of well-defined molecular catalysts capable of catalysing olefin metathesis are described in the literature. Several are commercially available in mg to g scale through the major chemicals retailers.[5] As a general tendency, the tolerance to impurities and functional groups usually increases in the order W < Mo < Ru. W catalysts can tolerate esters and amides as functional groups, Mo catalysts tolerate ketones in addition, and Ru catalysts can further tolerate alcohols and carboxylic acids. The most useful catalysts have some resemblance to the so-called Schrock, Grubbs and Hoveyda's metathesis catalysts that will be presented hereafter. The variations can merely be aimed to circumvent intellectual property issues or can indeed introduce effects that improve the performance of the catalyst or facilitate its extraction/recovery.

9.3.1 Ruthenium Catalysts with Well-defined Structures

Owing to their higher tolerance to functional groups and impurities, Ru complexes (Grubbs-type) are the most commonly used metathesis catalysts. Some Ru metathesis (pre-)catalysts are shown in Figure 9.4. They have an ylidene ligand, two chlorides in *trans* position, a bulky, strong sigma-donor ligand that stays coordinated throughout the catalytic cycle, and a stabilizing ligand, which is necessary to keep the pre-catalyst stable during synthesis and storage, but that will leave the coordination sphere of the metal to form the catalytically active species. These leaving ligands stay in the reaction mixture and can interfere with the stability and activity of these catalysts.

The so-called first-generation Grubbs' catalyst (**Ru-I**) is among the cheapest versions of commercial Ru catalysts, being the archetype of many variants. It is still the catalyst of choice for several important applications, but its efficiency is limited to alkyl-substituted double bonds with low steric constraints. Its activation requires the release of one PCy_3, which stays in the reaction mixture competing with the reactant for the vacant coordination site of the metal. **Ru-II** is known as a second-generation Grubbs' catalyst and features an N-heterocyclic carbene (NHC), in this example the 1,3-bis(2,4,6-trimethylphenyl)-4,5-dihydroimidazolyl-ylidene, which is a bulky, strongly sigma-donor ligand. The introduction of NHC ligands made the catalysts more reactive, allowing reaction with C=C substituted with electron-withdrawing groups such as acrylates and nitriles. The activation of the catalyst still depends on PCy_3 dissociation, but the resulting unsaturated Ru complex reacts faster with olefins than with PCy_3. Nevertheless, this ligand

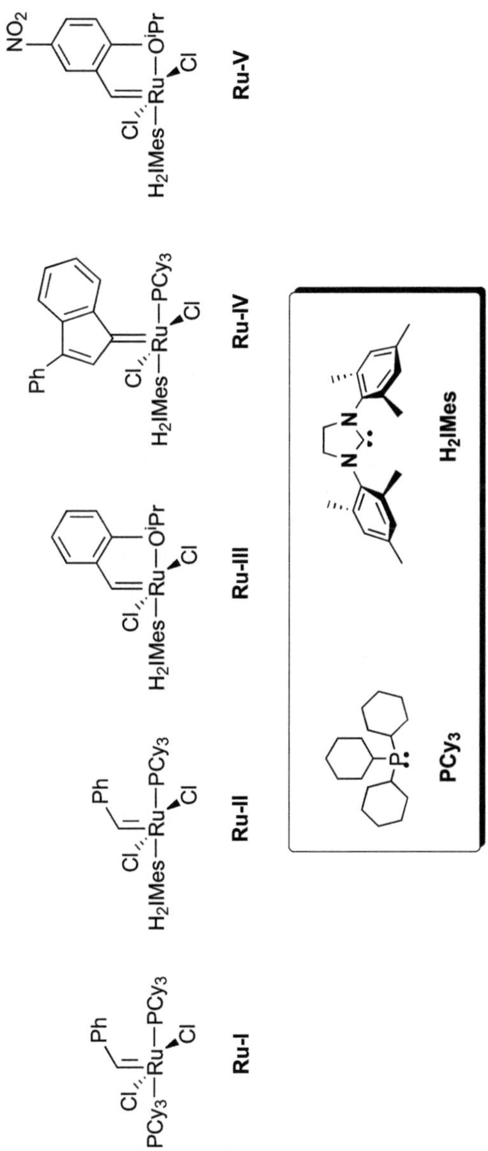

Figure 9.4 Grubbs' type metathesis catalysts.

may enable various decomposition pathways of the catalyst. The second-generation Hoveyda–Grubbs' catalyst (**Ru-III**) is currently among the most used metathesis catalysts because of its higher activity and improved stability, allowing a broader range of application. The alkylidene-ether moiety has both the function of initiator and stabilizing ligand. The unsaturated ether liberated in the initiation step does not contribute to catalyst decomposition pathways; rather, it may recapture the ruthenium active species forming a catalyst reservoir in the so-called "boomerang effect". Nevertheless, the price of **Ru-III** is considerably higher than that of **Ru-II**, from which it is synthesised. **Ru-IV** is considered similar to **Ru-II**, and **Ru-V** is similar to **Ru-III**, and they can perform better than their equivalents depending on the reactants and reaction conditions employed.

9.3.2 Tungsten and Molybdenum Catalysts of Well-defined Structures

The basic structure of Schrock's catalysts has tungsten or molybdenum in a high oxidation state as the central atom, a bulky imido ligand, two alkoxides, and one alkylidene as ligands. Complexes of the kind **Mo-I** are examples of highly active catalysts (Figure 9.5). The imido ligand is formally a four-electron donor and keeps the coordination number of the complex at four in a pseudo-tetrahedral geometry. There is no need for a ligand loss for initiation, which makes them very useful in specialty polymer synthesis *via* ROMP and ADMET. The bulkiness of the imido ligands helps to prevent the bimolecular deactivation pathways and the alkoxides containing electron-withdrawing substituents facilitate the olefin coordination by enhancing the Lewis acidity at the metal centre. Catalysts of the kind **Mo-II** have a chiral moiety and can be used in enantioselective reactions. Catalysts of the kind **Mo-III** have an alkoxide and a pyrrolide ligand. They are more robust than **Mo-I** or **Mo-II** and particularly good when *cis* (*Z*) selectivity is required in the C=C formation.

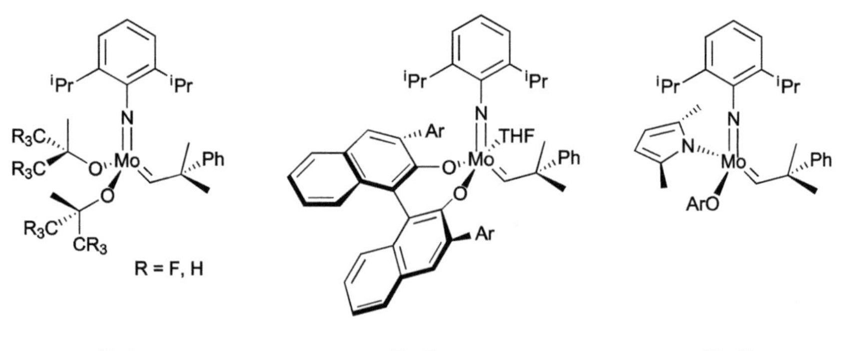

Figure 9.5 Schrock's type metathesis catalysts.

These complexes have the advantage of being highly active for metathesis, but are quite sensitive to impurities and need to be manipulated under a high-quality inert atmosphere. Tungsten complexes are more active, but their higher sensitivity to impurities and lower tolerance to functional groups in the substrate make the molybdenum ones more useful. Recently, versions stabilized with bidentate N-heterocyclic ligands were developed, but they require the addition of a Lewis acid to be activated.[10] A more practical approach is a formulation containing a pre-weighted molybdenum metathesis catalyst in paraffin tablets, which allow the storage and manipulation of the catalyst under more ordinary conditions, employing regular Schlenk techniques.[11] Owing to these improvements and their good selectivity in enantioselective (**Mo-II**) and *Z*-selective metathesis (**Mo-III**), more widespread use of these catalysts can be expected in the next future.

9.3.3 Molecular Catalyst Stability

In the scientific literature, the employment of 3–20 mol% (related to reactant) of metathesis catalyst is rather typical. This implies that the catalyst undergoes less than 30 cycles before its deactivation. For industrial applications, the percentage in mol of catalyst that can be used is much lower. Even for fine chemicals, 0.5 mol% of catalyst is a good guess for the upper limit. The retailing prices of the catalysts are not negligible, some may exceed GBP£400 per gram, but often the reactants themselves may have taken several synthetic steps before the metathesis, and for target-oriented synthesis there is a tendency to use an excess of catalyst to guarantee high yields. Nevertheless, a high catalyst concentration can accelerate its own decomposition paths and its products of decomposition may catalyse undesired reactions. Thus, a large increase in the amount of catalyst may lead to a modest increase in the yield.[12]

One of the main decomposition pathways of the catalyst is its bimolecular reaction to form an olefin and a metathesis-inactive complex (Figure 9.6). Thus, a high concentration of catalysts should be avoided.

The reactions involving terminal C=C lead to metal–methylidene intermediates ($M{=}CH_2$) that are particularly sensitive to the nucleophilic attack of adventitious Lewis bases such as amine impurities or even ligands originated from the pre-catalyst, such as PCy_3 in complexes **Ru-I** and **Ru-II**.

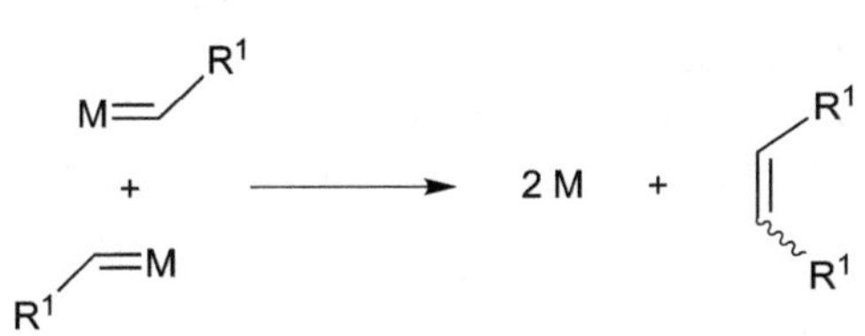

Figure 9.6 Bimolecular decomposition path of metathesis catalysts.

Water is deleterious to W or Mo catalysts. Impurities such as peroxides, Lewis bases (especially amines and phosphines), and molecular oxygen are deleterious even for Ru catalysts. Typical pre-treatments of the reactants to remove impurities are: passing through a neutral alumina pad (will remove peroxides), bubbling with inert gas (will remove molecular oxygen), or heating with magnesium silicate (*e.g.* Magnesol®) and silica (celite) (will destroy peroxides and will remove basic compounds such as amines).[13]

The decomposition products of the catalyst may lead to undesired consecutive reactions, such as isomerization, which is promoted by Ru–hydride species. Some isomerization products can act as inhibitors because they form Fischer-type carbenes that are quite unreactive for metathesis. To prevent isomerization, hydride quenchers such as quinones and dienes can be added to the reaction mixture.[14]

9.3.4 Catalyst Residue Removal

Residues of metals are problematic for some applications, particularly in the pharmaceutical industry. The amounts allowed by regulations are usually as low as <10 ppm in the final active pharmaceutical ingredient (API). As ruthenium-catalysed C=C metathesis has been employed both at the development stage and in industrial applications of many APIs, several methods of reducing ruthenium content to the limit level in the final products were applied and tested on an industrial scale. Normally they involve the addition of inexpensive polar ligands that will bind the catalyst and will allow its extraction along with other polar impurities. Examples of useful ligands are imidazole, 2-mercaptonicotinic acid (MNA), cysteine, and tris(hydroxymethyl)phosphine (THMP). Some of these ligands play an additional role of quenching the catalyst, which is useful to avoid loss in yield by consecutive reaction during, *e.g.*, the concentration of the reaction mixture to crystallise the product.[15] In an alternative approach, some catalysts may contain an attached polar group in one of their ligands, which allows their ready capture by filtration through a silica gel pad. The catalyst can be extracted from silica with a more polar solvent and used again.[16]

9.4 The Choice of Reaction Conditions

Screening for the best catalyst, while mandatory for industrial applications, is not always feasible in organic synthesis laboratories owing to the low availability of reactants or the unavailability of a catalyst set, *inter alia*. Searching in the literature for previous screening of similar reactions is obviously worthwhile. In the absence of these possibilities or information, **Ru-III** (and their variants) is a good first attempt. Nevertheless, the price of **Ru-III** may be significant even for lab-scale synthesis and is considerably higher than second-generation Grubbs' catalyst (**Ru-II**). Using the latter with enabling agents such as CuI or poly(4-vinylphenol)[12] is a way of increasing reactivity while employing a cheaper version of the catalyst. Using a large amount of catalyst added at once leads to a higher usage because the bimolecular mechanisms of deactivation are accelerated. Adding several smaller portions usually leads to better results in terms of yield *vs.* catalyst usage.

In industrial applications, besides the impact of the catalyst's cost on the final product, intellectual property issues, which are presently rather intricate, should also be considered.[5] The situation will improve since the patents of the most useful catalysts are going to expire in the next few years. The catalyst screening should be performed with reactants in the purity provided at industrial scale, or the purification costs should be taken into account in the process.

The temperature of the reaction typically ranges from 20 to 110 °C. In this range, the increase in temperature obviously makes the reaction faster but may lead to faster decomposition of the catalyst and selectivity loss owing to consecutive C=C isomerization. Metathesis is reversible and the thermodynamic equilibrium tends to be reached in detriment to the yield. The removal of volatile co-products can increase the yield for the desired product. Furthermore, some side products, such as ethylene, are detrimental to the stability of the catalyst because they result in metal–methylidene intermediates that are particularly sensitive to bimolecular and unimolecular decomposition pathways, as well as to attack by bases. Bubbling an inert gas in the solution or applying a static vacuum usually removes volatiles such as ethylene, propene and butenes.

Although it is possible to run metathesis in neat substrates, for most of the reactions, the use of a solvent is mandatory. The very efficient and once ubiquitous chlorinated solvents are now classified as hazardous or very hazardous and must be avoided in the context of green chemistry. For Ru catalysts, water, ethyl acetate, methyl-THF, diethylcarbonate and dimethylcarbonate are considered greener alternatives, but their efficiency is rather dependent on the substrate and catalyst.[17] Toluene has been used as a more general alternative: it is less problematic than the chlorinated ones, while still efficient; nevertheless, its production relies on petrochemicals. A promising alternative is the use of *p*-cymene, a solvent that can be obtained from renewable sources, and proved to be efficient in metathesis.[18]

9.5 Selected Examples of Metathesis in (Industrial) Organic Synthesis

There is undeniable interest in the application of metathesis in specialty polymer manufacture, including those of medical and pharmaceutical interest, employing both ROMP and ADMET. Because the mechanism is a living polymerization, it is possible to synthesise block copolymers and, for fast initiation catalysts, polymers with a narrow polydispersity can be obtained.[19] Related topics such as enyne metathesis[20] and alkyne metathesis,[21] although useful in organic synthesis, will not be treated in this chapter. Leading reviews on these subjects can be found elsewhere.[19–21] The examples in this chapter will focus on the applications of CM and RCM in organic synthesis. These sub-types have been integrated into the toolbox of organic chemists for some years[22] and have been enabling the synthesis of compounds that are difficult to reach by alternative synthetic routes. More specifically, the selected examples presented hereafter will focus on applications of industrial interest for pharmaceuticals or important applications in the sustainable chemistry context.

9.5.1 Cross Metathesis

In CM, reaching a high selectivity for the desired cross product is not straightforward. If both cross partners present the same reactivity for a given catalyst, the expected product distribution in a one-to-one molar mixture of reactants is the statistic: 50% of the desired CM product, along with 25% of the self-metathesis (SM) products of each reactant. If one of the reactants is cheap enough, the employment of a molar excess of this reactant will increase the yield of the cross product in relation to the more expensive reactant, leading, nevertheless, to the formation of large amounts of undesired SM co-product. The situation is better if the reactant in excess is not prone to SM with the catalyst employed: at length, more of the desired CM product will be formed with the complete consumption of the limiting reactant, and the reactant in excess can be recycled.[23]

An example of an application of CM in drug discovery is the synthesis of lysine sulfonamide derivatives **4**, which were tested for the treatment of HIV infection (Figure 9.7). Scientists from Merck have prepared a library of over a hundred of such derivatives employing the CM the sulfonamide with the general structure **1** with crotyl ketone derivatives **2**, yielding **3** in 90% yield. Some of the derivatives have shown IC_{50} of up to 0.005 nM.[16]

CM of olefins **5** and **6** was also used in the preparation of a synthetic intermediate **7** of aliskiren (**8**), an antihypertensive drug developed by Novartis and Speedel (Figure 9.8).[16] Although in this example the CM step seems not to be optimized, it brings out the potential for the use of CM in molecules containing chiral centres.

Compound **12** (paroxetine or Paxil) is a selective serotonin reuptake inhibitor, prepared from olefins **9** and **10**, *via* intermediate **11**. It is an API component of, *e.g.*, Seroxat (GlaxoSmithKline), commercialized in the United States. Although the original industrial route did not involve metathesis, an alternative route for its synthesis starts with CM, as depicted in Figure 9.9. The development of efficient new routes can make a competitive differential in the generic drug market when the API patent expires.[24]

The CM of bio-renewable propenylbenzenes **13** with commodity acrylates of type **14** (Figure 9.10) is a route to produce antioxidants of type **15**, useful in personal care formulations. Several cinnamates, including octyl methoxycinnamate, a major UV-A filter used in sunscreen lotions, were produced in up to 99% yield employing 0.5% of **Ru-III** as the catalyst.[25]

Perhaps the most striking current application of CM is in oleochemistry. Metathesis-based bio-refineries are producing olefins and unsaturated

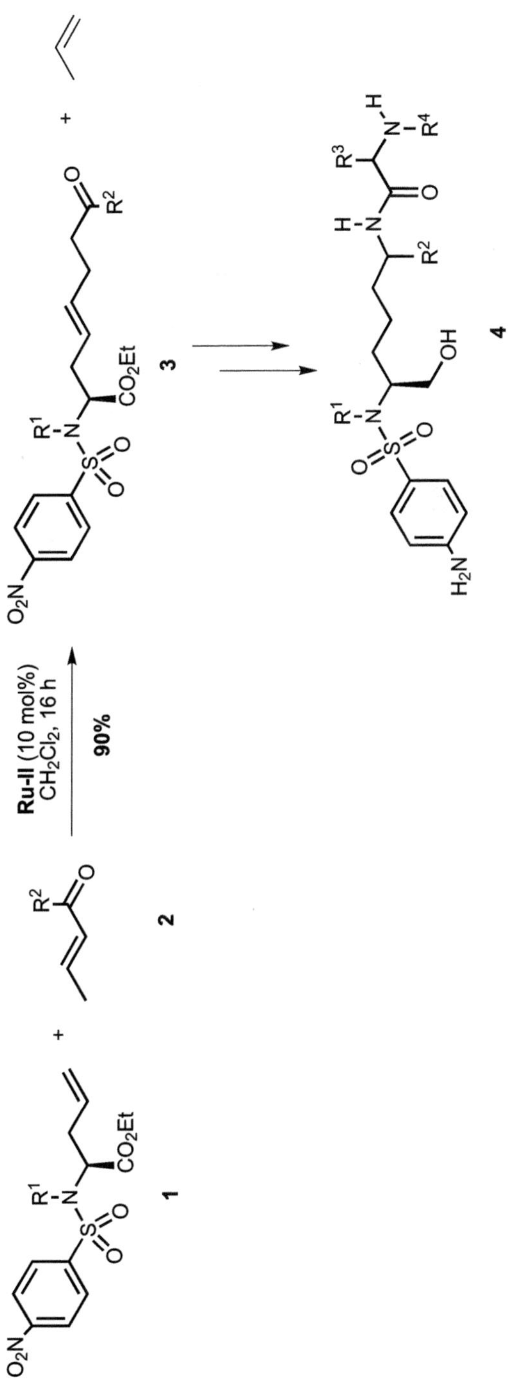

Figure 9.7 Synthesis of lysine sulfonamide derivatives **4**.

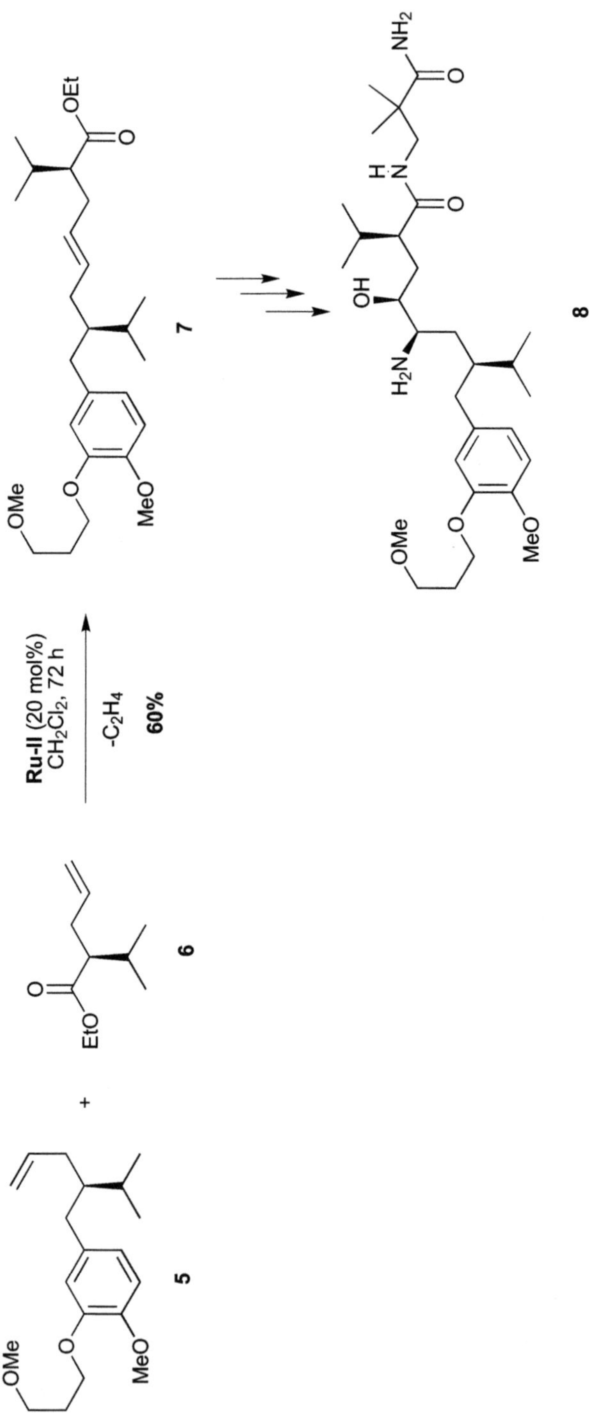

Figure 9.8 Synthetic route for aliskiren (**8**).

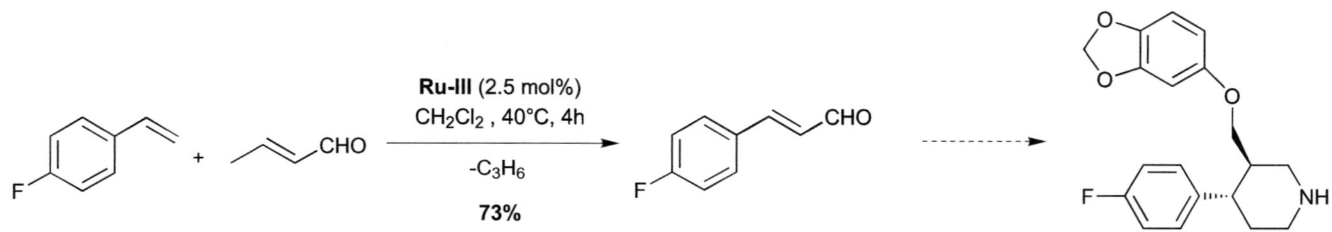

9 10 11 12

Figure 9.9 Synthesis of paroxetine (Paxil) (**12**).

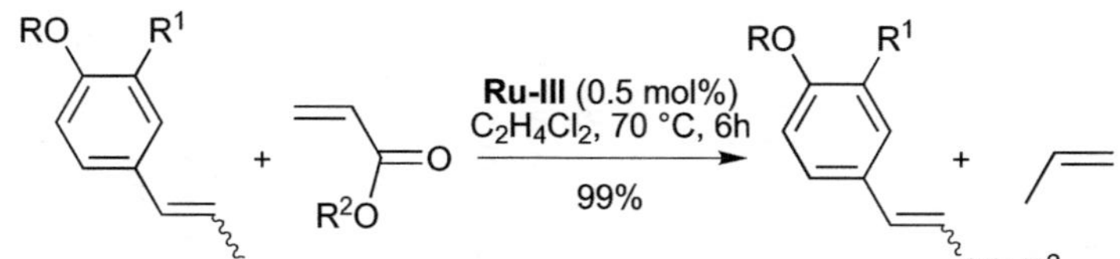

13 14 15

R, R^1		
a: Me, H	anethole	cinnamates
b: H , OMe	isoeugenol	ferulates
c: $-CH_2OCH_2-$	isosafrole	cinnamates

Figure 9.10 Synthesis of antioxidants from bio-renewables employing cross metathesis.

esters from palm oil in a 10^5 ton scale in Indonesia.[2] The classical heterogeneous metal oxide catalysts cannot be employed owing to their low tolerance to impurities and functional groups and thus the process employs well-defined molecular Ru catalysts. Natural triglycerides **16** have unsaturated moieties, such as oleic and linoleic chains. When they are treated with light olefins, such as ethylene, propylene or butenes, in the presence of a Ru metathesis catalyst, the products are olefins **17** and **18** and triglycerides of shorter chains **19** (Figure 9.11). The latter (**19**) can be transesterified into the omega-unsaturated ester **20**, having as side products saturated esters (**21**) and glycerol (**22**). The higher olefins (*e.g.* **10**) are drop-in products: they are identical or equivalent to the petrochemicals and can be directly integrated into their production chain as co-monomers for polyolefin and bases for high-performance lubricants, among other applications. The C_{10} unsaturated esters have been used as platforms for renewable polymers, surfactants and lubricants. As the price of this feedstock is relatively low, new applications in agrochemicals, flavour and fragrances, insect pheromones, and pharmaceuticals are expected.

9.5.2 Ring-closing Metathesis

RCM has been widely used for drug discovery and for the large-scale synthesis of various APIs. Many macrocyclic peptides of pharmaceutical interest, especially those for HCV NS3/4A serine protease inhibitors (Figure 9.12), have been synthesised employing RCM. Boehringer Ingelheim has pioneered the efforts towards scaling up the synthesis of ciluprevir (**23**), which, nevertheless, was halted in phase II trials. Other pharmaceutical companies, using similar approaches, have brought new drugs to the market. Vaniprevir (**24**) (Merck) and simeprevir (**25**) (Janssen Pharma) are synthesised industrially employing RCM in the critical step of their synthetic route.[5,15] It is worth mentioning that the synthesis of such macrocycles through alternative routes would usually increase the number of synthetic steps, making the process less interesting from the sustainability point of view.

Typical issues associated with the metathesis step are co-production of dimers and cyclooligomers of the dienes, isomerization of C=C, and epimerization. To prevent dimerization or cyclooligomerization, high dilutions of reactants (concentrations as low as 0.0025 M) are required in some cases. As a consequence, a large amount of solvent must be used and the cost of its recycling can be significant, in addition to the associated low productivity of the reactors. The problem can be minimized by the slow addition of the diene or by the installation of protecting groups, which induce such a diene conformation that RCM is favoured over oligomerization. Metal complex hydrides are formed by the decomposition of the catalyst in the course of the reaction and can promote the C=C isomerization. Quinones are used as hydride quenchers and their addition to the reaction mixture reduces the concurrent isomerization.[14]

In ciluprevir (**23**) development, significant improvements were made from the original scale-up synthesis to the optimized route (Figure 9.13).

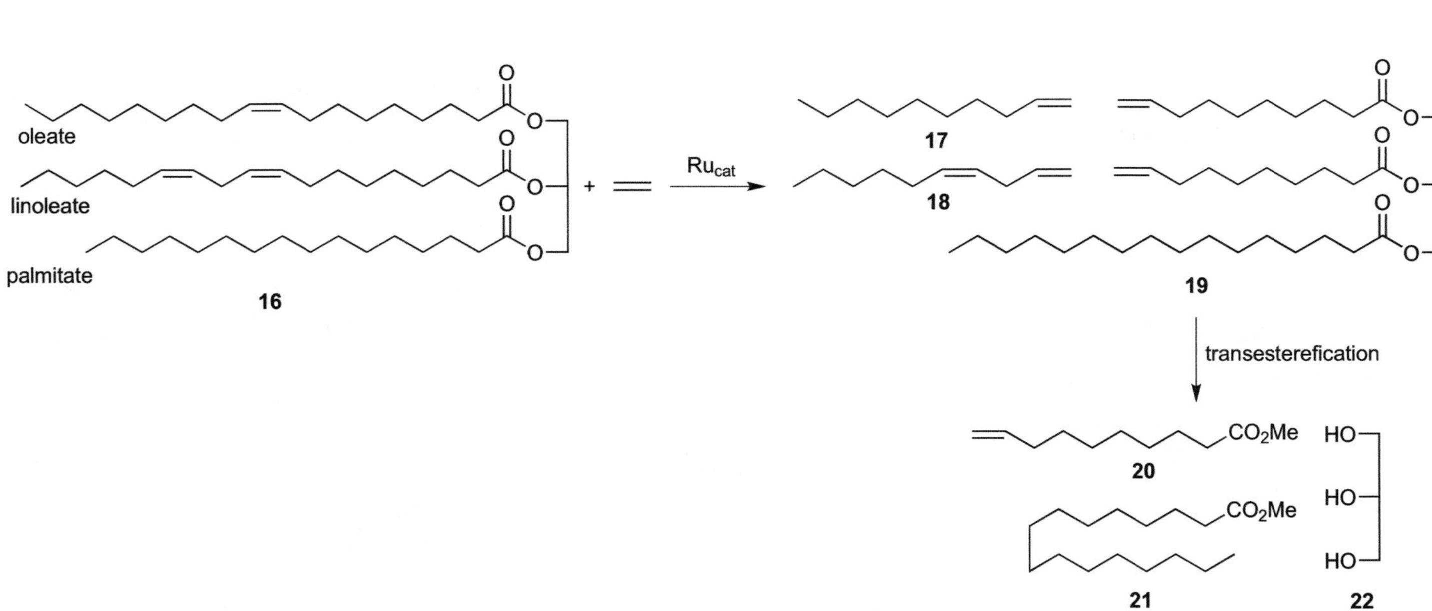

Figure 9.11 Ethenolysis of triacylglycerides.

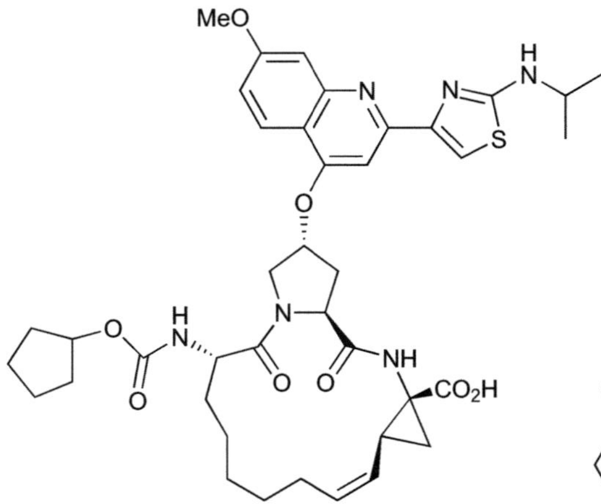

Ciluprevir (23)

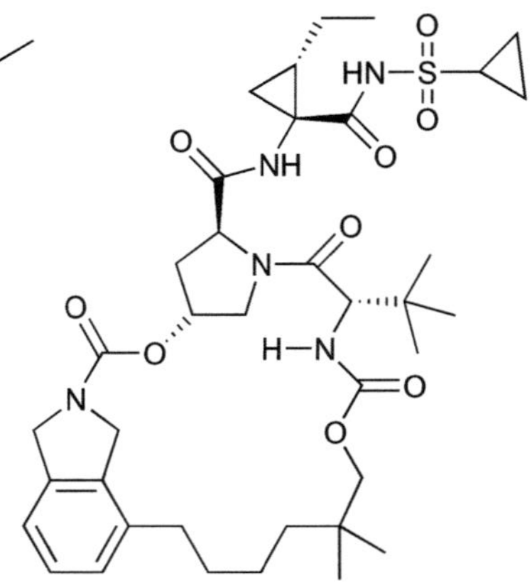

Vaniprevir (24)

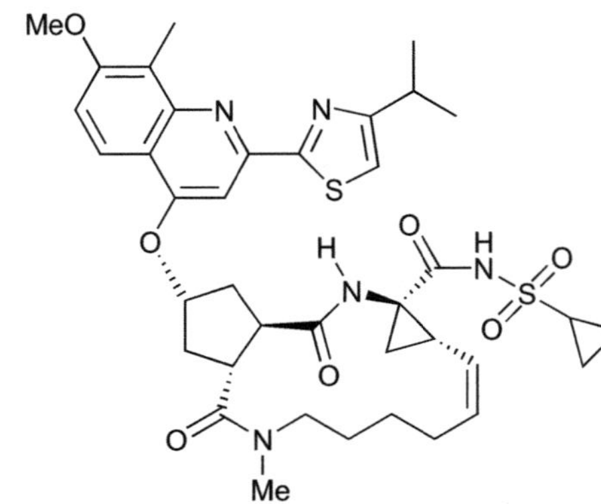

Simeprevir (25)

Figure 9.12 Macrocyclic peptides used as HCV NS3/4A serine protease inhibitors.

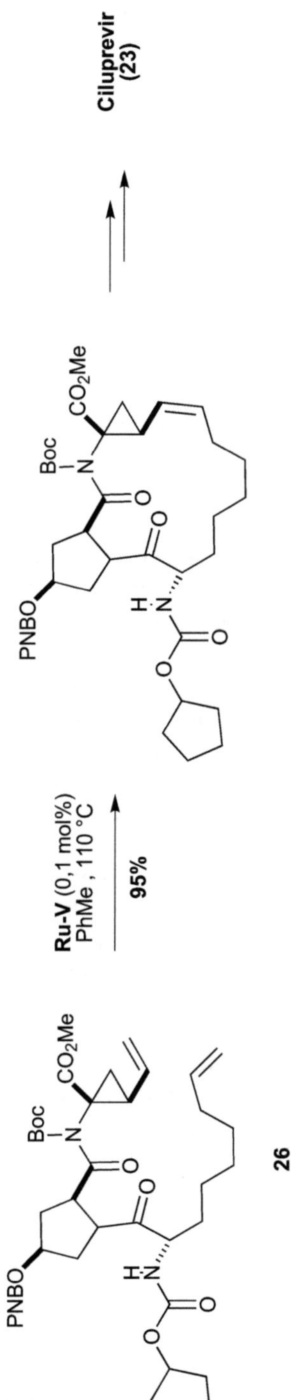

Figure 9.13 Synthesis of ciluprevir (23).

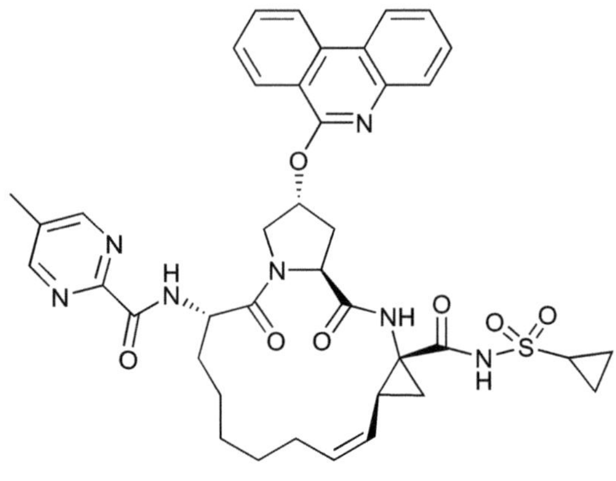

Paritaprevir (27)

Figure 9.14 Paritaprevir (**27**).

The percentage of Ru catalyst was reduced from 5 to 0.1 mol% when **Ru-I** was replaced by **Ru-V** (Figure 9.4), while the concentration of **26** was raised from 0.010 to 0.2 M by the employment of the N-protective *tert*-butyloxycarbonyl (BOC) group, which induced a conformation for **26** that favoured RCM.

Although ciluprevir did not reach the market, the analogous paritaprevir (**27**) (AbbVie) (Figure 9.14) is now approved for the treatment of hepatitis C virus infections in the USA, Canada, and Europe. The details about the synthesis of **27** are not disclosed, but considering that the macrocyclic ring is identical to ciluprevir (**23**), it is likely that the synthetic route involves also RCM.[5]

In the metathesis step of vaniprevir (**24**) synthesis, **Ru-III** was employed in 0.2 mol% at a quite high concentration of **28**. 2,6-Dichloroquinone (10%) was employed to prevent concurrent C=C isomerization (Figure 9.15).

Simeprevir (**25**) is a macrocyclic NS3 protease inhibitor, which can be synthesised employing RCM as a synthetic step (Figure 9.16). The amido group is protected *in situ* by the addition of chlorodifluoroacetic anhydride. The protecting chlorodifluoroacetyl group also favours the efficiency of RCM in **29** by the introduction of conformational restrictions that force the approximation of the two terminal double bonds. The reported initial conditions for this reaction employed a rather high catalyst loading (**Ru-IV** at 6 mol%) when dichloroethane was used as the solvent. Nevertheless, in refluxing toluene and using triethylammonium iodide as a promoter, it was possible to reduce the catalyst loading from 6 to 0.3%.[15]

9.6 Conclusion

With the development of the well-defined molecular catalysts, especially the ruthenium ones, olefin metathesis has been extended to C=C metathesis

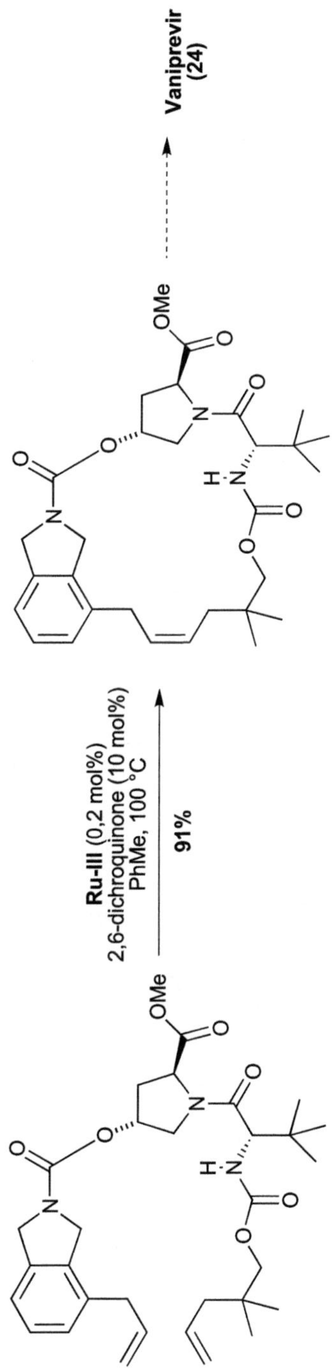

28

Figure 9.15 Synthesis of vaniprevir (24).

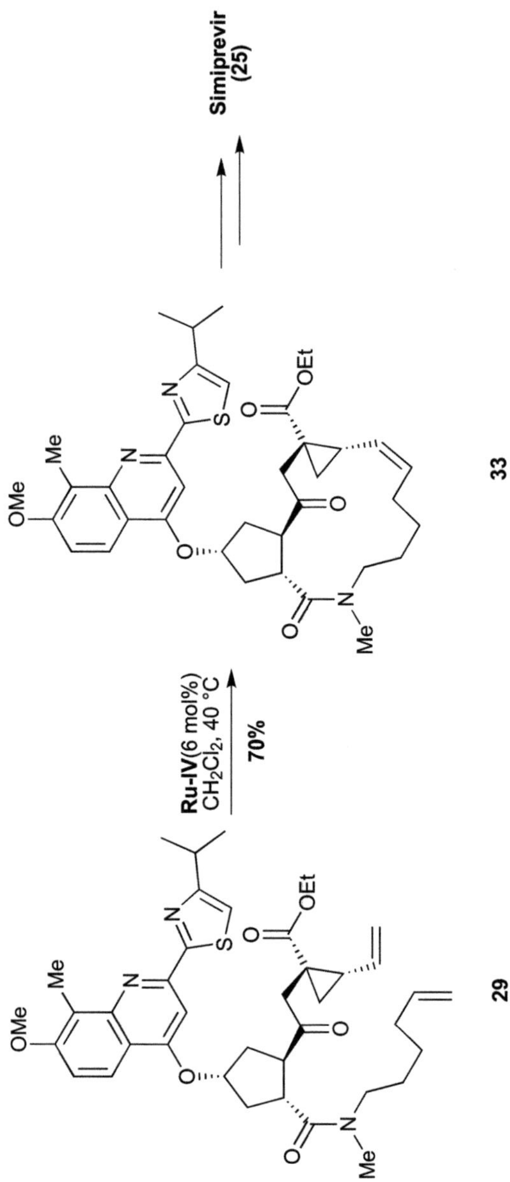

Figure 9.16 Synthesis of simeprevir (25).

and is now integrated into the toolbox of modern organic synthesis. This tool often shortens synthetic pathways and makes them more cost-effective. Various applications of metathesis in the synthesis of active pharmaceutical ingredients have been recently disclosed. It has also been very important for chemical diversity generation and drug discovery, and new applications in the industrial synthesis of APIs are expected. Besides the classical applications in petrochemicals, the new robust catalysts allowed the application of metathesis in bio-renewables such as plant oils. The amount of intermediate and fine chemicals presently generated in these bio-refineries exceeds 150 000 ton per year and with the pressure from society for greener processes, this value is likely to increase significantly. Thus, metathesis will occupy an even more important position in the chemical industry and should be taught at undergraduate and graduate levels for chemistry, chemical engineering, pharmaceutical sciences and other courses to which chemistry is a central area.

Acknowledgements

The authors thank the Brazilian funding agencies CNPq, CAPES and FAPEMIG for supporting scientific projects on the theme of this chapter.

References

1. J. C. Mol, *J. Mol. Catal. A: Chem.*, 2004, **213**, 39.
2. S. Chikkali and S. Mecking, *Angew. Chem., Int. Ed.*, 2012, **51**, 5802.
3. (a) D. E. Fogg and E. N. dos Santos, *Coord. Chem. Rev.*, 2004, **248**, 2365; (b) G. K. Zielinski and K. Grela, *Chem. – Eur. J.*, 2016, **22**, 9440.
4. https://www.nobelprize.org/nobel_prizes/chemistry/laureates/2005/ (last accessed March 2017).
5. C. S. Higman, J. A. Lummiss and D. E. Fogg, *Angew. Chem., Int. Ed.*, 2016, **55**, 3552.
6. J. L. Herisson and Y. Chauvin, *Makromol. Chem.*, 1971, **141**, 161.
7. S. Lwin and I. E. Wachs, *Acs Catal.*, 2014, **4**, 2505.
8. R. R. Schrock, *Angew. Chem., Int. Ed.*, 2006, **45**, 3748.
9. R. H. Grubbs, *Angew. Chem., Int. Ed.*, 2006, **45**, 3760.
10. J. Heppekausen and A. Furstner, *Angew. Chem., Int. Ed.*, 2011, **50**, 7829.
11. L. Ondi, G. M. Nagy, J. B. Czirok, A. Bucsai and G. E. Frater, *Org. Process Res. Dev.*, 2016, **20**, 1709.
12. A. G. Santos, G. A. Bailey, E. N. dos Santos and D. E. Fogg, *ACS Catal.*, 2017, **7**, 3181.
13. L. A. Ferreira and H. S. Schrekker, *Catal. Sci. Technol.*, 2016, **6**, 8138.
14. S. H. Hong, D. P. Sanders, C. W. Lee and R. H. Grubbs, *J. Am. Chem. Soc.*, 2005, **127**, 17160.
15. P. Wheeler, J. H. Phillips and R. L. Pederson, *Org. Process Res. Dev.*, 2016, **20**, 1182.
16. T. K. Olszewski, M. Figlus and M. Bieniek, *Chim. Oggi*, 2014, **32**, 22.

17. K. Skowerski, J. Bialecki, A. Tracz and T. K. Olszewski, *Green Chem.*, 2014, **16**, 1125.
18. A. V. Granato, A. G. Santos and E. N. dos Santos, *ChemSusChem*, 2017, **10**, 1832.
19. R. R. Schrock, *Acc. Chem. Res.*, 2014, **47**, 2457.
20. C. Fischmeister and C. Bruneau, *Beilstein J. Org. Chem.*, 2011, **7**, 156.
21. A. Furstner, *Angew. Chem., Int. Ed.*, 2013, **52**, 2794.
22. S. J. Meek, R. V. O'Brien, J. Llaveria, R. R. Schrock and A. H. Hoveyda, *Nature*, 2011, **471**, 461.
23. A. K. Chatterjee, T. L. Choi, D. P. Sanders and R. H. Grubbs, *J. Am. Chem. Soc.*, 2003, **125**, 11360.
24. R. L. Pederson, I. M. Fellows, T. A. Ung, H. Ishihara and S. P. Hajela, *Adv. Synth. Catal.*, 2002, **344**, 728.
25. J. A. M. Lummiss, K. C. Oliveira, A. M. T. Pranckevicius, A. G. Santos, E. N. dos Santos and D. E. Fogg, *J. Am. Chem. Soc.*, 2012, **134**, 18889.

CHAPTER 10

Tetravalent Boron-based Therapeutics

Q. MENG, M. WANG AND M. G. H. VICENTE*

Louisiana State University, Department of Chemistry, Baton Rouge, LA 70803, USA
*Email: vicente@lsu.edu

10.1 Introduction

A variety of boron compounds are found in nature, mainly in sedimentary rocks, seawater and coastal soils, and many others, for example structurally diverse organoboranes and polyhedral boron hydrides, can be readily synthesized in the laboratory. Boron-containing compounds have a wide range of applications, including in material and life sciences, medicine, energy and electronics.[1–5] In particular, the pharmacological uses of boron-containing compounds have been known for several decades, and many have been investigated in preclinical and clinical studies. Among these, one boron-based drug, bortezomib (**1**, Velcade®),[6] was approved for clinical use in 2003 for the treatment of multiple myeloma and non-Hodgkin's lymphoma, and another, tavaborole (**2**, Kerydin®),[7] was approved in 2014 as a topical anti-fungal agent for the treatment of onychomycosis (Figure 10.1).[8–12] Another important pharmaceutical application of boronated compounds is as boron delivery agents for the boron neutron capture therapy (BNCT) of cancers, particularly for high-grade brain tumors, such as glioblastoma multiforme, and for head and neck cancers.[13–15] BNCT is based on the high nuclear cross-section of non-radioactive ^{10}B nuclei, which upon irradiation with low energy thermal or epithermal neutrons produce high linear energy transfer

Green Chemistry Series No. 54
Sustainable Synthesis of Pharmaceuticals: Using Transition Metal Complexes as Catalysts
Edited by Mariette M. Pereira and Mário J. F. Calvete

Published by the Royal Society of Chemistry, www.rsc.org

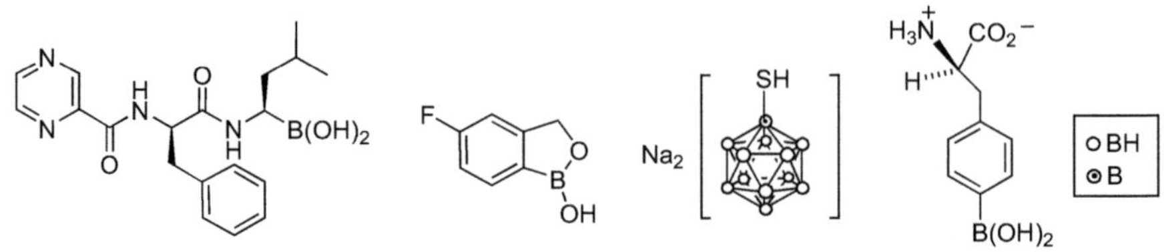

1: Bortezomib **2**: Tavaborole **3**: BSH **4**: BPA

Figure 10.1 Chemical structures of clinically approved boron drugs.

(high-LET) α particles and ^{7}Li nuclei, along with γ radiation and approximately 2.4 MeV of energy. The high-LET α-particles have less than 10 μm path lengths in tissue, which is about the diameter of one cell, so the damage is largely restricted to the ^{10}B-containing cells.[13–15] Two boron-containing compounds are approved for clinical use in BNCT, BSH ($Na_2B_{12}H_{11}SH$, **3**) and BPA (L-4-dihydroxy-borylphenylalanine, **4**), as shown in Figure 10.1.[16–18] Other boron-containing compounds under investigation as potential boron delivery agents for BNCT include boronated amino acids, peptides, monoclonal antibodies, nucleosides, carbohydrates, lipids and porphyrin derivatives.[14,18,19]

Boron has atomic number 5 and three valence electrons, and is classified as a metalloid with intermediate properties between metals and non-metals. Boron forms compounds with many other elements, for example with oxygen, nitrogen, hydrogen, phosphorus, the halogens, and with carbon. Trivalent (sp^2, trigonal planar) boron compounds, such as boric acid [$B(OH)_3$] and analogs, are Lewis acids, readily accepting a pair of electrons to form tetravalent compounds with tetrahedral (sp^3) geometry. In contrast to boric acid scaffolds, boron clusters, such as the polyhedral boron hydrides, are electron-deficient three-dimensional structures, in which boron is bound to five or six other atoms in the cluster. Among the boron clusters, derivatives of the dodecaborate anion ($B_{12}H_{12}^{2-}$) and the neutral icosahedral dicarbadodecaboranes ($C_2B_{10}H_{12}$) are the most used in BNCT and other medical applications due to their remarkably high photochemical, kinetic, and hydrolytic stabilities, high boron content, lipophilicity, and ease of derivatization. In this chapter, we review the synthesis and properties of different types of tetravalent boron therapeutics, with emphasis on boron dipyrromethene (abbreviated as BODIPY) derivatives and other naturally occurring and synthetic boron-containing compounds. Trivalent boron compounds, mainly boronic acid-based therapeutics, have been the subject of several recent reviews,[1,8,19–22] as have the boron cluster-based therapeutics,[23–26] and are therefore not a focus of this chapter.

10.2 Boron Dipyrromethenes

10.2.1 Structure and Properties

The basic structure and nomenclature of BODIPY (4,4-difluoro-4-bora-3a,4a-diaza-s-indacene) are shown in Figure 10.2. BODIPY dyes were first reported

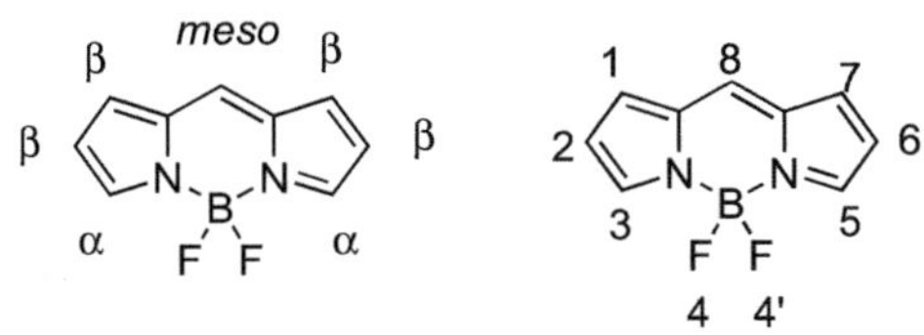

Figure 10.2 BODIPY structure and nomenclature.

in 1968 by Treibs and Kreuzer[27] but their popularity has increased spectacularly in the last two decades due to their many applications, including as labelling agents, chemical sensors, photosensitizers, fluorescent switches, and electroluminescent devices.[28–31] BODIPYs are readily prepared from the complexation of a dipyrromethene molecule with boron, usually accomplished using boron trifluoride diethyl etherate. The dipyrromethene is formed by linking the α-positions of two pyrrole rings *via* a sp^2 methylene carbon, called the *meso* bridge (C8). The complexation of the dipyrromethene with the BF_2 unit forms a rigid system and prevents the *cis*/*trans* isomerization of the dipyrromethene molecule. This structural rigidification allows the conjugation of the π-electron system along the carbon–nitrogen backbone, leading to unusually intense fluorescence quantum yields.[32] Other attractive BODIPY properties include excitation and emission wavelengths in the visible spectral region (>500 nm), large molar absorption coefficients (40 000–110 000 $M^{-1}\,cm^{-1}$), narrow emission bandwidths with high peak intensities, long excited singlet state lifetimes (1–10 ns), relatively high chemical and photochemical stability in both solution and solid states, versatile charge transfer properties, good solubility in most common solvents, and resistance toward aggregation in solution.[28,29] The replacement of C8 in the BODIPY with a nitrogen atom gives a so-called aza-BODIPY structure. In addition to the advantages of normal BODIPY molecules, aza-BODIPYs typically show *ca.* 90 nm red-shift in their absorption and emission bands.[33] Besides being promising agents for photodynamic therapy (PDT), aza-BODIPYs can also be utilized as luminescent chemosensors.[34–36]

The spectroscopic, chemical, and photophysical properties of BODIPY derivatives can be fine-tuned by regioselective functionalization at the peripheral positions on the BODIPY core, as shown in Figure 10.3,[37] or by total synthesis from functionalized pyrrole and *meso*-carbon units. The 2,6-positions bear the least positive charge, therefore they relatively easily undergo electrophilic aromatic substitution reactions, such as halogenation, nitration, sulfonation and formylation.[38–43] On the other hand, the α and β pyrrolic positions can undergo nucleophilic aromatic substitutions when functionalized with a leaving group, such as a halogen.[43–46] Knoevenagel condensation reactions can take place at the 3,5-positions, and subsequently at the 1,7-positions, when these are substituted with methyl groups by reaction with activated aldehydes.[47–49] The 1,2,3,5,6,7-positions can also be regioselectively functionalized *via* Pd(0)-catalyzed cross-coupling reactions,

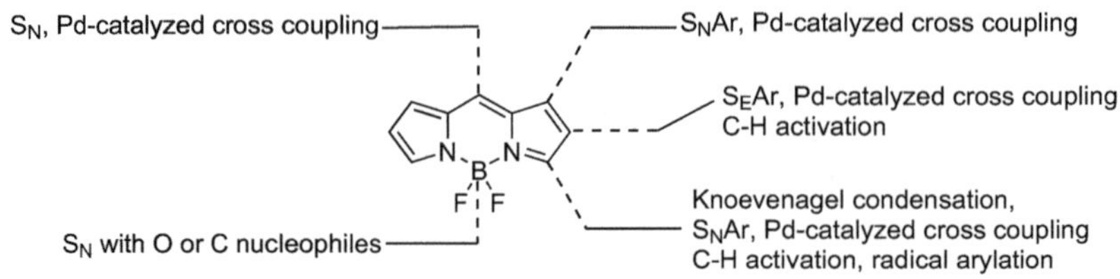

Figure 10.3 Overview of BODIPY functionalization.

such as Suzuki, Stille, Sonogashira and Heck couplings, on halogenated BODIPYs.[43–46,50–56] Functionalization at the *meso*-8-position is often achieved *via* nucleophilic substitutions, or by Pd(0)-catalyzed cross-coupling reactions (see Chapter 8) on 8-halogenated BODIPYs.[43–46,53,55–57] Catalytic amounts of $Pd(PPh_3)_4$ or $Pd(PPh_3)_2Cl_2$ are employed as the most common catalysts among the cross-coupling reactions at the carbon 1,2,3,5,6,7,8-positions, and most catalysts can be recycled after the cross-coupling reactions *via* filtration. In some cases, $Pd(dppf)Cl_2$ has been reported as a highly efficient catalyst for the functionalization of 3,5-diiodo-BODIPYs,[54] and $Pd(PCy_3)G_2$ for Stille coupling of 2,6-dichloro-BODIPYs.[43,55] Toluene and DMF are the most popular solvents used in the cross-coupling reactions owing to their relatively high boiling points, although THF used as co-solvent was shown to significantly increase the yields of Suzuki cross-coupling reactions on 3,5-diiodo-BODIPYs.[54]

Boron functionalizations by substitution of the fluorides, mainly with oxygen and carbon nucleophiles, are accomplished by reaction with strong nucleophiles (such as alkoxides or Grignard reagents) or with milder nucleophiles in the presence of a Lewis acid, such as $AlCl_3$ or $SnCl_4$,[58–67] despite the need for conscientious use for environmental reasons.

10.2.2 Application in PDT

PDT is a minimally invasive type of treatment for several forms of cancers and pre-cancerous conditions, age-related macular degeneration, and other diseases.[68–72] PDT is a tri-modal cancer therapy that combines three key components: a photosensitizer (PS), light of appropriate wavelength, and molecular oxygen. Upon light irradiation, the PS is activated and transfers the absorbed energy to ground state oxygen generating highly reactive singlet oxygen ($^1O_2$) and other reactive oxygen species (ROS), which cause cell death.[68,69] Desirable properties of PDT photosensitizers include: (1) preferential accumulation in target tissue; (2) low dark toxicity; (3) high phototoxicity; (4) absorption within the photodynamic therapeutic window (650–800 nm) with high extinction coefficients, (5) high chemical, kinetic and photostability, and (6) high quantum yield of PS triplet state ($\Phi_T > 0.4$), long triplet state lifetimes ($\tau_T \sim 1$ μs) and energy ($E_T \geq 95$ kJ mol^{-1}), for efficient energy transfer to triplet molecular oxygen and high singlet oxygen

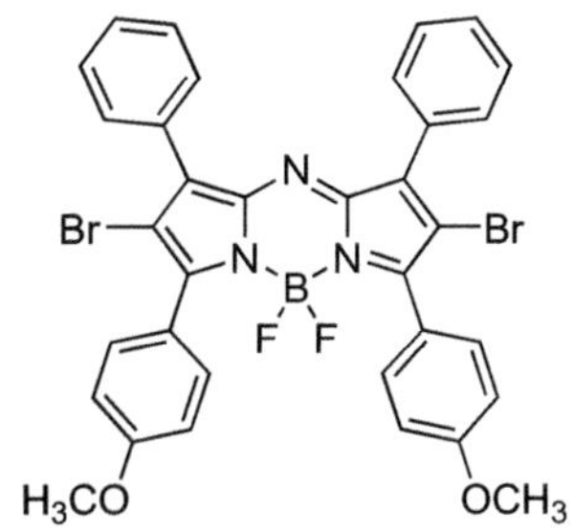

5: ADPM06

Figure 10.4 Chemical structure of aza-BODIPY photosensitizer ADPM06.[88]

quantum yields.[73,74] Current PDT anticancer agent Photofrin® is an FDA-approved porphyrin-based compound for the treatment of melanoma, esophageal, digestive tract, genitourinary tract, and lung cancers. Other PDT photosensitizers under clinical investigation are tetrapyrrolic-based macrocycles, including chlorins, bacteriochlorins, and phthalocyanines.[75–80] Although effective in PDT, Photofrin® has some drawbacks in that it absorbs only weakly in the red region of the spectrum and it is a mixture of compounds (porphyrin monomers and oligomers) with prolonged retention times in tissues, which often causes patient photosensitivity for several weeks after treatment.[81] Therefore, in addition to tetrapyrrolic macrocycles, other PS with strong absorptions within the PDT therapeutic window and favorable photophysical and tumor-targeting properties are being explored for application in PDT, including phenothiazinium-based photosensitizers[82–84] and BODIPYs.[31,85]

In 2002, O'Shea *et al.* reported for the first time the synthesis of 2,6-dibrominated aza-BODIPYs and their investigation as PDT photosensitizers.[86] In particular, compound **5** (ADPM06), shown in Figure 10.4, bearing *para*-methoxyphenyl groups at the 3,5-positions showed a *ca.* 30 nm bathochromic shift in the absorption band compared to the corresponding phenyl derivative and increased singlet oxygen quantum yield. In preclinical animal studies using dynamic PET-MRI (positron emission tomography-magnetic resonance imaging), aza-BODIPY **5** was shown to be a promising PDT photosensitizer with a vascular-targeting mechanism of action, causing 71% tumor ablation in mice bearing mammary tumors at a drug dose of 2 $mg\,kg^{-1}$ and light dose of 150 $J\,cm^{-2}$.[87,88]

10.2.3 Structure Modifications

Chemical modifications of BODIPYs are necessary for the design of effective PS, not only to convert these usually highly fluorescent boron compounds to PS with efficient intersystem crossing ability from the singlet to the triplet excited state by the heavy atom effect,[38,89,90] but also to extend their π-systems for excitation within the PDT therapeutic window.

10.2.3.1 Halogenation Reactions

The incorporation of heavy halogen atoms, such as bromine or iodine, onto the BODIPY platform can generate useful PS for PDT applications owing to enhanced intersystem crossing by the heavy atom effect. In particular, the introduction of bromine or iodine atoms at the 2,6-positions of the BODIPY tends to favor intersystem crossing and singlet oxygen generation.[31,38,85,88,91] Two methods exist for introduction of halogen atoms on the BODIPY platform, either by direct halogenation reactions[89,92] or by total synthesis from halogenated precursors.[90,93,94]

Bromination of BODIPYs at the 2,6-positions can be achieved using *N*-bromosuccinimide (NBS) or bromine, as shown in Scheme 10.1. For example, 1,3,5,7-tetramethyl-8-phenyl-BODIPY was brominated with NBS in refluxing CCl_4, using AIBN as the initiator (Scheme 10.1a).[43,95,96] On the other hand, 1,3,5,7,8-pentamethyl-BODIPY was brominated at room temperature using bromine in dichloromethane (DCM) or benzene (Scheme 10.1b) in 93% yield.[97] 8-Phenyl-BODIPY was also brominated in high yield under the same

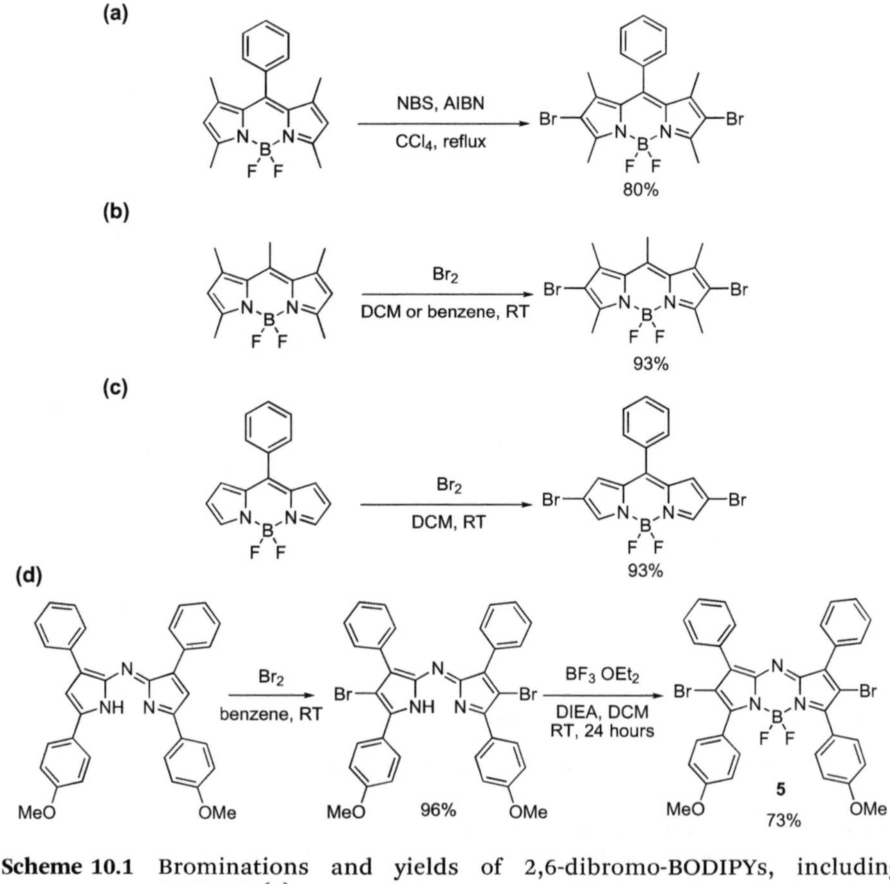

Scheme 10.1 Brominations and yields of 2,6-dibromo-BODIPYs, including ADPM06 (5).

conditions, using three equivalents of bromine (Scheme 10.1c).[92] Similar bromination conditions were employed in the synthesis of photosensitizer ADPM06 (**5**) by performing the bromination before complexation with boron (Scheme 10.1d).[98] Although fluorinated and chlorinated solvents, as well as benzene, have been mostly employed in bromination reactions due to enhanced BODIPY solubility in these solvents, alternative greener solvents, including methanol, acetonitrile and THF, have also been recently used to prepare brominated BODIPYs in fair yields.[99,100]

The iodine atom is heavier than the bromine atom, therefore the incorporation of iodine at the 2,6-positions of BODIPYs is thought to further enhance intersystem crossing and singlet oxygen generation. The common iodination reagents, *N*-iodosuccinimide (NIS) and I_2, are less efficient than the corresponding brominating reagents, therefore I_2 and HIO_3 in refluxing ethanol are often employed to iodinate BODIPYs (Scheme 10.2). Nagano *et al.* reported that *meso*-free alkylated BODIPYs were iodinated using I_2 and HIO_3 to give the corresponding 2,6-diiodo-BODIPYs without further iodination at the *meso*-position (Scheme 10.2a).[38] Under similar conditions, Ortiz *et al.* poly-iodinated an 8-aryl-BODIPY (Scheme 10.2b), and showed that the di-, tri- and tetra-iodo-BODIPYs were all potential PS with single oxygen quantum yields ranging from 0.83 to 0.87.[89] NIS in DCM at room

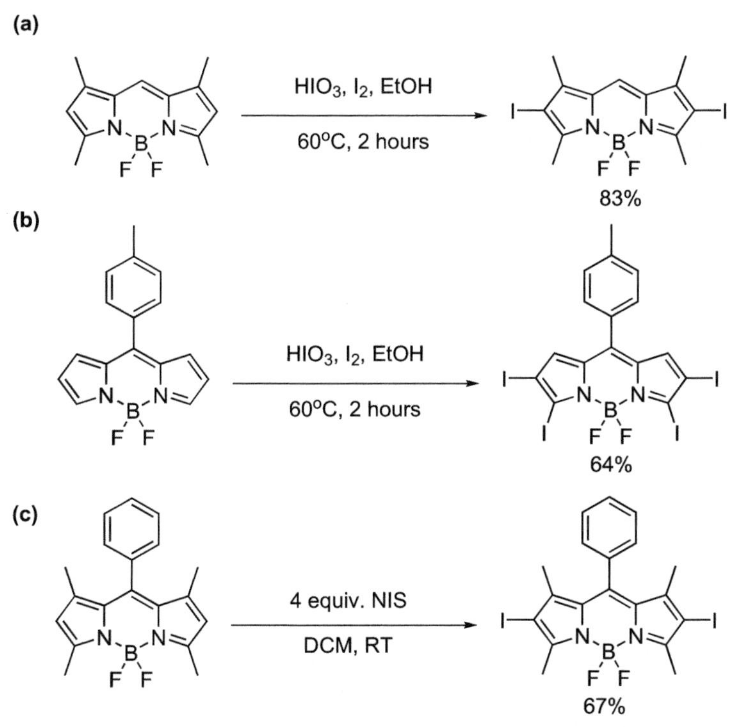

Scheme 10.2 Iodination conditions to give 2,6-diiodo-BODIPYs.

temperature can also be used to iodinate the BODIPY's 2,6-positions in good yields (Scheme 10.2c).[101]

10.2.3.2 Knoevenagel Reactions

Knoevenagel condensation reactions are one the most common methods used to generate π-extended BODIPYs with large bathochromic shifts of the absorption and emission bands, of about 100 nm per styryl group. The starting materials, methylated BODIPYs and functionalized aryl aldehydes, are readily synthesized in the laboratory or available from commercial sources. The number of styryl units incorporated *via* the Knoevenagel reactions can be controlled using different stoichiometry of reagents and reaction times. For alkylated BODIPYs, the reaction occurs usually first at the 3,5-methyl groups, followed by the 1,7-methyls.[49,102,103] However, when formyl groups are present at the 6- and 2,6-positions, the Knoevenagel reactions occur first at the 1,7-positions.[104]

Akkaya *et al.* reported the synthesis of three 2,6-dibromo-BODIPYs symmetrically substituted with styryl units at the 3,5-positions. One promising PS is BODIPY **6** (Figure 10.5) containing nine polyethylene glycol (PEG) groups for enhanced solubility in aqueous solution. This compound was found to have an $EC_{50} \sim 200$ nM in K562 cells, upon LED irradiation at 2.5 mW cm^{-2} fluence rate for 4 hours.[105] Ng and co-workers reported the synthesis of a series of symmetric and asymmetric distyryl-BODIPYs with iodine in place of bromine at the 2,6-positions, including compounds **7** and **8** (Figure 10.5).[106,107] Some of the BODIPYs were highly phototoxic, with EC_{50} values of 7 and 15 nM for **7** and **8**, respectively, in HT29 cells, and preferential subcellular localization in the endoplasmic reticulum (ER) and the lysosomes. More recently, Ng and co-workers reported folate receptor-targeted distyryl BODIPY **9** synthesized using "click chemistry" (Figure 10.5). BODIPY **9** was highly phototoxic to KB cells, with an EC_{50} of 0.06 μM and localized subcellularly in the ER and lysosomes.[108]

Our group reported the synthesis and investigation of 11 2,6-diiodo-BODIPYs bearing different 8-aryl groups, including 3,5- and 3,4-dimethoxyphenyl, pentafluorophenyl, thienyl and bisthienyl groups.[109] All BODIPYs, with one exception, showed no dark cytotoxicity ($IC_{50} > 400$ μM) toward HEp2 cells, and the most phototoxic ($IC_{50} = 4$–7.5 μM) were BODIPYs **10** and **11** bearing *meso*-dimethoxyphenyl groups (Figure 10.6). Based on these studies, a series of BODIPYs with extended π-systems from Knoevenagel condensations, bearing mono- and distyryl groups containing indolyl, pyrrolyl, thienyl and tri(ethylene glycol)phenyl units, was investigated.[103] Among these, the mono-styryl-BODIPYs were found to be phototoxic ($IC_{50} = 2$–15 μM at 1.5 J cm^{-2} in HEp2 cells) while the distyryl-BODIPYs had low or no phototoxicity ($IC_{50} > 100$ mM at 1.5 J cm^{-2}) toward HEp2 cells. The most promising PDT photosensitizer of the series was BODIPY **12**, which was also found to produce ROS in murine hepatoma cells and to localize preferentially in the cell ER and mitochondria.

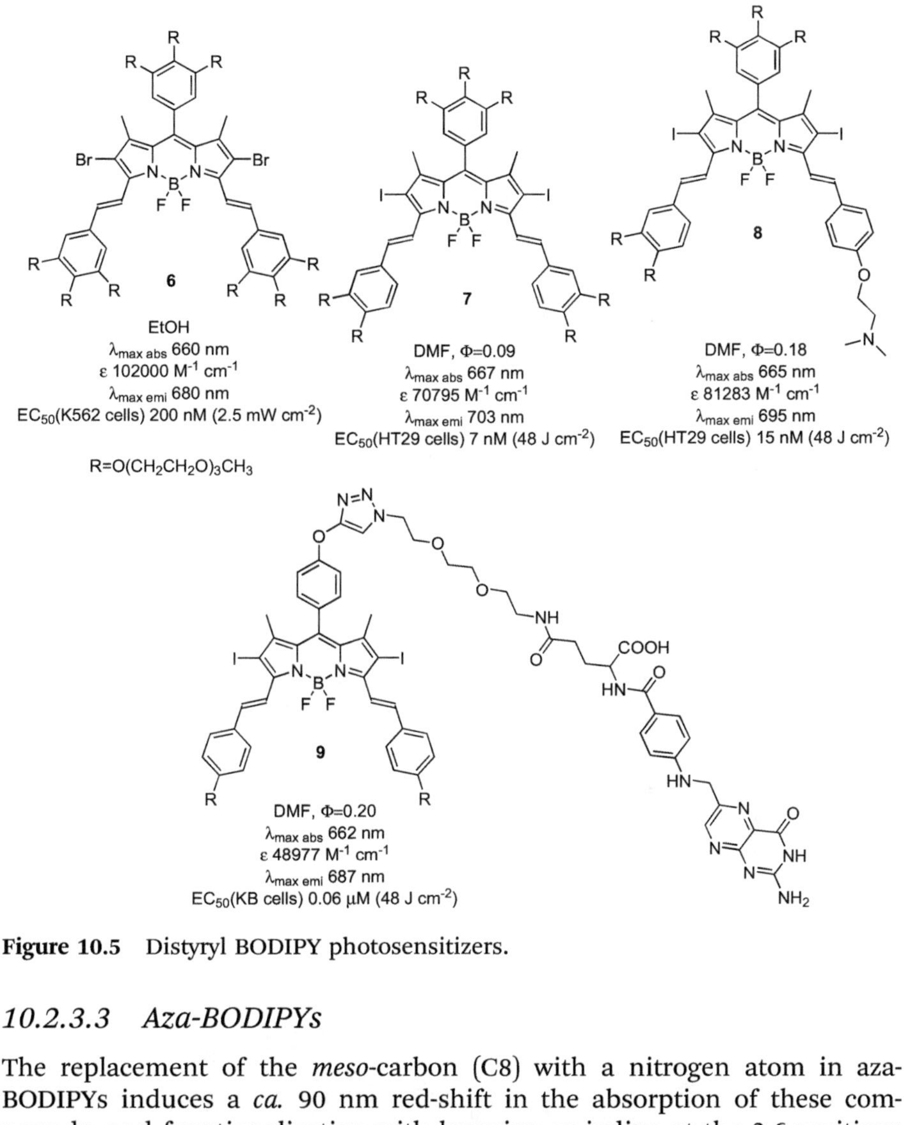

Figure 10.5 Distyryl BODIPY photosensitizers.

10.2.3.3 Aza-BODIPYs

The replacement of the *meso*-carbon (C8) with a nitrogen atom in aza-BODIPYs induces a *ca.* 90 nm red-shift in the absorption of these compounds, and functionalization with bromine or iodine at the 2,6-positions further red-shifts the absorption by another 20–30 nm. Therefore, aza-BODIPYs typically absorb within the PDT therapeutic window and are promising PSs.[98,110]

O'Shea *et al.* compared a series of aza-BODIPYs bearing different aryl units at the 1,3,5,7-positions and found that the presence of *para*-methoxyphenyl groups enhanced the singlet oxygen quantum yields. The most promising PDT photosensitizer of the series, ADPM06 (**5**), showed high photostability, triplet state quantum yield of 72%, singlet oxygen quantum yield of 74%, and it was investigated both *in vitro* and *in vivo* studies.[86–88,91] High phototoxicity was observed using light doses of 8 J cm^{-2} in a large number of

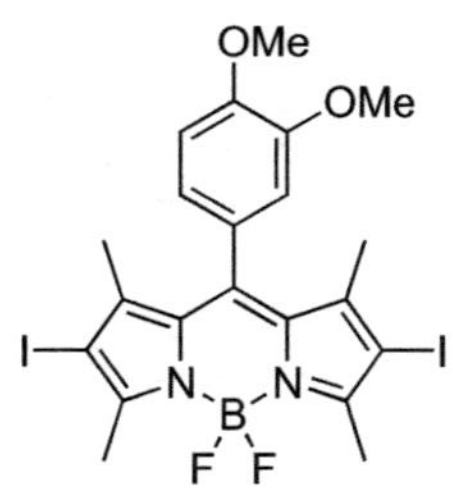

10

THF, Φ=0.04
$\lambda_{max\ abs}$ 532 nm
log ε 4.88M^{-1} cm^{-1}
$\lambda_{max\ emi}$ 546 nm
IC_{50}(HEp2 cells) 7.5 μM (1.5 J cm^{-2})

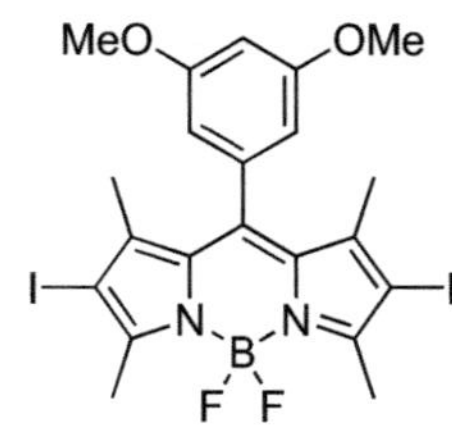

11

THF, Φ=0.05
$\lambda_{max\ abs}$ 534 nm
log ε 4.94M^{-1} cm^{-1}
$\lambda_{max\ emi}$ 547 nm
IC_{50}(HEp2 cells) 4.0 μM (1.5 J cm^{-2})

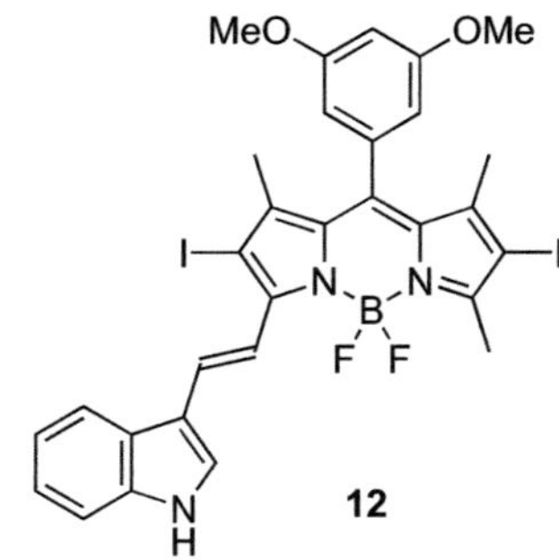

12

DMSO, Φ=0.17
$\lambda_{max\ abs}$ 636 nm
ε 29512 M^{-1} cm^{-1}
$\lambda_{max\ emi}$ 676 nm
IC_{50}(HEp2 cells) 2.0 μM (1.5 J cm^{-2})

Figure 10.6 BODIPY photosensitizers **10–12**.

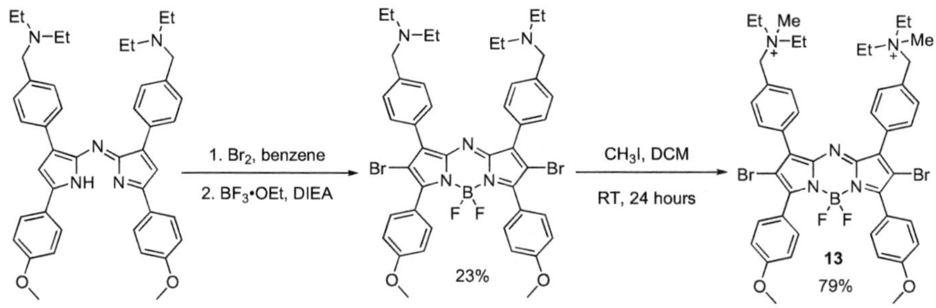

Scheme 10.3 Preparation of cationic aza-BODIPY **13** with antimicrobial activity.

human tumor cell lines, including some drug-resistant cell lines. In *in vivo* studies, ADPM06 was found to rapidly clear both from the skin and internal organs, while retaining in the tumor, and induced 71% tumor ablation at 2 mg kg^{-1} drug dose and 150 J cm^{-2} light dose, comparable to the biological efficacy of clinically approved PDT agents in mice xenograft models. Radiolabelling with ^{18}F at the boron atom,– allowed the application of a combined PET/MRI imaging modality to investigate the vascular-targeting response to therapy, and the results demonstrated that ADMP06 is a promising photosensitizer for Phase I clinical trials.

O'Shea *et al.* also reported the synthesis of di-cationic aza-BODIPY **13** as shown in Scheme 10.3, and investigated its antibacterial activity. This compound was rapidly taken up by both Gram-positive and Gram-negative bacterial strains, as well as by pathogenic yeasts, while minimal uptake was observed in a human cell line (MDA-MB-231). In a study using 5 μg mL^{-1} of **13** under 75 J cm^{-2} light dose, a broad spectrum pathogen response was found, with 3.6 and 5.7 log_{10} decreases for Gram-negative bacterium *E. coli* and the pathogenic yeast *Candida albicans*, respectively.[111]

10.2.3.4 *Aryl-fusion*

In addition to halogenation, Knoevenagel reactions, and replacement of C8 with a nitrogen atom, BODIPYs with fused aromatic rings have been proposed as promising photosensitizers for PDT. Thienyl-fused and furanyl-fused BODIPYs are accessible from the corresponding aryl-fused pyrrole precursors, and the bromination reactions are usually accomplished in the presence of excess bromine.[112–114] For example, the di- and tetra-brominated aryl-fused BODIPYs **14** and **15** (Figure 10.7) showed favorable singlet oxygen quantum yields by comparison with a reference PS compound, (5-(4-methoxyphenyl)-10,15,20-tetraphenyl-21,23-dithiaporphyrin), a core modified porphyrin (CMP). A 1.2-fold rate of oxidation of DPBF (1,3-diphenylisobenzofuran) by **14** was observed, while a 0.5-fold rate for **15** was found, compared to that of CMP. Higher photostabilities of BODIPY **14** and **15** were also reported by comparing with mTHPC, a second-generation PDT

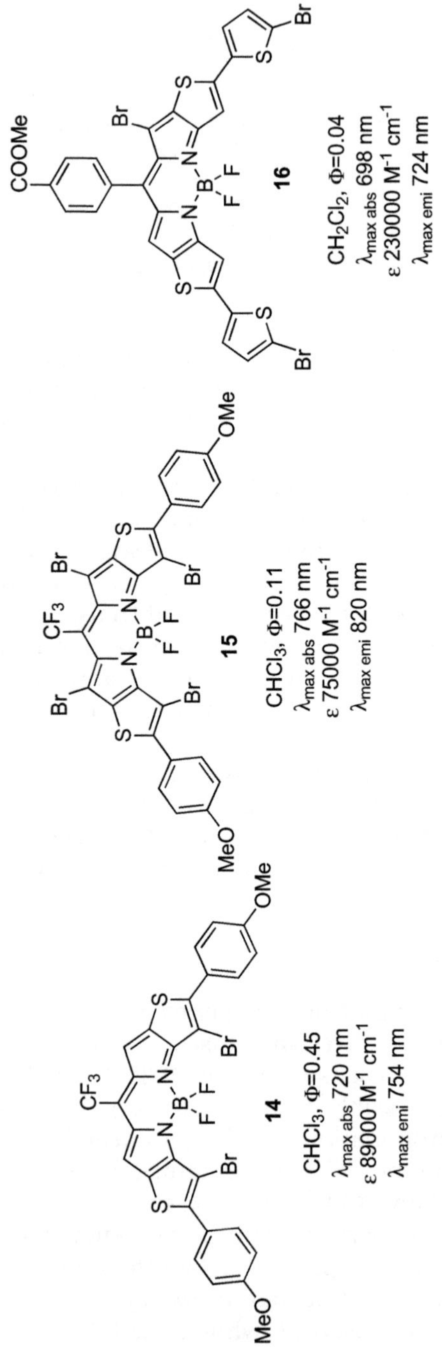

Figure 10.7 Thienyl-fused BODIPYs for PDT applications.

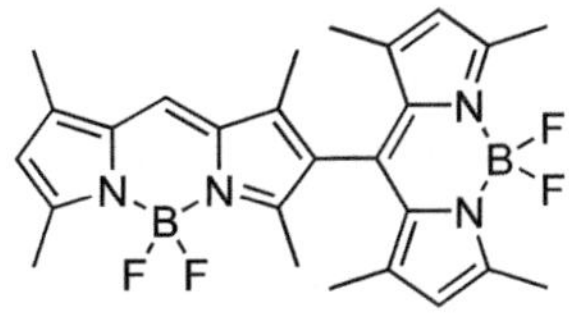

17

$CHCl_3$, Φ=0.03
$\lambda_{max\ abs}$ 514 nm
ε 119400 M^{-1} cm^{-1}
$\lambda_{max\ emi}$ 527 nm

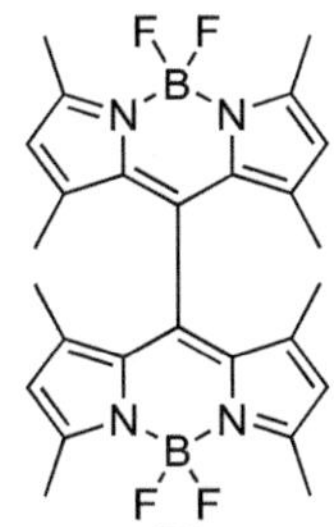

18

$CHCl_3$, Φ=0.31
$\lambda_{max\ abs}$ 515 nm
$\lambda_{max\ emi}$ 588 nm

Figure 10.8 BODIPY dimers as promising photosensitizers.

photosensitizer.[114] Shen and co-workers reported similar BODIPYs bearing *meso*-aryl units in place of the trifluoromethyl group and upon bromination with NBS in THF the unsymmetric BODIPY **16** bearing three bromines was obtained. This BODIPY was found to have high phototoxicity toward HeLa cells (IC_{50} = 7.12 μM) upon irradiation with a 635 nm laser beam with power density of 400 mW cm^{-2}.[100]

Halogen-free BODIPY dimers with enhanced intersystem crossing have also been proposed as PDT photosensitizers. BODIPY dimers connected at different positions, including *meso*–*meso*,[115] *meso*–β,[116] α–α,[117] and β–β[118] directly linked BODIPYs, have been investigated. Among these, the *meso*–β and *meso*–*meso* linked dimers are the most promising photosensitizers, for example, compounds **17** and **18** (Figure 10.8) with singlet oxygen quantum yields of 0.51 and 0.46, respectively. However, these BODIPY dimers do not absorb within the PDT therapeutic window and therefore require further functionalization, for example *via* Knoevenagel condensations.

BODIPYs have been conjugated to porphyrins and other molecules for PDT applications.[119–124] For example, Smith *et al.* reported the synthesis of β-fused porphyrin-BODIPY systems with red-shifted absorption and emission spectra,[119] while D'Souza *et al.* installed a C_{60} unit at the boron position of a halogen-free tetraphenyl aza-BODIPY,[120] and Zhao *et al.* synthesized mono-, di- and tri-C_{60}-containing styryl-BODIPYs.[121] Among these, the tri-C_{60}-styryl BODIPY showed a long-lived triplet excited state (τ_T = 123.2 μs). Other porphyrin-BODIPY conjugates have been synthesized *via* substitution reactions on the boron center with porphyrinbenzyloxyl groups,[122] by cross-coupling reactions[123] or *via* "click chemistry".[124] However, no singlet oxygen quantum yields nor PDT activity have been reported for these compounds.

10.2.4 Application in BNCT

BNCT is a binary anticancer therapy based on the $^{10}B(n,\alpha)^{7}Li$ capture reaction, particularly promising for the treatment of high-grade brain tumors due to the much deeper penetration of epithermal neutron beams compared with light used in PDT. However, treating diseases of the brain is challenging, predominantly as a result of the blood–brain barrier (BBB), which prevents most drugs from penetrating into the brain tumor cells or other targeted tissue.[18,23,125–129] Many physicochemical properties of drugs influence their BBB permeability, including molecular weight, lipophilic character, size, polar surface area, and charge.[130–132] Since BODIPYs have low molecular weights and good permeability across cellular membranes, they were recently investigated by our group as boron delivery vehicles for BNCT. Carborane clusters were introduced *via* Pd(0)-catalyzed Suzuki cross-couplings or nucleophilic substitution reactions on halogenated BODIPY precursors.[133,134] A series of four carboranyl-BODIPYs (**19–22**; Figure 10.9) was initially synthesized and their BBB permeability evaluated using the human brain endothelial cell line hCMEC/D3 as a BBB model. Among these, only BODIPY **22** showed higher permeability ($Pe = 38.75 \pm 1.12 \times 10^{-6}$ cm s^{-1}) than lucifer yellow, a marker for low BBB permeability.[133] Based on these results, we synthesized another series of carboranyl-BODIPYs **23-28** (Figure 10.9) with

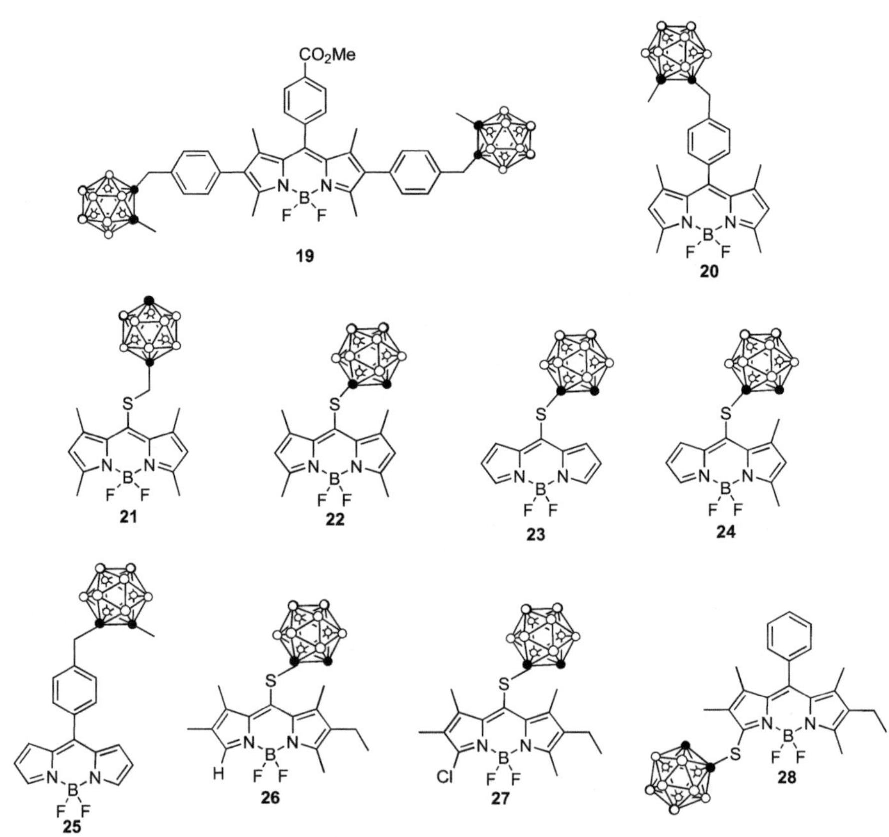

Figure 10.9 Carborane-functionalized BODIPYs.

molecular weights in the range 266–527 Da and octanol–water partition coefficients, log *P*, in the range 1.5–2.7. All BODIPYs had low cytotoxicity toward human glioma T98G cells and were taken up by these cells to different extents, accumulating preferentially in the ER. Among this series of compounds, BODIPYs **22, 23, 24, 26,** and **27** showed higher BBB permeability than lucifer yellow, with **23** having the highest Pe value ($164 \pm 3 \times 10^{-6}$ cm s^{-1}), in part due to its low molecular weight and favorable hydrophobic character (log $P = 1.50$).[134] These BODIPYs with high boron content are promising boron delivery agents for BNCT, as 15–30 μg ^{10}B g $tumor^{-1}$, or approximately 10^9 atoms tumor $cell^{-1}$, is estimated to be required for effective therapy, depending on its tumor distribution.[13–15]

10.3 Naturally Occurring Tetravalent Boron Therapeutics

10.3.1 Boromycin

Boromycin (**29**) is a polyether-macrolide antibiotic, initially isolated from the *Streptomyces* strain in 1967.[135] It is notable for being the first natural product found to contain boron. The anionic tetravalent boron in boromycin and related naturally occurring boron therapeutics is able to complex metal cations, such as K^+ or Na^+ (Figure 10.10). Boromycin was found to kill Gram-positive bacteria through binding at the cytoplasmic membrane.[136] In 1996, boromycin was also found to have strong anti-human immunodeficiency virus (HIV) ability, which is probably due to the inhibition of replication of HIV strains.[137] The IC_{50} values of boromycin toward HIV-1 (LAV-1) and HIV-2 (LAV-2) are 0.008 and 0.007 μM, respectively. The anti-HIV activity and cytotoxicity of boromycin in several cell lines are summarized in Table 10.1.[137]

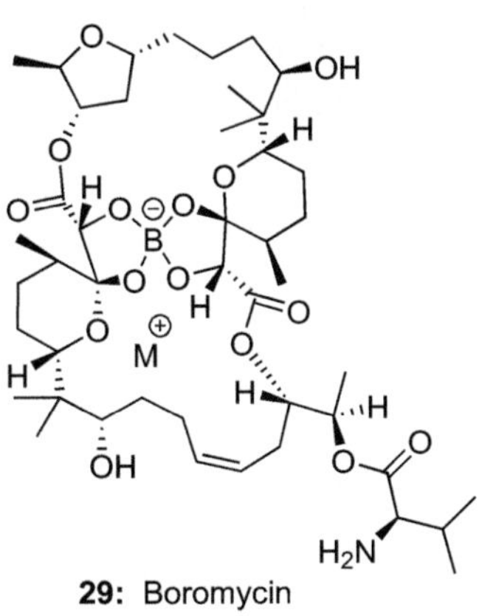

29: Boromycin

Figure 10.10 Structure of boromycin.

Table 10.1 Anti-HIV activity and cytotoxicity of boromycin.

Virus	Strain	Cell	Assay	Day of analysis	Anti-HIV activity IC_{50} (μM)	Anti-HIV activity IC_{90} (μM)	Cytotoxicity CC_{50} (μM)
HIV-1	LAV-1	MT-4	CPE	6	0.008	0.027	0.22
	RF	MT-4	CPE	5	0.11	0.18	0.17
	KK-1	MT-4	CPE	6	0.14	0.31	0.22
	KK-1	PBMC	MT-4/CPE	10	<0.1	<0.1	4.3
	KK-1	PBMC	RT active	10	1.8	4.8	4.3
HIV-2	LAV-2	MT-4	CPE	6	0.007	0.019	0.22

10.3.2 Aplasmomycin A, B and C

In 1976 the antibiotic aplasmomycin A (**30**, Figure 10.11) was extracted from a strain of *Streptomyces griseus* isolated from shallow sea mud in Sagami Bay, Japan.[138] The boromycin-like structure with a boron atom in the center was determined by X-ray crystallography. This compound was found to inhibit the growth of Gram-positive bacteria, including mycobacteria *in vitro* and plasmodia *in vivo*. Aplasmomycins B (**31**) and C (**32**)[139] were the minor products isolated under the same culture conditions. Aplasmomycin A and aplasmomycin B have similar ability to transport K^+, while aplasmomycin C and deboroaplasmomycin have much decreased K^+ transport ability and also lower biological activity compared with aplasmomycins A and B. The antibacterial activity of the asplasmomycins is summarized in Table 10.2.[139]

10.3.3 Tartrolons

Through the screening of myxobacteria, antibacterial metabolites tartrolon A_1, A_2, A_3 and B were isolated from Gram-negative eubacteria, *Sorangium cellulosum* strain So ce678.[140] The boron-containing macrodiolide structure, similar to boromycin and aplasmomycin, was characterized by X-ray crystallography.[141] The first total synthesis of tartrolons B (**33**) was published by Berger and Muzler in 1999.[142] Tartrolon C (**34**), was isolated from a *Streptomyces* species and it has a similar structure to tartrolon B, except for an extra oxygen atom in each monomeric subunit (Figure 10.12).[142] This compound is active against beet armyworm and tobacco budworm, with a minimum emergent larvicide concentration of 125 ppm for each insect.[143]

10.3.4 Borophycin

The potent cytotoxin borophycin (**35**, Figure 10.13) was isolated from the lipophilic extraction of a marine strain of the blue-green alga *Nostoc linckia.*[144] The structure, determined by NMR and X-ray crystallography, is made up of two identical halves, similar to the antibiotics boromycin and aplasmomycin A (Figure 10.14).[145,146] All three compounds are acetate-derived polyketides with the same boron-binding C_1–C_7 mode in the center of the symmetric

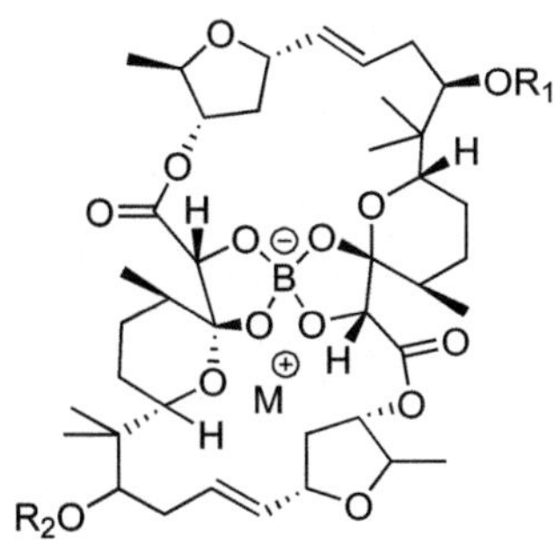

30: Aplasmomycin A $R_1=R_2=H$
31: Aplasmomycin B $R_1=Ac, R_2=H$
32: Aplasmomycin C $R_1=R_2=Ac$

Figure 10.11 Structures of aplasmomycin A, B and C.

Table 10.2 Biological activity of asplasmomycins A, B and C, using the diameter of inhibition zone with aplasmomycin (500 μg ml^{-1}) by disk diffusion method.

Compounds	*Bacillius subtilis* PCI 219	*Staphylococcus aureus* FDA 209P	*Staphylococcus aureus* Smith	*Bacillus subtilis* PCI 219
Aplasmomycin A	21.5	22.0	20.5	18.5
Aplasmomycin B	22.0	22.0	20.5	18.5
Aplasmomycin C	11.0	10.5	+	0
Deboroaplasmomycin	+	+	0	0

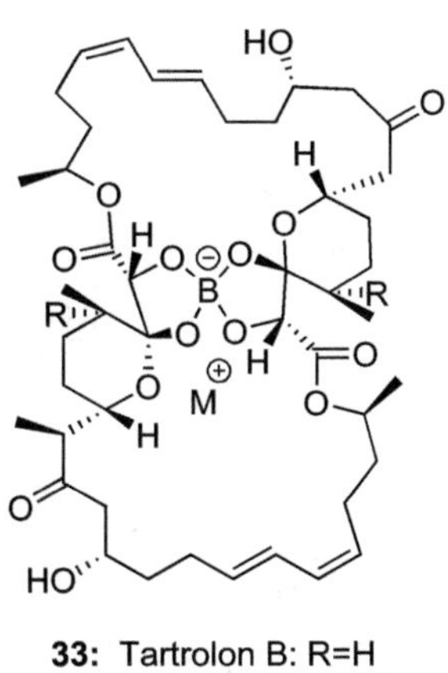

33: Tartrolon B: R=H
34: Tartrolon C: R=OH

Figure 10.12 Structures of tartrolons B and C.

structure. However, the biosynthesis of borophycin differs from those of boromycin and aplasmomycin. Phosphoglycerate or phosphoenolpyruvate are used in the biosynthesis of boromycin and aplasmomycin, while acetate and

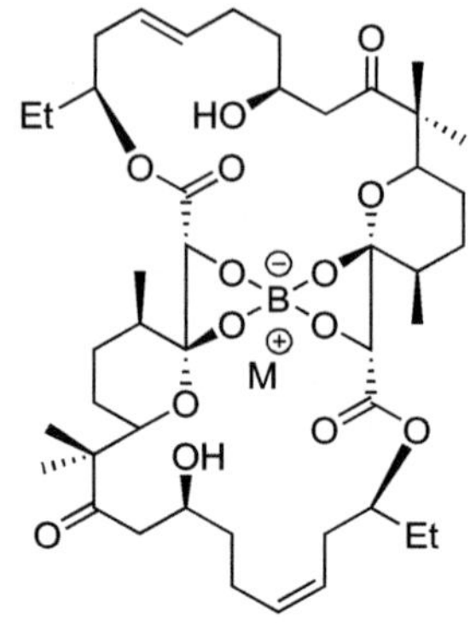

35: Borophycin

Figure 10.13 Structure of borophycin.

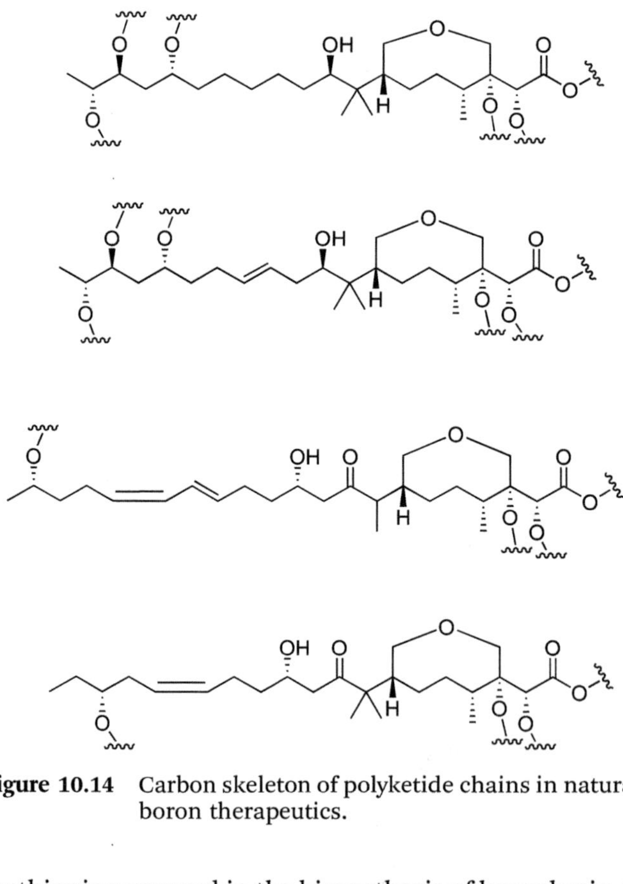

29: Boromycin

30: Aplasmomycin A

33: Tartrolon B

35: Borophycin

Figure 10.14 Carbon skeleton of polyketide chains in naturally occurring tetravalent boron therapeutics.

methionine are used in the biosynthesis of borophycin. Borophycin was found to have antitumor activity toward cancer cell lines LoVo [minimum inhibitory concentration (MIC) 0.066 μg mL^{-1}] and KB (MIC 3.3 μg mL^{-1}).[144]

10.4 Naturally-inspired Boron Therapeutics

10.4.1 Borinate Ester Derivatives

Based on borinic acid picolinate esters, several boron-containing antibacterial agents were synthesized (Scheme 10.4) and the MIC obtained by screening against Gram-positive and Gram-negative bacteria.[147] THF was the solvent used for the preparation of the aryl borinic acid intermediates, and ethanol–water mixtures for the final picolinate ester derivatives. Compound **36g**, designated AN0128, was identified to have the most potent antibacterial and anti-inflammatory activities. AN0128 could significantly reduce the formation of an inflammatory infiltrate and reduce bone loss, measured histologically and by micro-CT.[148] Table 10.3 summarizes the biological activity of borinate esters **36a–q** bearing different R^1 and R^2 groups.

10.4.2 Boronate Ester Derivatives

In 2003, Westcott *et al.* developed an expedite synthesis of novel boronate esters **37–42** bearing heterocyclic ring systems using ethylenediamine derivatives.[149] The syntheses of boronate esters were mainly conducted in diethyl ether or dichloromethane solutions. The compounds in Figure 10.15 were found to show considerable antifungal activity against *Aspergillus niger* and *Aspergillus flavus*, and moderate antibacterial activity against *Bacillus cereus*.[149] The pinacol group in these compounds is believed to enhance their biological activity relative to boronic acid.

10.4.3 Boroxazolidones

Boroxazolidones are a type of boron-containing compound obtained by reacting α-amino acids with boranes or borinates, as shown in Scheme 10.5. Through condensation with an aldehyde, the corresponding Schiff base

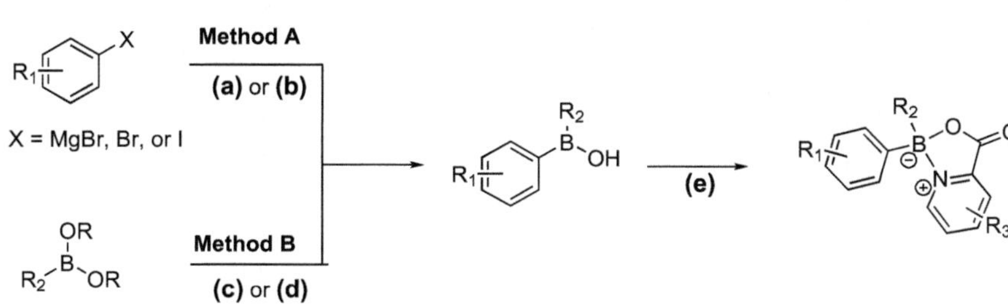

Scheme 10.4 Synthesis of borinate esters. Reaction conditions: (a) $B(OMe)_3$ (0.5 equiv), THF, 0 °C to rt (when X = MgBr); (b) BuLi, $B(OMe)_3$ (0.5 equiv.), THF, −78 °C to rt (when X = Br); (c) (when X = MgBr), THF, 0 °C to rt; (d) (when X = Br), BuLi, THF, −78 °C to rt; (e) R_3-picolinic acid, EtOH, H_2O, reflux.

Table 10.3 Minimum inhibitory concentration (MIC, l g mL^{-1}) for borinic esters **36**.

Compound	R[1]	R[2]	*Staphylococcus aureus*	*Staphylococcus epidermidis*	*Propionibacterium acnes*	*Bacillus subtilis*	*Haemophilus influenzae*
Erythromycin			0.5	0.15	0.1	0.1	4
36a	3-Cl	3-Cl–Ph	≤0.125	8	10	16	16
36b	4-Cl	4-Cl–Ph	4	1	1	1	16
36c	3-Cl	Pyridin-3-yl	16	32	nt[a]	nt[a]	32
36d	4-Cl	Pyridin-3-yl	64	32	nt[a]	nt[a]	16
36e	4-Cl	2-Cl-pyridin-5-yl	32	32	nt[a]	nt[a]	32
36f	3-Cl	Thiophen-3-yl	32	32	10	16	32
36g	3-Cl-4-Me	3-Cl-4-Me–Ph	1	0.5	0.3	1	>64
36h	3-Cl-4-Me	4-Me	32	32	10	16	32
36i	3-Cl-4-Me	Phenethyl	0.5	1	1	1	>64
36j	3-F	3-F–Ph	>64	>64	>100	>64	>64
36k	3-Cl	3-SMe–Ph	8	8	3	4	>64
36l	3-Cl	3-SMe–Ph	8	8	3	4	>64
36m	3-Cl-4-F	3-Cl-4-F–Ph	1	8	3	8	4
36n	3-Cl-4-OEt	3-Cl-4-OEt–Ph	2	2	1	2	>64
36o	3-Cl-4-NMe_2	3-Cl-4-NMe_2–Ph	32	32	nt[a]	64	>64
36p	3-Cl-4-Me	4-Me–Ph	4	2	3	2	>64
36q	4-Cl-2-Me	4-Cl-2-Me–Ph	4	2	0.3	0.5	16

[a]nt: not tested.

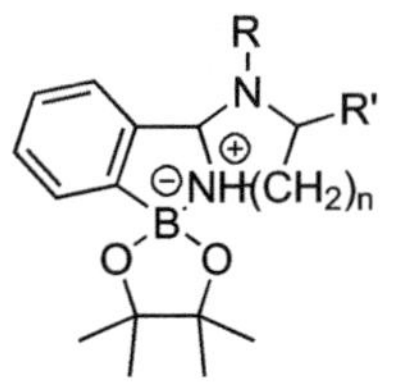

Figure 10.15 Structure of boronate esters with biological activity.

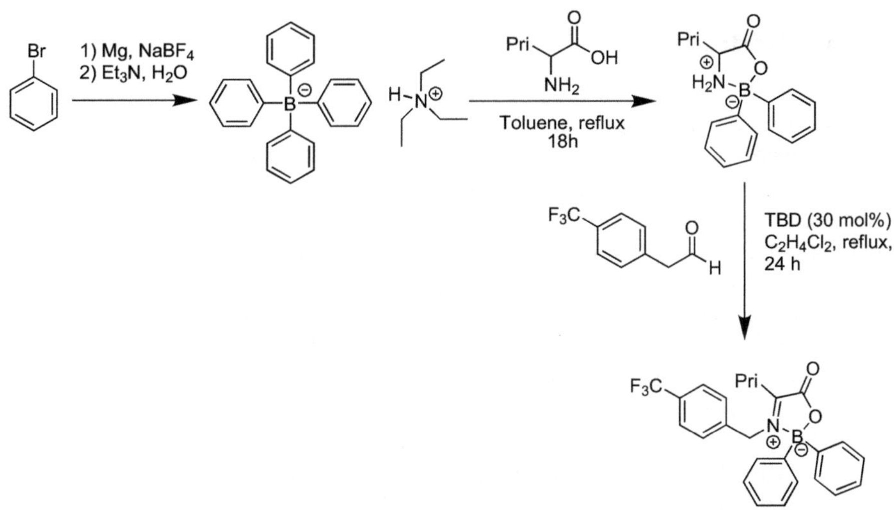

43

Scheme 10.5 Synthesis of boroxazolidone **43**.

compounds show moderate cytotoxicity against colorectal adenocarcinoma cells, with the most active compound **43** in this series having an IC_{50} of 76 μM.[150] A brief study of the effect of the same compound against human brain astrocytoma cells revealed an IC_{50} of 268 μM.

10.4.4 Arylspiroborates

A series of arylspiroborate compounds was prepared from 2,3-dihydroxy-naphthalene, as shown in Scheme 10.6, and characterized by Geier *et al.*[151] The solvents used in this case were metal hydroxide aqueous solutions, including LiOH, NaOH and $Ca(OH)_2$. All seven compounds **44–50** showed moderate antifungal activity against *C. albicans* and *Mycobacterium tuberculosis*, with the alkali metal salts being the most active compounds. However, none of the compounds showed appreciable antibacterial activity against *Pseudomonas aeruginosa* or *Staphylococcus aureus*. Through further work, these boron compounds have the potential to treat tuberculosis.

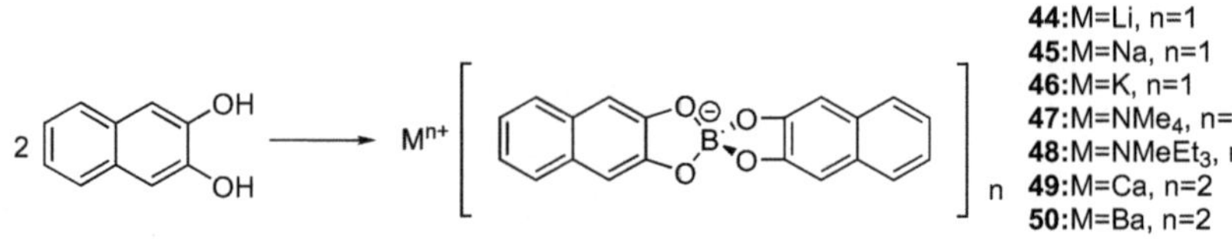

Scheme 10.6 Synthesis of metal and ammonium arylspiroborates. Reaction conditions: 1 equiv. $B(OH)_3$, MOH, H_2O, RT, or 0.5 equiv. $B(OH)_3$, $M(OH)_2$, H_2O, RT.

10.5 Conclusion

Boron-containing drugs have been investigated for several decades and continue to be developed as therapeutic agents with anticancer, antiviral, antibacterial, antifungal, and other biological activities. Boron-based drugs bortezomib (Velcade®) and tavaborole (Kerydin®) are clinically approved for the treatment of multiple myeloma and of onychomycosis of the toenails, respectively, and many others are in different phases of preclinical and clinical development. In this chapter, we review the main classes of tetravalent boron-based therapeutics currently under investigation, with emphasis on BODIPY derivatives, and on naturally occurring and inspired compounds. The number of boron-containing therapeutics is expected to continue to increase as new synthetic compounds, such as those based on BODIPYs and related systems, continue to be investigated for potential applications in diagnosis as well as in the treatment of cancer and other diseases, including *via* PDT and BNCT. Although benzene and chlorinated solvents, such as dichloromethane, have been the most commonly employed in BODIPY syntheses and functionalizations, alternative greener solvents and co-solvents, including toluene, methanol, acetonitrile, and THF, are currently used. On the other hand, boron compounds derived from borinic and boronic acids are generally synthesized in aqueous conditions and can potentially display potent antifungal and/or antibacterial activities.

References

1. B. C. Das, P. Thapa, R. Karki, C. Schinke, S. Das, S. Kambhampati, S. K. Banerjee, P. Van Veldhuizen, A. Verma, L. M. Weiss and T. Evans, *Future Med. Chem.*, 2013, **5**, 653–676.
2. P. Hunter, *EMBO Rep.*, 2009, **10**, 125–128.
3. T. Řezanka and K. Sigler, *Phytochemistry*, 2008, **69**, 585–606.
4. M. M. Smedskjaer, J. C. Mauro, R. E. Youngman, C. L. Hogue, M. Potuzak and Y. Yue, *J. Phys. Chem. B*, 2011, **115**, 12930–12946.
5. M. Richold, *Biol. Trace Elem. Res.*, 1998, **66**, 121–129.
6. J. Adams, M. Behnke, S. Chen, A. A. Cruickshank, L. R. Dick, L. Grenier, J. M. Klunder, Y.-T. Ma, L. Plamondon and R. L. Stein, *Bioorg. Med. Chem. Lett.*, 1998, **8**, 333–338.

7. F. L. Rock, W. Mao, A. Yaremchuk, M. Tukalo, T. Crépin, H. Zhou, Y.-K. Zhang, V. Hernandez, T. Akama, S. J. Baker, J. J. Plattner, L. Shapiro, S. A. Martinis, S. J. Benkovic, S. Cusack and M. R. K. Alley, *Science*, 2007, **316**, 1759–1761.
8. S. J. Baker, C. Z. Ding, T. Akama, Y.-K. Zhang, V. Hernandez and Y. Xia, *Future Med. Chem.*, 2009, **1**, 1275–1288.
9. L. Ciani and S. Ristori, *Expert Opin. Drug Discovery*, 2012, **7**, 1017–1027.
10. W. Yang, X. Gao and B. Wang, *Med. Res. Rev.*, 2003, **23**, 346–368.
11. E. Genin, M. Reboud-Ravaux and J. Vidal, *Curr. Top. Med. Chem.*, 2010, **10**, 232–256.
12. M. Benderdour, T. Bui-Van, A. Dicko and F. Belleville, *J. Trace Elem. Med. Biol.*, 1998, **12**, 2–7.
13. A. H. Soloway, W. Tjarks, B. A. Barnum, F.-G. Rong, R. F. Barth, I. M. Codogni and J. G. Wilson, *Chem. Rev.*, 1998, **98**, 1515–1562.
14. R. F. Barth, J. A. Coderre, M. G. H. Vicente and T. E. Blue, *Clin. Cancer Res.*, 2005, **11**, 3987–4002.
15. R. F. Barth, M. G. H. Vicente, O. K. Harling, W. Kiger, K. J. Riley, P. J. Binns, F. M. Wagner, M. Suzuki, T. Aihara, I. Kato and S. Kawabata, *Radiat. Oncol.*, 2012, **7**, 146.
16. L. Kankaanranta, T. Seppälä, H. Koivunoro, K. Saarilahti, T. Atula, J. Collan, E. Salli, M. Kortesniemi, J. Uusi-Simola, P. Välimäki, A. Mäkitie, M. Seppänen, H. Minn, H. Revitzer, M. Kouri, P. Kotiluoto, T. Seren, I. Auterinen, S. Savolainen and H. Joensuu, *Int. J. Radiat. Oncol., Biol., Phys.*, 2012, **82**, e67–e75.
17. J. W. Hopewell, T. Gorlia, L. Pellettieri, V. Giusti, B. Stenstam and K. Sköld, *Appl. Radiat. Isot.*, 2011, **69**, 1737–1740.
18. M. Sibrian-Vazquez and M. G. H. Vicente, Boron tumor-delivery for BNCT: Recent Developments and Perspectives, in *Boron Science: New Technologies & Applications*, ed. N. S. Hosmane, CRC Press, 2011, pp. 203–232.
19. A. Adamczyk-Woźniak, K. M. Borys and A. Sporzyński, *Chem. Rev.*, 2015, **115**, 5224–5247.
20. V. M. Dembitsky, A. A. Al Quntar and M. Srebnik, *Chem. Rev.*, 2011, **111**, 209–237.
21. S. J. Baker, J. W. Tomsho and S. J. Benkovic, *Chem. Soc. Rev.*, 2011, **40**, 4279–4285.
22. P. C. Trippier and C. McGuigan, *MedChemComm*, 2010, **1**, 183–198.
23. S. Xuan and M. G. H. Vicente, Recent Development of Boron Delivery agents for Boron Neutron Capture Therapy, in *Medicinal Chemistry of Boron Compounds*, ed. C. Vinas and E. Hey-Hawkins, Wiley Publishers, in press.
24. M. F. Hawthorne, *Comments Inorg. Chem.*, 2010, **31**, 153–163.
25. M. Scholz and E. Hey-Hawkins, *Chem. Rev.*, 2011, **111**, 7035–7062.
26. Z. J. Lesnikowski, New opportunities in boron chemistry for medical applications. Boron tumor-delivery for BNCT: Recent Developments and Perspectives, in *Boron Science: New Technologies & Applications*, ed. N. S. Hosmane, CRC Press, 2011, pp. 3–19.

27. A. Treibs and F.-H. Kreuzer, *Justus Liebigs Ann. Chem.*, 1968, **718**, 208–223.
28. A. Loudet and K. Burgess, *Chem. Rev.*, 2007, **107**, 4891–4932.
29. G. Ulrich, R. Ziessel and A. Harriman, *Angew. Chem., Int. Ed.*, 2008, **47**, 1184–1201.
30. N. Boens, V. Leen and W. Dehaen, *Chem. Soc. Rev.*, 2012, **41**, 1130–1172.
31. A. Kamkaew, S. H. Lim, H. B. Lee, L. V. Kiew, L. Y. Chung and K. Burgess, *Chem. Soc. Rev.*, 2013, **42**, 77–88.
32. A. B. Descalzo, H.-J. Xu, Z. Shen and K. Rurack, *Ann. N. Y. Acad. Sci.*, 2008, **1130**, 164–171.
33. V. F. Donyagina, S. Shimizu, N. Kobayashi and E. A. Lukyanets, *Tetrahedron Lett.*, 2008, **49**, 6152–6154.
34. M. M. Salim, E. A. Owens, T. Gao, J. H. Lee, H. Hyun, H. S. Choi and M. Henary, *Analyst*, 2014, **139**, 4862–4873.
35. M. J. Hall, L. T. Allen and D. F. O'Shea, *Org. Biomol. Chem.*, 2006, **4**, 776–780.
36. S. O. McDonnell and D. F. O'Shea, *Org. Lett.*, 2006, **8**, 3493–3496.
37. N. Boens, B. Verbelen and W. Dehaen, *Eur. J. Org. Chem.*, 2015, **30**, 6577–6595.
38. T. Yogo, Y. Urano, Y. Ishitsuka, F. Maniwa and T. Nagano, *J. Am. Chem. Soc.*, 2005, **127**, 12162–12163.
39. M. Shah, K. Thangaraj, M.-L. Soong, L. T. Wolford, J. H. Boyer, I. R. Politzer and T. G. Pavlopoulos, *Heteroat. Chem.*, 1990, **1**, 389–399.
40. T. G. Pavlopoulos, J. H. Boyer and G. Sathyamoorthi, *Appl. Opt.*, 1998, **37**, 7797–7800.
41. L. Li, B. Nguyen and K. Burgess, *Bioorg. Med. Chem. Lett.*, 2008, **18**, 3112–3116.
42. L. Jiao, C. Yu, J. Li, Z. Wang, M. Wu and E. Hao, *J. Org. Chem.*, 2009, **74**, 7525–7528.
43. N. Zhao, S. Xuan, F. R. Fronczek, K. M. Smith and M. G. H. Vicente, *J. Org. Chem.*, 2015, **80**, 8377–8383.
44. H. Wang, M. G. H. Vicente, F. R. Fronczek and K. M. Smith, *Chem. – Eur. J.*, 2014, **20**, 5064–5074.
45. N. Zhao, M. G. H. Vicente, F. R. Fronczek and K. M. Smith, *Chem. – Eur. J.*, 2015, **21**, 6181–6192.
46. N. Zhao, S. Xuan, F. R. Fronczek, K. M. Smith and M. G. H. Vicente, *J. Org. Chem.*, 2017, **82**, 3880–3885.
47. E. Deniz, G. C. Isbasar, O. A. Bozdemir, L. T. Yildirim, A. Siemiarczuk and E. U. Akkaya, *Org. Lett.*, 2008, **10**, 3401–3403.
48. O. Buyukcakir, O. A. Bozdemir, S. Kolemen, S. Erbas and E. U. Akkaya, *Org. Lett.*, 2009, **11**, 4644–4647.
49. T. Uppal, N. V. S. D. K. Bhupathiraju and M. G. H. Vicente, *Tetrahedron*, 2013, **69**, 4687–4693.
50. T. Rohand, M. Baruah, W. Qin, N. Boens and W. Dehaen, *Chem. Commun.*, 2006, 266–268.

51. T. Rohand, W. Qin, N. Boens and W. Dehaen, *Eur. J. Org. Chem.*, 2006, 4658–4663.
52. L. Jiao, C. Yu, T. Uppal, M. Liu, Y. Li, Y. Zhou, E. Hao, X. Hu and M. G. H. Vicente, *Org. Biomol. Chem.*, 2010, **8**, 2517–2519.
53. H. Wang, F. R. Fronczek, M. G. H. Vicente and K. M. Smith, *J. Org. Chem.*, 2014, **79**, 10342–10352.
54. Q. Meng, F. R. Fronczek and M. G. H. Vicente, *New J. Chem.*, 2016, 5740–5751.
55. N. Zhao, S. Xuan, B. Byrd, F. R. Fronczek, K. M. Smith and M. G. H. Vicente, *Org. Biomol. Chem.*, 2016, **14**, 6184–6188.
56. S. Xuan, N. Zhao, X. Ke, Z. Zhou, F. R. Fronczek, K. M. Kadish, K. M. Smith and M. G. H. Vicente, *J. Org. Chem.*, 2017, **82**, 2545–2557.
57. V. Leen, P. Yuan, L. Wang, N. Boens and W. Dehaen, *Org. Lett.*, 2012, **14**, 6150–6153.
58. S. L. Niu, G. Ulrich, R. Ziessel, A. Kiss, P.-Y. Renard and A. Romieu, *Org. Lett.*, 2009, **11**, 2049–2052.
59. C. Goze, G. Ulrich, L. J. Mallon, B. D. Allen, A. Harriman and R. Ziessel, *J. Am. Chem. Soc.*, 2006, **128**, 10231–10239.
60. C. Goze, G. Ulrich and R. Ziessel, *Org. Lett.*, 2006, **8**, 4445–4448.
61. S. Goeb and R. Ziessel, *Org. Lett.*, 2007, **9**, 737–740.
62. L. Bonardi, G. Ulrich and R. Ziessel, *Org. Lett.*, 2008, **10**, 2183–2186.
63. G. Ulrich, S. Goeb, A. De Nicola, P. Retailleau and R. Ziessel, *J. Org. Chem.*, 2011, **76**, 4489–4505.
64. G. Ulrich, C. Goze, M. Guardigli, A. Roda and R. Ziessel, *Angew. Chem., Int. Ed.*, 2005, **44**, 3694–3698.
65. C. Goze, G. Ulrich and R. Ziessel, *J. Org. Chem.*, 2007, **72**, 313–322.
66. A. L. Nguyen, P. Bobadova-Parvanova, M. Hopfinger, F. R. Fronczek, K. M. Smith and M. G. H. Vicente, *Inorg. Chem.*, 2015, **54**, 3228–3236.
67. A. L. Nguyen, M. Wang, P. Bobadova-Parvanova, Q. Do, Z. Zhou, F. R. Fronczek, K. M. Smith and M. G. H. Vicente, *J. Porphyrins Phthalocyanines*, 2016, **20**, 1409–1419.
68. B. W. Henderson and T. J. Dougherty, *Photochem. Photobiol.*, 1992, **55**, 145–157.
69. T. J. Dougherty, C. J. Gomer, B. W. Henderson, G. Jori, D. Kessel, M. Korbelik, J. Moan and Q. Peng, *J. Natl. Cancer Inst.*, 1998, **90**, 889–905.
70. T. J. Dougherty, *J. Clin. Laser Med. Surg.*, 2002, **20**, 3–7.
71. M. A. MacCormack, *Semin. Cutaneous Med. Surg.*, 2008, **27**, 52–62.
72. Q. Peng and K. Berg, *Photochem. Photobiol. Sci.*, 2007, **6**, 1233–1233.
73. M. J. Garland, C. M. Cassidy, D. Woolfson and R. F. Donnelly, *Future Med. Chem.*, 2009, **1**, 667–691.
74. R. R. Allison and C. H. Sibata, *Photodiagn. Photodyn. Ther.*, 2010, **7**, 61–75.
75. W. M. Sharman, C. M. Allen and J. E. van Lier, *Drug Discovery Today*, 1999, **4**, 507–517.

76. M. F. Grahn, A. McGuinness, R. Benzie, R. Boyle, M. L. de Jode, M. G. Dilkes, B. Abbas and N. S. Williams, *J. Photochem. Photobiol., B*, 1997, **37**, 261–266.
77. H. A. Isakau, M. V. Parkhats, V. N. Knyukshto, B. M. Dzhagarov, E. P. Petrov and P. T. Petrov, *J. Photochem. Photobiol., B*, 2008, **92**, 165–174.
78. B. Aveline, T. Hasan and R. W. Redmond, *Photochem. Photobiol.*, 1994, **59**, 328–335.
79. E. D. Sternberg, D. Dolphin and C. Brückner, *Tetrahedron*, 1998, **54**, 4151–4202.
80. E. S. Nyman and P. H. Hynninen, *J. Photochem. Photobiol., B*, 2004, **73**, 1–28.
81. A. Hajri, S. Wack, C. Meyer, M. K. Smith, C. Leberquier, M. Kedinger and M. Aprahamian, *Photochem. Photobiol.*, 2002, **75**, 140–148.
82. M. Wainwright, D. A. Phoenix, L. Rice, S. M. Burrow and J. Waring, *J. Photochem. Photobiol., B*, 1997, **40**, 233–239.
83. M. Wainwright and R. M. Giddens, *Dyes Pigm.*, 2003, **57**, 245–257.
84. M. Wainwright, *Photodiagn. Photodyn. Ther.*, 2005, **2**, 263–272.
85. S. G. Awuah and Y. You, *RSC Adv.*, 2012, **2**, 11169–11183.
86. J. Killoran, L. Allen, J. F. Gallagher, W. M. Gallagher and D. F. O'Shea, *Chem. Commun.*, 2002, 1862–1863.
87. A. T. Byrne, A. E. O'Connor, M. Hall, J. Murtagh, K. O'Neill, K. M. Curran, K. Mongrain, J. A. Rousseau, R. Lecomte, S. McGee, J. J. Callanan, D. F. O'Shea and W. M. Gallagher, *Br. J. Cancer*, 2009, **101**, 1565–1573.
88. A. E. O'Connor, M. M. McGee, Y. Likar, V. Ponomarev, J. J. Callanan, D. F. O'Shea, A. T. Byrne and W. M. Gallagher, *Int. J. Cancer*, 2012, **130**, 705–715.
89. M. J. Ortiz, A. R. Agarrabeitia, G. Duran-Sampedro, J. B. Prieto, T. A. Lopez, W. A. Massad, H. A. Montejano, N. A. García and I. L. Arbeloa, *Tetrahedron*, 2012, **68**, 1153–1162.
90. S. H. Lim, C. Thivierge, P. Nowak-Sliwinska, J. Han, H. van den Bergh, G. Wagnières, K. Burgess and H. B. Lee, *J. Med. Chem.*, 2010, **53**, 2865–2874.
91. W. M. Gallagher, L. T. Allen, C. O'Shea, T. Kenna, M. Hall, A. Gorman, J. Killoran and D. F. O'Shea, *Br. J. Cancer*, 2005, **92**, 1702–1710.
92. L. Jiao, W. Pang, J. Zhou, Y. Wei, X. Mu, G. Bai and E. Hao, *J. Org. Chem.*, 2011, **76**, 9988–9996.
93. A. Burghart, H. Kim, M. B. Welch, L. H. Thoresen, J. Reibenspies, K. Burgess, F. Bergström and L. B. A. Johansson, *J. Org. Chem.*, 1999, **64**, 7813–7819.
94. C.-W. Wan, A. Burghart, J. Chen, F. Bergström, L. B. A. Johansson, M. F. Wolford, T. G. Kim, M. R. Topp, R. M. Hochstrasser and K. Burgess, *Chem. – Eur. J.*, 2003, **9**, 4430–4441.
95. Z. Dost, S. Atilgan and E. U. Akkaya, *Tetrahedron*, 2006, **62**, 8484–8488.
96. A. Cihaner and F. Algı, *React. Funct. Polym.*, 2009, **69**, 62–67.

97. Y. Hayashi, S. Yamaguchi, W. Y. Cha, D. Kim and H. Shinokubo, *Org. Lett.*, 2011, **13**, 2992–2995.
98. A. Gorman, J. Killoran, C. O'Shea, T. Kenna, W. M. Gallagher and D. F. O'Shea, *J. Am. Chem. Soc.*, 2004, **126**, 10619–10631.
99. S. Zhu, J. Zhang, G. Vegesna, A. Tiwari, F.-T. Luo, M. Zeller, R. Luck, H. Li, S. Green and H. Liu, *RSC Adv.*, 2012, **2**, 404–407.
100. Y. Yang, Q. Guo, H. Chen, Z. Zhou, Z. Guo and Z. Shen, *Chem. Commun.*, 2013, 3940–3942.
101. G.-L. Fu, H. Pan, Y.-H. Zhao and C.-H. Zhao, *Org. Biomol. Chem.*, 2011, **9**, 8141–8146.
102. C. Yu, Y. Xu, L. Jiao, J. Zhou, Z. Wang and E. Hao, *Chem. - Eur. J.*, 2012, **18**, 6437–6442.
103. J. H. Gibbs, Z. Zhou, D. Kessel, F. R. Fronczek, S. Pakhomova and M. G. H. Vicente, *J. Photochem. Photobiol., B*, 2015, **145**, 35–47.
104. L. Wang, J.-W. Wang, A.-J. Cui, X.-X. Cai, Y. Wan, Q. Chen, M.-Y. He and W. Zhang, *RSC Adv.*, 2013, **3**, 9219–9222.
105. S. Atilgan, Z. Ekmekci, A. L. Dogan, D. Guc and E. U. Akkaya, *Chem. Commun.*, 2006, 4398–4400.
106. H. He, P.-C. Lo, S.-L. Yeung, W.-P. Fong and D. K. P. Ng, *J. Med. Chem.*, 2011, **54**, 3097–3102.
107. H. He, P.-C. Lo, S.-L. Yeung, W.-P. Fong and D. K. P. Ng, *Chem. Commun.*, 2011, **47**, 4748–4750.
108. M. R. Ke, S.-L. Yeung, D. K. P. Ng, W.-P. Fong and P.-C. Lo, *J. Med. Chem.*, 2013, **56**, 8475–8483.
109. J. H. Gibbs, L. T. Robins, Z. Zhou, P. Bobadova-Parvanova, M. Cottam, G. T. McCandless, F. R. Fronczek and M. G. H. Vicente, *Bioorg. Med. Chem.*, 2013, **21**, 5770–5781.
110. P. Batat, M. Cantuel, G. Jonusauskas, L. Scarpantonio, A. Palma, D. F. O'Shea and N. D. McClenaghan, *J. Phys. Chem. A*, 2011, **115**, 14034–14039.
111. D. O. Frimannsson, M. Grossi, J. Murtagh, F. Paradisi and D. F. O'Shea, *J. Med. Chem.*, 2010, **53**, 7337–7343.
112. K. Umezawa, Y. Nakamura, H. Makino, D. Citterio and K. Suzuki, *J. Am. Chem. Soc.*, 2008, **130**, 1550–1551.
113. K. Umezawa, A. Matsui, Y. Nakamura, D. Citterio and K. Suzuki, *Chem. - Eur. J.*, 2009, **15**, 1096–1106.
114. S. G. Awuah, J. Polreis, V. Biradar and Y. You, *Org. Lett.*, 2011, **13**, 3884–3887.
115. O. A. Bozdemir, S. Erbas-Cakmak, O. O. Ekiz, A. Dana and E. U. Akkaya, *Angew. Chem., Int. Ed.*, 2011, **50**, 10907–10912.
116. W. Pang, X.-F. Zhang, J. Zhou, C. Yu, E. Hao and L. Jiao, *Chem. Commun.*, 2012, 5437–5439.
117. A. B. Nepomnyashchii, M. Bröring, J. Ahrens and A. J. Bard, *J. Am. Chem. Soc.*, 2011, **133**, 19498–19504.
118. A. B. Nepomnyashchii, M. Bröring, J. Ahrens and A. J. Bard, *J. Am. Chem. Soc.*, 2011, **133**, 8633–8645.

119. K. Tan, L. Jaquinod, R. Paolesse, S. Nardis, C. Di Natale, A. Di Carlo, L. Prodi, M. Montalti, N. Zaccheroni and K. M. Smith, *Tetrahedron*, 2004, **60**, 1099–1106.
120. A. N. Amin, M. E. El-Khouly, N. K. Subbaiyan, M. E. Zandler, S. Fukuzumi and F. D'Souza, *Chem. Commun.*, 2012, 206–208.
121. L. Huang, X. Yu, W. Wu and J. Zhao, *Org. Lett.*, 2012, **14**, 2594–2597.
122. B. Brizet, A. Eggenspiller, C. P. Gros, J.-M. Barbe, C. Goze, F. Denat and P. D. Harvey, *J. Org. Chem.*, 2012, **77**, 3646.
123. S. Pascal, L. Bucher, N. Desbois, C. Bucher, C. Andraud and C. P. Gros, *Chem. – Eur. J.*, 2016, **22**, 4971.
124. A. Eggenspiller, A. Takai, M. E. El-Khouly, K. Ohkubo, C. P. Gros, C. Bernhard, C. Goze, F. Denat, J.-M. Barbe and S. Fukuzumi, *J. Phys. Chem. A*, 2012, **116**, 3889.
125. W. M. Pardridge, *NeuroRx*, 2005, **2**, 3–14.
126. J. F. Deeken and W. Löscher, *Clin. Cancer Res.*, 2007, **13**, 1663–1674.
127. J. M. Provenzale, S. Mukundan and M. Dewhirst, *Am. J. Roentgenol.*, 2005, **185**, 763–767.
128. E. Roda, S. Nion, G. Bernocchi and T. Coccini, *Brain Res.*, 2014, **1583**, 34–44.
129. I. B. Sivaev and V. V. Bregadze, *Eur. J. Inorg. Chem.*, 2009, 1433–1450.
130. H. E. Vries, J. Kuiper, A. G. Boer, T. J. C. V. Berkel and D. D. Breimer, *Pharmacol. Rev.*, 1997, **49**, 143–156.
131. A. M. Palmer and M. S. Alavijeh, *Drug Discovery Today*, 2012, **17**, 1068–1078.
132. M. D. Habgood, D. J. Begley and N. J. Abbott, *Cell. Mol. Neurobiol.*, 2000, **20**, 231–253.
133. J. H. Gibbs, H. Wang, N. V. S. D. K. Bhupathiraju, F. R. Fronczek, K. M. Smith and M. G. H. Vicente, *J. Organomet. Chem.*, 2015, **798**, 209–213.
134. S. Xuan, N. Zhao, Z. Zhou, F. R. Fronczek and M. G. H. Vicente, *J. Med. Chem.*, 2016, **59**, 2109–2117.
135. R. Hütter, W. Keller-Schien, F. Knüsel, V. Prelog, G. C. Rodgers, P. Suter, G. Vogel, W. Voser and H. Zähner, *Helv. Chim. Acta*, 1967, **50**, 1533–1539.
136. W. Pache and H. Zähner, *Arch. Mikrobiol.*, 1969, **67**, 156–165.
137. J. Kohno, T. Kawahata, T. Otake, M. Morimoto, H. Mori, N. Ueba, M. Nishio, A. Kinumaki, S. Komatsubara and K. Kawashima, *Biosci., Biotechnol., Biochem.*, 1996, **60**, 1036–1037.
138. Y. Okami, T. Okazaki, T. Kitahara and H. Umezawa, *J. Antibiot.*, 1976, **29**, 1019–1025.
139. K. Sato, T. Okazari, K. Maeda and Y. Okami, *J. Antibiot.*, 1978, **31**, 632–635.
140. D. Schummer, H. Irschik, H. Reichenbach and G. Höfle, *Liebigs Ann. Chem.*, 1994, 283–289.
141. D. Schummer, D. Schomburg, H. Irschik, H. Reichenbach and G. Höfle, *Liebigs Ann.*, 1996, 965–969.

142. M. Berger and J. Mulzer, *J. Am. Chem. Soc.*, 1999, **121**, 8393–8394.
143. P. Lewer, E. L. Chapin, P. R. Graupner, J. R. Gilbert and C. Peacock, *J. Nat. Prod.*, 2003, **66**, 143–145.
144. T. Hemscheidt, M. P. Puglisi, L. K. Larsen, G. M. Patterson, R. E. Moore, J. L. Rios and J. Clardy, *J. Org. Chem.*, 1994, **59**, 3467–3471.
145. V. M. Dembitsky, R. Smoum, A. A. Al-Quntar, H. A. Ali, I. Pergament and M. Srebnik, *Plant Sci.*, 2002, **163**, 931–942.
146. V. M. Dembitsky, A. A. Al-Quntar and M. Srebnik, *Chem. Rev.*, 2010, **111**, 209–237.
147. S. J. Baker, T. Akama, Y.-K. Zhang, V. Sauro, C. Pandit, R. Singh, M. Kully, J. Khan, J. J. Plattner, S. J. Benkovic, V. Lee and K. R. Maples, *Bioorg. Med. Chem. Lett.*, 2006, **16**, 5963–5967.
148. Q. Luan, T. Desta, L. Chehab, V. J. Sanders, J. Plattner and D. T. Graves, *J. Dent. Res.*, 2008, **87**, 148–152.
149. A. M. Irving, C. M. Vogels, L. G. Nikolcheva, J. P. Edwards, X.-F. He, M. G. Hamilton, M. O. Baerlocher, F. J. Baerlocher, A. Decken and S. A. Westcott, *New J. Chem.*, 2003, **27**, 1419–1424.
150. J. Raunio, J. Mannoja, T. Nguyen, N. Ahmad, N. M. Kemppainen, R. G. Franzén, M. Kandhavelu and N. R. Candeias, *RSC Adv.*, 2017, **7**, 20620–20627.
151. M. J. Geier, E. G. Bowes, G. M. Lee, H. Li, T. O'Neill, A. Flewelling, C. M. Vogels, A. Decken, C. A. Gray and S. A. Westcott, *Heterocycl. Chem.*, 2013, **24**, 116–123.

Subject Index